Progress in Mathematics

3 CORE

Progress in Mathematics

Pupils' Book 1G
1G Mental Tests and Phase Tests
1G Copy Masters
Pupils' Book 2G
2G Mental Tests and Phase Tests
2G Copy Masters
Pupils' Book 3G
3G Mental Tests and Phase Tests
3G Copy Masters

Pupils' Book 1C
1C Mental Tests and Phase Tests
1C Copy Masters
Pupils' Book 2C
2C Mental Tests and Phase Tests
2C Copy Masters
Pupils' Book 3C
3C Mental Tests and Phase Tests
3C Copy Masters

4th and 5th-year volumes are in preparation, as is a third 'layer' — the 'E' ('extension') books.

Progress in Mathematics

3 CORE

Les Murray BA
Senior Teacher and Head of Mathematics, Garstang County High School

Stanley Thornes (Publishers) Ltd

First published in 1986 by:
Stanley Thornes (Publishers) Ltd
Old Station Drive
Leckhampton, CHELTENHAM GL53 ODN
England

British Library Cataloguing in Publication Data

Murray, L.
Progress in mathematics.
Book 3C
1. Mathematics—Examinations, questions, etc.
I. Title
510′.76 QA43

ISBN 0–85950–181–7

Typeset by KEYTEC, Bridport, Dorset
Printed and bound in Great Britain at The Bath Press, Avon

Preface

This book contains numerous, carefully graded questions. It is not intended to be worked through from cover to cover; the teacher should be selective in the use of exercises and questions. Worked examples have only been included where desirable to allow greater freedom for a teacher to choose his or her own methods.

Although most chapter headings in this book, as in previous books, have been arranged in parallel with those in the G series, I do not believe it is educationally desirable that all chapters should run parallel. The main difference in the 3rd-year books is the omission from book 3C of the three later chapters on trigonometry, vectors and matrices. In my opinion, these chapters are not needed at this level, and if it is felt by any teacher that a few pupils ought to be introduced to those topics then book 3G can easily be used with the pupils concerned. For the same reason, Chapter 10 of book 3G on 'Brackets and Factorising' has been replaced by 'Decimals and Vulgar Fractions', while Chapter 11 on 'Loci' has been replaced by 'Circumference of a Circle'.

Revision sections have been inserted at intervals, and once again photocopy masters are available to the teacher for exercises in which pupils may benefit by their provision. Exercises for which masters are available have been labelled **M**.

The completion of this book has been dependent on the valued help and advice given to me by many people, in particular by Mrs Alice Dickson of Garstang County High School, who has carefully and painstakingly worked through the whole text and has provided the answers as well as giving welcome advice; and to Mr J. Britton, Head of Mathematics at Copthall School, London, for his most useful comments. My thanks also go to the staff and pupils of Garstang County High School for their interest and co-operation while writing has been in progress; to Casio Electronics and Texas Instruments for the use of calculators; and to all those who have helped in any way.

Les Murray
1986

Acknowledgements

The author and publishers are grateful to the following:

Vallardei Editore Graficke for the photograph of the Matterhorn on p.10

British Railways Board for the timetable on p.81

McGraw-Hill Book Company for the illustration of Pythagoras on p.296

Contents

1 Sets

Exercise 1

1. Set W shows the set of whole numbers that are less than 10. List the odd numbers that are in set W.

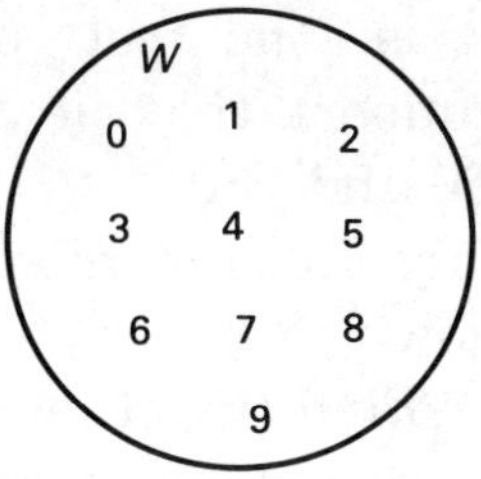

2. List the whole numbers that are in set P.

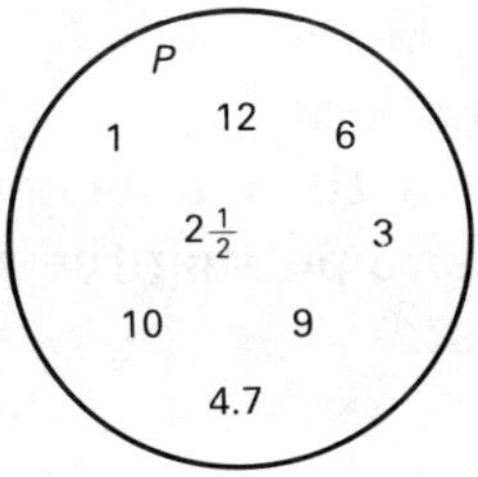

3. From set G, list:
   (*a*) the numbers that are less than 6,
   (*b*) the numbers that have one digit,
   (*c*) the even numbers,
   (*d*) the numbers that divide exactly by 5,
   (*e*) the numbers that are greater than 4 and less than 10.

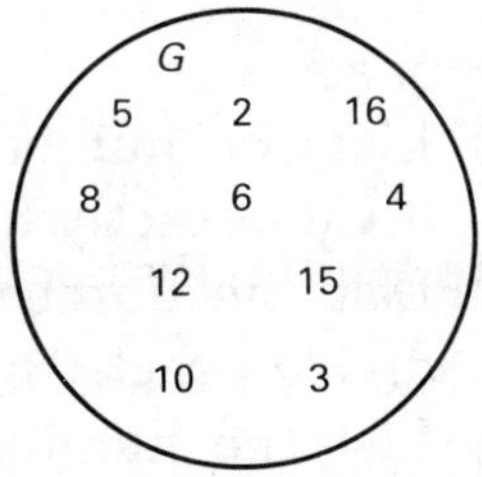

4. D shows a set of 2-digit numbers.

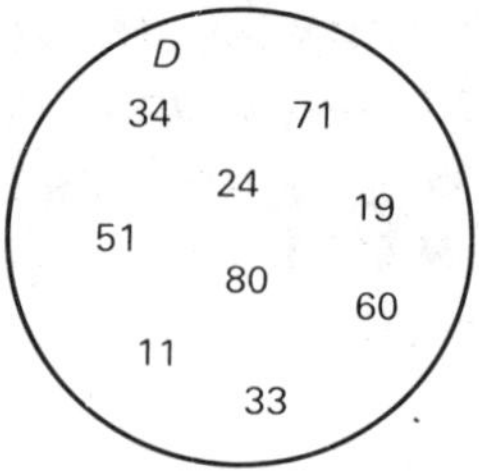

(*a*) List the numbers in set D whose digits add up to 6.

(*b*) Find two more 2-digit numbers that are not in set D whose digits add up to 6.

5. Copy set V.
V is the set of even numbers that lie between 160 and 180.
Only 3 members are shown.

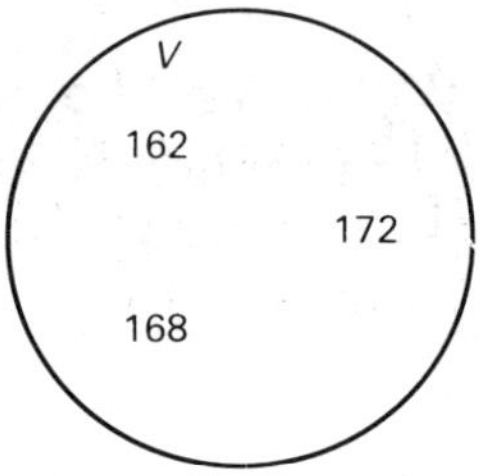

(*a*) Find the other members and write them on your copy.

(*b*) How many members does set V have?

6. T is a set of 3-digit numbers.
Write the numbers in order of size putting the smallest first.

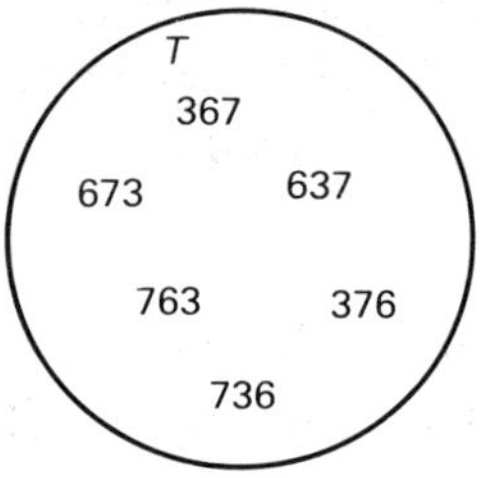

Exercise 2

1. From set A:

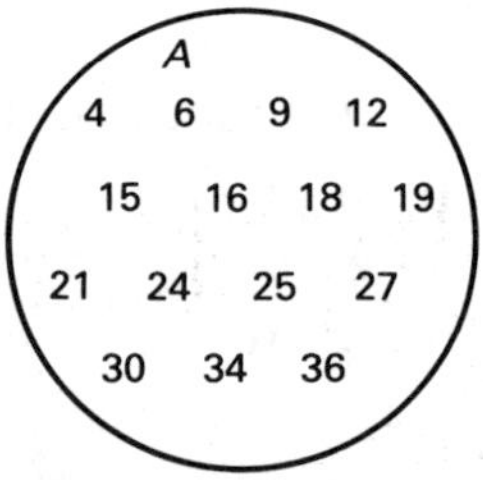

(*a*) List the numbers that divide exactly by 2.

(*b*) List the numbers that divide exactly by 3.

(*c*) List the numbers that divide exactly by 2 and also by 3.

(*d*) Describe the numbers that have been listed in the answer to part (*c*).

2. (*a*) From set *B*, list the factors of 24.
(*b*) Write three factors of 24 that are not given in set *B*.

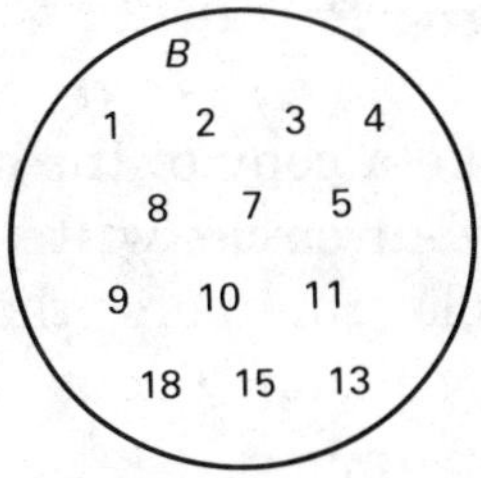

3. (*a*) From set *C*, list the multiples of 9.
(*b*) Write two multiples of 9 that are not in set *C*.
(*c*) How many multiples of 6 are in set *C*?

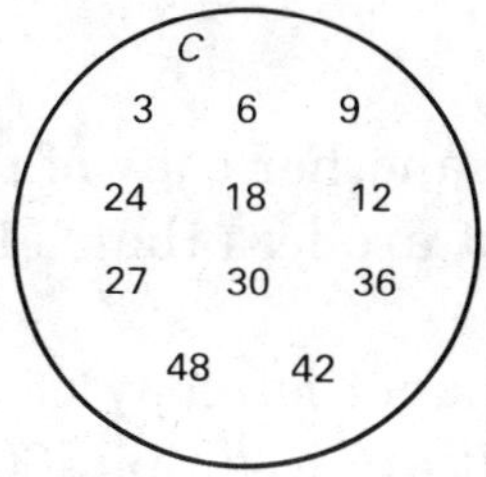

4. *P* is the set of prime numbers that are less than 20. Two members are missing. Copy the diagram and fill in the missing two prime numbers.

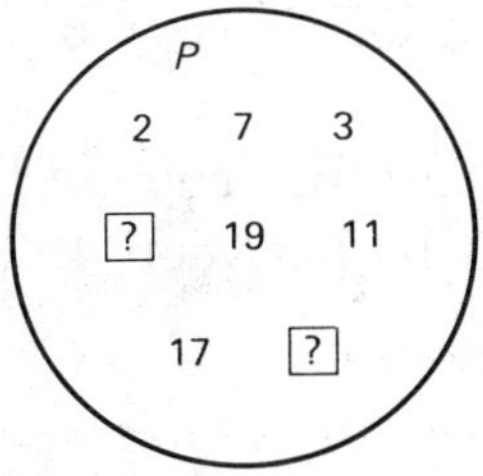

5. From set *X*, list:
(*a*) the factors of 36 that are less than 10,
(*b*) the multiples of 4,
(*c*) the prime numbers that are bigger than 20.

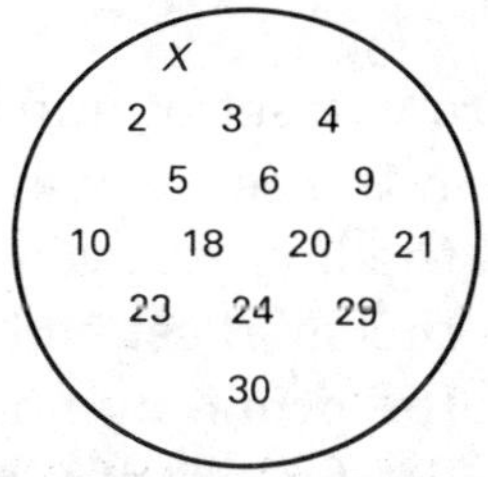

6. *Y* is the set of factors of 30. Two of the factors are missing.
Find them.

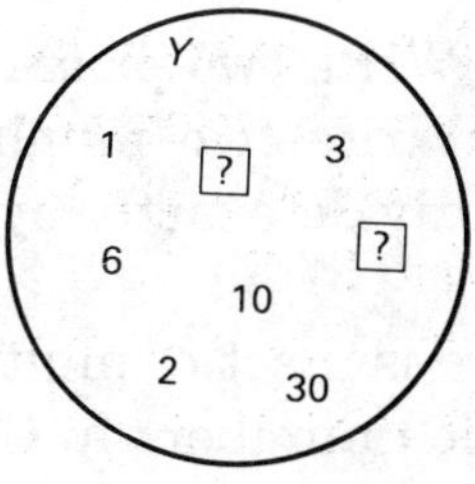

Exercise 3

1. Make a copy of the circle. In your circle, write all the whole numbers that are bigger than 6 but smaller than 15.

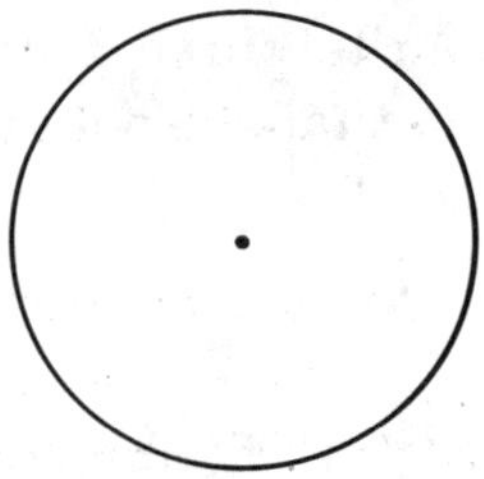

2. On another copy of the circle, write all the whole numbers that are less than 11.

3. On another copy of the circle, write all the multiples of 8 that are less than 60.

4. On another copy of the circle, write all the factors of 32.

5. On another copy of the circle, write all the prime numbers that lie between 40 and 50.

Exercise 4

1. Here is a set of numbers: $\{2, 3, 5, 6, 8, 12, 19, 21\}$.
From the set above, write:
(*a*) the largest even number,
(*b*) the smallest odd number,
(*c*) the prime numbers,
(*d*) the factors of 60,
(*e*) the multiples of 3.

2. (*a*) Write two prime numbers that lie between 30 and 40.
(*b*) Write two numbers that lie between 40 and 50 and divide exactly by 7.

3. Here is a set of multiples of 5: $\{15, 20, 35, 45, 70, 90\}$.
Which numbers in the set are also multiples of 10?

4. Write the set of multiples of 7 that are less than 60.

5. Write 3 different prime numbers in which the units digit is 1.

6. The set of numbers that end in 5 are all exactly divisible by 5.
(*a*) Are the numbers that end in 4 always exactly divisible by 4?
(*b*) Are the numbers that end in 2 always exactly divisible by 2?
(*c*) Are the numbers that end in 6 always exactly divisible by 6?

Exercise 5

A In each of the following sequences, find the next two numbers:

1. 1, 5, 9, 13, 17, [?], [?]
2. 5, 10, 20, 40, [?], [?]
3. 1, 3, 6, 10, 15, [?], [?]
4. 1, 4, 9, 16, 25, [?], [?]
5. 31, 28, 25, 22, [?], [?]

B The following sets of numbers form sequences. Write each set of numbers in order of size, smallest to largest, then find the next two numbers in the sequence:

1. 10, 1, 7, 4, 13
2. 3.5, 2, 1.5, 3, 2.5
3. 20, 2, 14, 8
4. 6, 24, 12, 3
5. $2\frac{1}{4}$, $1\frac{3}{4}$, 3, 2, $2\frac{3}{4}$, $2\frac{1}{2}$

Exercise 6

1. 9 is a square number:

Set S contains some more square numbers.
Find them. For each one, draw its dot pattern.

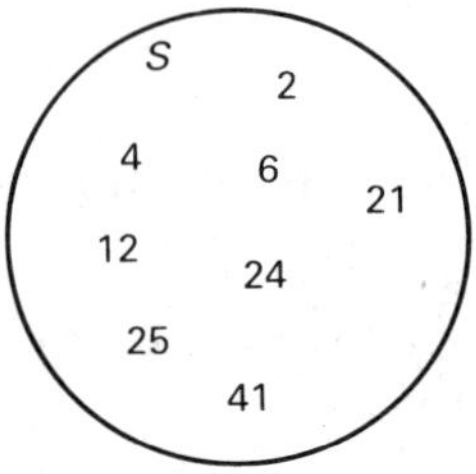

2. 10 is a triangular number:

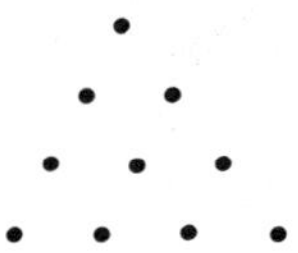

Set T contains some more triangular numbers. Find them. For each one, draw its dot pattern.

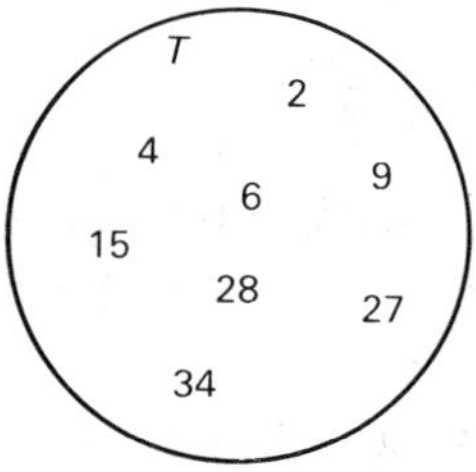

3 Find a number that lies between 30 and 40 that is both a square number and a triangular number. Draw both dot patterns.

4. (*a*) Some of the numbers in set R are rectangular numbers. List them in order of size, largest first.

(*b*) Draw a dot pattern to show the third largest rectangular number in set R.

(*c*) What are the non-rectangular numbers called?

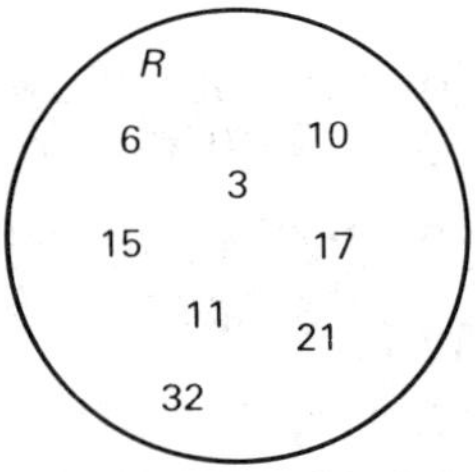

2 Approximations and Estimations

Exercise 1

1. Is 56 closer to 50 or to 60?

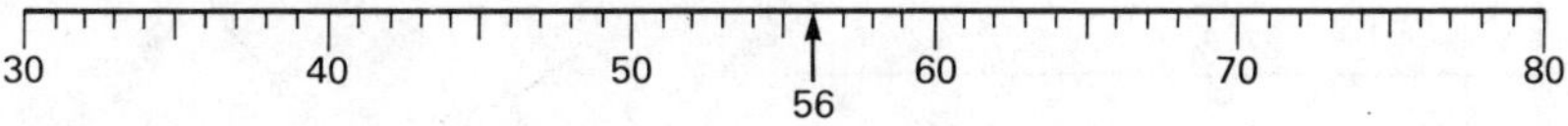

2. Is 430 closer to 400 or to 500?
3. Is 689 closer to 600 or to 700?
4. Is 7200 closer to 7000 or to 8000?
5. Is 2700 closer to 2000 or to 3000?
6. Is 8029 closer to 8000 or to 9000?
7. Is 280 closer to 200 or to 300?
8. Is 3280 closer to 3000 or to 4000?
9. Is 1762 closer to 1000 or to 2000?
10. Is 9834 closer to 9000 or to 10 000?

Exercise 2

A Round these numbers to the nearest ten:

1. 46	**3.** 23	**5.** 77	**7.** 31	**9.** 55
2. 87	**4.** 94	**6.** 19	**8.** 62	**10.** 98

B Round these numbers to the nearest hundred:

1. 260	**3.** 420	**5.** 130	**7.** 876	**9.** 680
2. 790	**4.** 673	**6.** 545	**8.** 310	**10.** 975

C Round these numbers to the nearest thousand:

1. 5100	**3.** 6700	**5.** 4380	**7.** 7060	**9.** 2541
2. 3300	**4.** 8260	**6.** 1920	**8.** 3726	**10.** 9864

Exercise 3

A Give the lengths of the following lines correct to the nearest centimetre:

1.

2.

3.

4.

5.

6.

7.

8.

B Give each of these correct to the nearest centimetre:

1. 2.9 cm	**4.** 6.6 cm	**7.** 1.8 cm	**10.** 15.4 cm
2. 4.7 cm	**5.** 3.2 cm	**8.** 6.1 cm	**11.** 28.5 cm
3. 7.9 cm	**6.** 5.4 cm	**9.** 5.8 cm	**12.** 9.7 cm

C Give each of these correct to the nearest kilometre:

1. 3.3 km	**4.** 1.4 km	**7.** 4.5 km	**10** 62.7 km
2. 5.9 km	**5.** 7.6 km	**8.** 31.6 km	**11.** 5.49 km
3. 8.7 km	**6.** 2.2 km	**9.** 78.3 km	**12.** 9.76 km

D Write these decimals correct to the nearest whole number:

1. 6.8	**4.** 15.6	**7.** 7.93	**10.** 65.49
2. 8.2	**5.** 42.1	**8.** 4.18	**11.** 3.867
3. 1.9	**6.** 2.87	**9.** 54.84	**12.** 99.75

Exercise 4

1. Round £8645 to the nearest thousand pounds.
2. Round £392 to the nearest hundred pounds.
3. Round 482 m to the nearest hundred metres.
4. Round 53 ℓ to the nearest ten litres.
5. Round 92 mm to the nearest ten millimetres.
6. Round 679 yd to the nearest hundred yards.
7. Round £5498 to the nearest thousand pounds.
8. Round 408 g to the nearest hundred grams.

Exercise 5

1. (*a*) Round 586 to the nearest hundred.
   (*b*) Round 586 to the nearest ten.

2. (*a*) Round 6284 to the nearest thousand.
   (*b*) Round 6284 to the nearest hundred.
   (*c*) Round 6284 to the nearest ten.

3. (*a*) Round 3756 to the nearest thousand.
   (*b*) Round 3756 to the nearest hundred.
   (*c*) Round 3756 to the nearest ten.

4. (*a*) Round 47 259 to the nearest ten thousand.
   (*b*) Round 47 259 to the nearest thousand.
   (*c*) Round 47 259 to the nearest hundred.
   (*d*) Round 47 259 to the nearest ten.

Exercise 6

1. A battery is 4.9 cm long. Write its length to the nearest centimetre.

2. A car cost £7450. What is that to the nearest thousand pounds?

3. There were 823 pupils in school. Write that number to the nearest hundred.

4. There were 676 mℓ of water in a jug. Write the amount correct to the nearest hundred millilitres.

5. There were 7.8 ℓ of water in a bucket. Write the amount correct to the nearest litre.

6. A table is 73.1 cm high. Write its height correct to the nearest centimetre.

7. The Matterhorn (on the border of Switzerland and Italy) is 4477 m high. Write its height correct to the nearest hundred metres.

8. A bar of soap has a mass of 142 g. Write its mass correct to the nearest ten grams.

9. Mrs Green's house cost £28 850. How much was that correct to the nearest thousand pounds?

10. A 5 p piece weighs 5.6552 g. Write its mass correct to the nearest gram.

Exercise 7

Write each of the following numbers correct to 1 significant figure:

1. 47	**6.** 3.84	**11.** 146	**16.** 58.05
2. 81	**7.** 265	**12.** 398.2	**17.** 475
3. 19.3	**8.** 26.5	**13.** 41.7	**18.** 9.29
4. 86.9	**9.** 2.65	**14.** 69.76	**19.** 9.75
5. 7.8	**10.** 338.7	**15.** 1.81	**20.** 97.5

Exercise 8

Estimate answers to the following. Work with 1 significant figure:

1. 48×82	**6.** 36×541	**11.** 23×881
2. 79×61	**7.** 74×76	**12.** 4206×37
3. 92×57	**8.** 740×760	**13.** 3162×89
4. 186×63	**9.** 68×72	**14.** 98×636
5. 408×284	**10.** 680×72	**15.** 470×112

Exercise 9

Write whether or not each answer is reasonable:

1. A box holds 12 cartons of milk. A shop has 62 boxes. The manager decided that they had just less than 600 cartons of milk.

2. Mrs Ray drove 43 km each day in travelling to and from work. She worked a 5-day week. She said she travelled over 200 km a week.

3. Mr Kerai earned £200 per week. He said that was more than £10 000 a year.

4. A car mileometer showed 43 296 miles at the beginning of the year and 52 830 miles at the end of the year. The car travelled over 10 000 miles during the year.

Exercise 10

In this exercise, work with one significant figure.

1. A salesperson drove about 210 km each day for 5 days. Estimate the total distance travelled.

2. A box holds 30 tins of soup. Estimate the number of tins in 290 boxes.

3. A box holds 12 bottles of sauce. Estimate the number of bottles in 285 boxes.

4. There are 2720 bottles of shampoo. Estimate the number of boxes needed to store the shampoo if each box holds 18 bottles.

5. Eight people weigh 65 kg, 69 kg, 61 kg, 59 kg, 62 kg, 56 kg, 72 kg and 68 kg. Estimate their total mass.

6. One dining room suite cost £935 while a cheaper one cost £639.95. Estimate the difference in cost.

7. Helena worked for 18 h at £4.15 an hour. Estimate how much she earned.

8. A jar of jam weighs 678 g including the jar. Estimate the mass of 21 jars of jam. Give your answer in kilograms.

Exercise 11

For each question, write which calculation gives the greater value:

1. 57×38 or 60×40
2. 61×51 or 60×50
3. 74×84 or 70×80
4. 22×13 or 20×10
5. 86×75 or 90×80
6. 17×19 or 20×20
7. 632×44 or 600×40
8. 767×59 or 800×60
9. 81×407 or 80×400
10. 123×14 or 100×10
11. 869×35 or 900×40
12. 276×590 or 300×600

3 Number and Number Patterns

Exercise 1

A Write in figures:

1. forty-two
2. seventy-eight
3. six hundred
4. nine hundred and seventy
5. two hundred and five
6. three hundred and eighty-one
7. four thousand
8. seven thousand one hundred
9. two thousand eight hundred and sixteen
10. nine thousand six hundred and seventy
11. five thousand seven hundred and nine
12. one thousand and one
13. eight thousand and fourteen
14. six thousand and thirty
15. fifty-three thousand
16. ninety thousand
17. seventy-one thousand four hundred
18. forty thousand and ninety-three
19. twenty-five thousand and sixty
20. one million

B Write in words:

1. 49
2. 86
3. 200
4. 530
5. 402
6. 612
7. 9000
8. 1593
9. 2905
10. 7061
11. 63 070
12. 3 000 000

Exercise 2

For each sentence, write the given number in figures.

1. The present cost twenty-six pounds.

2. There were thirty-two pupils in the class.

3. There are one hundred and sixty-three pupils in the third year.

4. There are seven hundred and ninety pupils in school.

5. Three hundred and forty-five people went to the concert.

6. Six thousand four hundred people live in our town.

7. There were eighteen thousand, six hundred and seven spectators at the hockey match.

8. There were thirty-four thousand, eight hundred and seventy spectators at the football match.

Exercise 3

Answer the following:

1. Which is the smallest number of 492, 208, 763, 194 and 326?

2. Which is the larger:
   (*a*) 491 or 728?
   (*b*) 3102 or 3097?
   (*c*) 23 094 or 19 678?

3. Write these numbers in order of size putting the smallest first:

   869, 47, 832, 907, 98, 1102, 1012

4. (*a*) Is 285 greater than 258?
(*b*) Is 109 less than 111?
(*c*) Is 486 less than 459?
(*d*) Is 2064 greater than 1997?

5. Copy each sentence and fill in the missing words 'is greater than' or 'is less than':

(*a*) 19 [?] 17
(*b*) 184 [?] 199
(*c*) 209 [?] 186
(*d*) 1043 [?] 992
(*e*) 3981 [?] 4023
(*f*) 15 064 [?] 20 000

6. Using each of the digits 2, 6 and 9 once only in each number, write as many different 3-digit numbers as you can. Give these numbers in order of size starting with the smallest first.

7. Using each of the digits 1, 3, 4 and 7 once only in each number, write:
(*a*) the largest number that can be made,
(*b*) the largest even number that can be made,
(*c*) the smallest 4-digit even number that can be made,
(*d*) the largest 2-digit number that can be made,
(*e*) all the 3-digit numbers that can be made, giving them in order of size with the largest first.

8. Using four of the digits 8, 6, 4, 3 and 1 write:
(*a*) the smallest 4-digit number that can be made,
(*b*) the smallest 4-digit odd number that can be made.

Exercise 4

Copy the following and fill in the missing operation, +, −, × or ÷, to make each answer correct:

1. 4 [?] 4 = 8
2. 6 [?] 2 = 12
3. 12 [?] 2 = 6
4. 16 [?] 4 = 4

5. 14 [?] 7 = 7

6. 20 [?] 5 = 15

7. 2 [?] 4 = 6

8. 8 [?] 8 = 16

9. 3 [?] 3 = 9

10. 6 [?] 4 = 24

11. 12 [?] 6 = 6

12. 24 [?] 8 = 3

13. 16 [?] 8 = 8

14. 28 [?] 4 = 7

15. 40 [?] 5 = 8

16. 7 [?] 2 = 9

17. 5 [?] 3 = 15

18. 18 [?] 2 = 16

19. 6 [?] 2 = 2 [?] 4

20. 30 [?] 5 = 5 [?] 5

21. 8 [?] 4 = 24 [?] 2

22. 7 [?] 4 = 24 [?] 4

23. 14 [?] 2 = 15 [?] 8

24. 18 [?] 6 = 9 [?] 6

Exercise 5

1. The graph shows the number of packets of crisps sold each day at a shop. Answer the following from the graph:

(*a*) How many packets of crisps were sold on Thursday?

(*b*) On which day were the fewest packets of crisps sold?

(*c*) How many more packets were sold on Friday than on Tuesday?

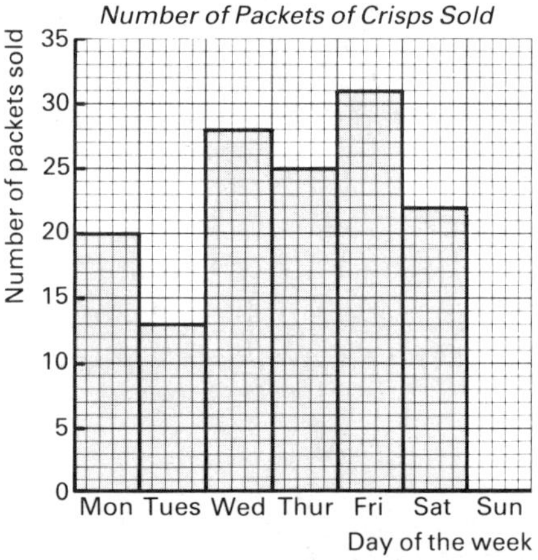

(*d*) Find the total number of packets of crisps sold on Monday, Tuesday and Wednesday.

(*e*) Find the total number of packets of crisps sold during the week.

(*f*) If the crisps are delivered in boxes with 48 packets in each box, how many boxes were needed for the week?

2. In the following football matches, 3 points are awarded for a win, 1 for a draw and 0 for losing. Copy and complete the table:

Number of games played	Number of wins	Number of draws	Number of games lost	Total points
9	6		1	20
10		3	2	18
15	9	4		
	7	1	4	
18	5		7	
25		9	4	
24	14		5	
20	10			34
	5		7	17
16		4		25

Exercise 6

A Carry out these calculations:

1.
$$\begin{array}{r} 3746 \\ +2517 \\ \hline \end{array}$$

2.
$$\begin{array}{r} 275 \\ 486 \\ +138 \\ \hline \end{array}$$

3.
$$\begin{array}{r} 5814 \\ 263 \\ +8025 \\ \hline \end{array}$$

4.
$$\begin{array}{r} 3426 \\ -1254 \\ \hline \end{array}$$

5.
$$\begin{array}{r} 8302 \\ -3147 \\ \hline \end{array}$$

6.
$$\begin{array}{r} 7024 \\ -2578 \\ \hline \end{array}$$

7. $\begin{array}{r} 347 \\ \times \quad 6 \\ \hline \\ \hline \end{array}$

8. $\begin{array}{r} 2068 \\ \times \quad 9 \\ \hline \\ \hline \end{array}$

9. $\begin{array}{r} 6103 \\ \times \quad 2 \\ \hline \\ \hline \end{array}$

10. $4\overline{)672}$

11. $3\overline{)8025}$

12. $9\overline{)3168}$

B Copy and complete:

1.

		4	7	
	×	2	4	
	9	4	0	20 × 47
+	?	?	?	4 × 47
1	?	?	?	

2.

		5	8	
	×	1	9	
	?	8	0	10 × 58
+	?	?	2	9 × 58
?	?	?	2	

C Set out in columns then work out:

1. 5613 + 2996
2. 8325 − 1708
3. 9002 − 4576
4. 349 × 4
5. 7146 ÷ 6
6. 38 × 21

Exercise 7

Use a calculator for this exercise:

a	b	c	d	e	f	g	h	i	j	k	l	m	n	o	p	q	r	s	t	u	v	w	x	y	z
1	2	3	4	5	6	7	8	9	10	11	12	13	14	15	16	17	18	19	20	21	22	23	24	25	26

From the table,

d = 4, o = 15, g = 7

2 × 4   2 × 15   2 × 7

8   30   14

dog = 8, 30, 14 in code (using × 2 as the *encoder*)

1. Use $\times 2$ as the encoder to write these words in code:
(*a*) bull (*b*) house (*c*) train

2. *Decode* these (that is, find the words) if the encoder was $\times 2$:
(*a*) 28, 42, 26, 4, 10, 36
(*b*) 38, 34, 42, 2, 36, 10
(*c*) 26, 2, 40, 16, 38

3. Encode these words (use $\times 3$ as the encoder):
(*a*) sweet (*b*) triangle (*c*) orange

4. Decode these (the encoder was $\times 3$):
(*a*) 24, 15, 72, 3, 21, 45, 42
(*b*) 6, 54, 45, 69, 42
(*c*) 60, 24, 15, 27, 54

5. Decode these:
(*a*) 60, 45, 70, 25 (encoder = $\times 5$)
(*b*) 72, 135, 162, 81, 234, 135, 126, 180, 9, 108 (encoder = $\times 9$).
(*c*) 12, 36, 72, 12, 48, 20 (encoder = $\times 4$)

6. Encode these:
(*a*) octagon (encoder = $\times 6$)
(*b*) rectangle (encoder = $\times 8$)
(*c*) straight (encoder = $\times 7$)

7. Try to decode the following (the decoders have not been given):
(*a*) 4, 72, 12
(*b*) 3, 42, 21, 36, 15
(*c*) 96, 6, 108, 6, 72, 72, 30, 72
(*d*) 30, 4, 40, 42, 38, 10
(*e*) 162, 72, 135, 117, 18, 189, 171
(*f*) 160, 144, 8, 128, 40, 208, 72, 168, 104
(*g*) 45, 95, 75, 95, 15, 25, 60, 25, 95
(*h*) 112, 35, 126, 112, 35, 98, 28, 63, 21, 147, 84, 7, 126

Exercise 8

1. Find the value of $87 + 46$.

2. Evaluate 6×57.

3. Choose the correct answer to 42×29:
A 462 B 1218 C 174 D 818 E 1209

4. Work out $7326 + 1847$.

5. Find the difference between 93 and 54.

6. What number, when subtracted from 85, gives 47?

7. A postman delivered 59 letters on one road and 76 letters on another. How many is that altogether?

8. I. Ball had scored 68 runs at cricket. How many more runs did he need to make a century?

9. A box holds a dozen eggs.
(*a*) How many eggs will there be in 7 boxes?
(*b*) How many eggs will there be in 20 boxes?
(*c*) How many boxes are needed for 72 eggs?
(*d*) How many boxes are needed for 360 eggs?

10. A car park has 4 floors. Each has space for 68 cars. How many cars are there when the car park is full?

11. Eight people share 143 sweets equally amongst themselves. How many will each get? How many will be left?

12. A box holds 40 cards. How many packs of a dozen cards can be made up from three boxes?

Exercise 9

Answer these as quickly as possible:

1. $7 + 9 + 3$
2. $46 + 38 + 54$
3. $31 + 47 - 31$
4. $2 \times 9 \times 5$
5. $25 \times 3 \times 4$
6. $38 \times 7 \div 7$
7. $9 + 7 = \boxed{?} + 9$
8. $36 + 59 = \boxed{?} + 36$
9. $4 \times 8 = \boxed{?} \times 4$
10. $7 \times 6 = 6 \times \boxed{?}$
11. If $36 + 45 = 81$, find $45 + 36$.
12. If $63 + 29 = 92$, find $92 - 29$.
13. If $6 \times 13 = 78$, find 13×6.
14. If $7 \times 29 = 203$, find $203 \div 7$.
15. If $21 \times 45 = 945$, find $945 \div 21$.

Exercise 10

Use a calculator throughout this exercise. Estimate each answer as a check.

1. Find the sum of 3869 and 2184.
2. Find the product of 51 and 67.
3. Find the difference between 2063 and 1845.
4. Find a number that is 491 more than 326.
5. There are 1082 pupils in a school.
   If 593 of these are girls, how many boys are there?
6. Evaluate $2183 \div 37$.
7. 9 coaches each carry 53 passengers.
   How many passengers is that altogether?

8. The North Sea has an area of 575 300 km^2 while the Red Sea has an area of 437 700 km^2. How much bigger is the North Sea?

9. There are 24 h in a day. How many hours are there in a year (365 days)?

10. 160 pencils are put into packs of a dozen. How many packs can be made up? How many pencils are left over?

Exercise 11

Ask a friend to:

(*a*)	Key in his or her age on a calculator, e.g.	13
(*b*)	Multiply it by 5	65
(*c*)	Add 12	77
(*d*)	Multiply by 4	308
(*e*)	Subtract 4	304
(*f*)	Multiply by 5	1520
(*g*)	Add the number of the month of his or her birth (e.g. May = 5)	1525
(*h*)	Multiply by 25	38 125
(*i*)	Add 18	38 143
(*j*)	Multiply by 4	152 572
(*k*)	Add the date of his or her birthday (e.g. 27 May, so add 27)	152 599
(*l*)	Subtract 50	152 549
(*m*)	Now take the calculator from your friend. Subtract 22 022.	130 527

Look at the display. The last two digits give the day of the birthday. The next two digits give the month. The first digit or digits give the age.

DON'T FORGET to press [=] at each stage.

Brackets and Order of Calculations

Exercise 12

Find the value of:

1. (*a*) $5 \times 3 + 6$
   (*b*) $6 + 5 \times 3$
2. (*a*) $24 \div 2 + 4$
   (*b*) $4 + 24 \div 2$
3. $5 \times 10 - 6$
4. $3 + 3 \times 3$
5. $26 - 5 \times 4$
6. $10 - 3 \times 2$
7. $10 - 2 \times 3$
8. $8 + 20 \div 4$
9. $32 \div 4 + 12$
10. $38 - 14 \div 2$
11. $40 \div 10 + 10$
12. $17 - 5 \times 3$

Exercise 13

Use a calculator to help you to answer these:

1. $5 \times 8 + 21$
2. $21 + 5 \times 8$
3. $45 - 36 \div 9$
4. $60 - 12 \times 3$
5. $3 \times 14 - 12$
6. $125 \div 5 - 19$
7. $23 + 11 \times 7$
8. $84 \div 3 + 13$
9. $100 - 6 \times 15$
10. $13 \times 8 - 27$
11. $39 + 56 \div 8$
12. $47 + 49 \times 7$

The use of brackets can change the order in which calculations are carried out.
Note that $15 - 8 - 3 \quad - 4$ (Check it.)
However, $15 - (8 - 3) = 15 - 5 = 10$ (The calculation inside the brackets must be worked out first.)

Exercise 14

Work out the following:

e.g. (*a*) $6 \times (4 + 3)$
$= 6 \times 7$
$= \underline{\underline{42}}$

(*b*) $6 \times 4 + 3$
$= 24 + 3$
$= \underline{\underline{27}}$

1. (*a*) $9 - (6 - 2)$
   (*b*) $9 - 6 - 2$
2. (*a*) $4 \times (2 + 6)$
   (*b*) $4 \times 2 + 6$
3. $3 \times (8 - 4)$
4. $(9 - 3) \times 4$
5. $(10 + 8) \div 6$
6. (*a*) $(5 + 3) \times 7$
   (*b*) $5 + 3 \times 7$
7. (*a*) $24 \div (4 + 2)$
   (*b*) $24 \div 4 + 2$
8. $24 \div (8 - 6)$
9. $(30 - 10) \div 5$
10. $(18 - 12) \times 8$
11. (*a*) $23 - (12 + 5)$
    (*b*) $23 - 12 + 5$
12. (*a*) $24 \div (6 \div 2)$
    (*b*) $24 \div 6 \div 2$
13. $5 \times (9 - 3)$
14. (*a*) $(30 - 9) \div 3$
    (*b*) $30 - 9 \div 3$
15. $20 \div (11 - 7)$
16. $(15 + 1) \div 8$
17. $9 \times (6 + 4)$
18. $(11 + 4) \times 6$
19. $8 \times (14 - 6)$
20. (*a*) $(23 - 7) \times 2$
    (*b*) $23 - 7 \times 2$

Exercise 15

Find the value of the following:

e.g. $(3 + 2) \times (7 - 3)$
$= 5 \times 4$
$= \underline{\underline{20}}$

1. $(9 - 4) \times (8 - 5)$
2. $(6 - 3) \times (1 + 9)$
3. $(10 + 8) - (3 + 11)$
4. $(12 + 15) \div (12 - 3)$
5. $(5 - 2) \times (6 + 4)$
6. $(10 - 4) \div (9 - 7)$
7. $(9 + 3) \times (9 - 3)$
8. $(38 - 10) \div (4 + 3)$
9. $(9 + 10) - (6 - 2)$
10. $(15 - 2) - (12 - 9)$

Exercise 16

Copy each statement. If it is correct, leave it. If it is wrong, insert brackets to make it correct:

1. $4 + 3 \times 2 = 10$
2. $4 + 3 \times 2 = 14$
3. $7 - 4 \times 6 = 18$
4. $12 + 4 \div 2 = 8$
5. $20 - 8 \div 2 = 16$
6. $6 \times 8 - 2 = 36$
7. $7 \times 3 + 8 = 77$
8. $36 \div 9 - 3 = 6$
9. $4 \times 7 + 2 = 30$
10. $2 \times 8 - 4 = 8$

Number Patterns

Exercise 17

1. The dot pattern shows the square number, 16. Draw a dot pattern to show the square number 25.

2. The dot pattern shows the triangular number 10. Draw a dot pattern to show the triangular number 15.

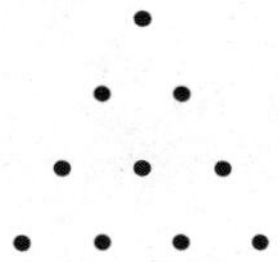

3. The dot pattern shows the rectangular number 15. Draw two different dot patterns to show the rectangular number 18.

Exercise 18

Copy these number patterns and give the next three steps:

1.
$9 \times 11 - 99$
$9 \times 21 = 189$
$9 \times 31 = 279$
$9 \times 41 = 369$
and so on

2.
$9 \times 11 = 99$
$9 \times 22 = 198$
$9 \times 33 = 297$
$9 \times 44 = 396$
and so on

3.
$1 \times 9 = 9$
$21 \times 9 = 189$
$321 \times 9 = 2889$
$4321 \times 9 = 38889$
and so on

4.
$5 \times 9 = 45$
$55 \times 9 = 495$
$555 \times 9 = 4995$
$5555 \times 9 = 49995$
and so on

5. $1 \times 8 + 1 = 9$
$12 \times 8 + 2 = 98$
$123 \times 8 + 3 = 987$
$1234 \times 8 + 4 = 9876$
and so on

6. $1 \times 5 + 1 = 6$
$2 \times 5 + 1 = 11$
$3 \times 5 + 1 = 16$
$4 \times 5 + 1 = 21$
and so on

Exercise 19

Copy and complete the number pattern.

Write what you notice about the answers.

21	−	12	=	9
31	−	13	=	?
41	−	14	=	?
51	−	15	=	?
?	−	?	=	?
?	−	?	=	?
?	−	?	=	?
91	−	19	=	72

Directed Numbers

Exercise 20

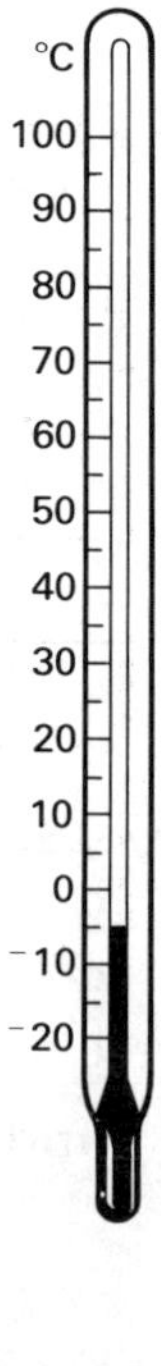

1. What temperature is shown on the thermometer?

2. Which is colder, $^{-}4$ °C or 8 °C?

3. Which is warmer, $^{-}7$ °C or $^{-}8$ °C?

4. Which is the lowest temperature out of:
$^{-}3$ °C, $^{-}5$ °C, 2 °C, $^{-}1$ °C and $^{-}2$ °C?

5. Write these temperatures in order from coldest to hottest:
14 °C, $^{-}6$ °C, 12 °C, 2 °C, $^{-}10$ °C, $^{-}16$ °C, 0 °C

6. Is $^{-}9$ °C warmer than $^{-}3$ °C?

7. Here is a list of temperatures:

−3 °C	4 °C	17 °C	−9 °C	29 °C	12 °C
15 °C	−12 °C	−8 °C	32 °C	−1 °C	0 °C

(*a*) Which is the third highest temperature in the list?
(*b*) Which is the third lowest temperature in the list?
(*c*) From the list, which temperatures lie between −10 °C and 10 °C?
(*d*) Write the temperatures in the list that are less than −2 °C.

Exercise 21

1. What is the difference between the two temperatures marked on the sketch of a thermometer?

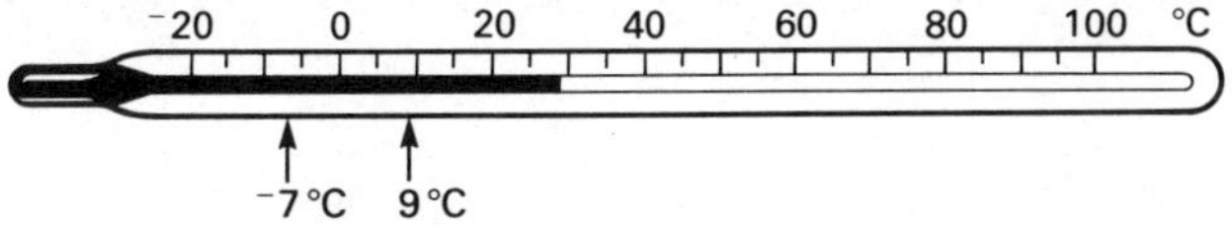

2. Copy and complete the following table for the given thermometer:

	Temperature shown	Temperature change (°C)	New temperature (°C)
(*a*)	?	up 24	?
(*b*)	?	down 12	?
(*c*)	?	down 40	?
(*d*)	?	up ?	31
(*e*)	?	down ?	14
(*f*)	?	down ?	−2

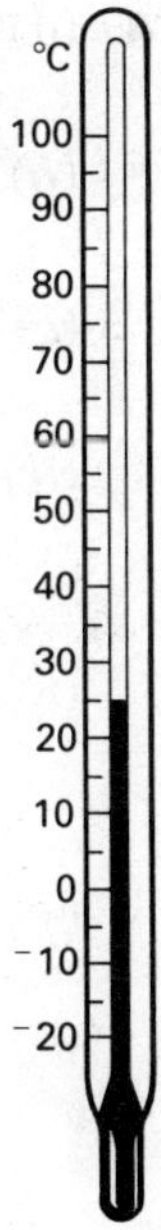

3. Copy and complete the following table for the given thermometer:

	Temperature shown	Temperature change (°C)	New temperature (°C)
(*a*)	?	up 4	?
(*b*)	?	up 18	?
(*c*)	?	down 6	?
(*d*)	?	up ?	8
(*e*)	?	up ?	⁻1
(*f*)	?	down ?	⁻16

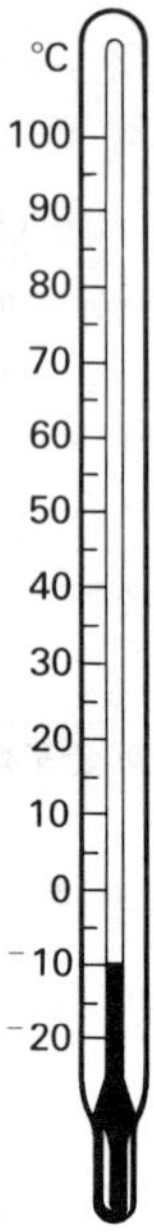

4. (*a*) What temperature, in degrees Celsius, is shown on the thermometer?
(*b*) If the temperature shown is increased by 8 °C, what would the new temperature be?
(*c*) If the temperature shown went down by 9 °C, what would the new temperature be?
(*d*) Through how many degrees Celsius would the temperature shown need to go up to reach 42 °C?
(*e*) Through how many degrees Celsius would the temperature shown need to drop to reach 17 °C?
(*f*) Through how many degrees Celsius would the temperature shown need to drop to reach freezing point?

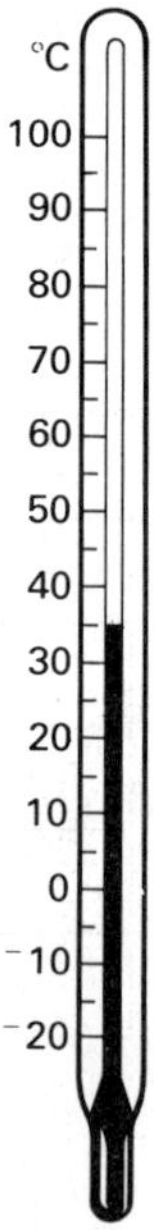

5. At 6 o'clock in the morning the temperature was $^{-}7$ °C. By noon it was 5 °C. By how much had the temperature risen?

6. At 7 o'clock in the morning the temperature was $^{-}3$ °C. It had gone up by 8 °C by noon. What was the temperature at noon?

7. At 4 o'clock in the afternoon the temperature was 9 °C. By midnight it had fallen 13 °C. What was the temperature at midnight?

8. Copy and complete the table:

Minimum temperature (°C)	Maximum temperature (°C)	Change in temperature (°C)
$^{-}4$	5	9
5	12	
$^{-}3$	9	
$^{-}6$		10
	3	11
$^{-}18$	$^{-}4$	

9. At Tummel Bridge, Tayside, on 9 May 1978 the temperature varied between $^{-}7$ °C and 22 °C. What is the difference between the two temperatures?

10. When Viking 1 landed on Mars on 20 July 1976 it measured the temperature to be $^{-}86$ °C just after dawn and $^{-}31$ °C near noon. What is the difference between these two temperatures?

4 Money

Earning Money

People work to earn money.
Weekly earnings are called *wages*.

Although wages are paid weekly, they are often worked out as an amount paid per hour called the *basic hourly rate* (or *basic rate*).

Wages = hours worked × hourly rate

Exercise 1 **M**

A Copy and complete the table:

	Hours worked	Hourly rate	Wage
1.	4	£3	
2.	8	£5	
3.	6	£4	
4.	10	£7	
5.	7	£6	
6.	4	£2.50	
7.	7	£3.50	
8.	5	£4.50	
9.	20	£6	
10.	40	£4	

B **1.** Mrs Adams earned £4 an hour. How much does she earn in a day if she works 8 h?

2. Mr Baker earned £5 an hour. How much does he earn if he works 7 h?

3. Mr Conway earned £7 an hour. How much does he earn if he works 9 h?

4. Mrs Dawson earned £2.50 an hour for 10 h. Find her total earnings.

5. Mr Edwards earned £6 an hour for 36 h work.
Find his total wage.

6. Mr Ford's hourly rate was £4.
Find his total wage for 38 h work.

7. How much did Mrs Grant earn in 40 h at £7 an hour?

8. How much did Mrs Hill earn in 35 h at £3 an hour?

Exercise 2

A Copy and complete the table:

	Hours worked	Hourly rate	Wage
1.	8	£3.50	
2.	9	£3.20	
3.	10	£1.90	
4.	7	£2.80	
5.	20	£4.50	
6.	40	£5.10	
7.	25	£4.60	
8.	28	£3.70	
9.	34	£6.20	
10.	38	£2.60	

B **1.** How much did Mr Isles earn if he worked for 6 h at £7.50 an hour?

2. Connie earned £2.50 an hour. Find her earnings for 7 h work.

3. Warren worked for 9 h at £2.40 an hour. Find his total earnings.

4. Mrs Jaffrey earned £5.20 an hour. Find her earnings for 8 h work.

5. Mr Keegan worked a 40-hour week. Calculate his wage if his hourly rate was £3.90.

6. Sarah Lewis earned £2.70 an hour. How much would she earn in 20 h?

7. Mrs McCallum worked for 36 h at an hourly rate of £4.10. Calculate her earnings.

8. Mr Naylor worked a 35-hour week. Calculate his weekly earnings if his basic hourly rate was £5.30.

Exercise 3 **M**

A ready reckoner can be used to find weekly earnings.

A Copy and complete this ready reckoner:

Rate	Number of hours worked				
	36	37	38	39	40
£2	£72	£74	£76		£80
£2.50	£90	£92.50	£95	£97.50	
£3		£111	£114	£117	
£3.50	£126	£129.50	£133	£136.50	£140
£4	£144	£148		£156	£160
£4.50	£162	£166.50	£171		£180
£5		£185		£195	
£5.50	£198		£209	£214.50	
£6	£216	£222			£240

B Use the ready reckoner on the previous page to find the weekly earnings of someone who works:

1. 40 h at £4.50 an hour
2. 36 h at £5.50 an hour
3. 38 h at £3 an hour
4. 37 h at £5 an hour
5. 39 h at £3.50 an hour
6. 36 h at £4 an hour
7. 39 h at £2.50 an hour
8. 38 h at £5.50 an hour
9. 37 h at £4.50 an hour
10. 39 h at £4 an hour

The wages that are earned in a week are called the *gross weekly wages* (or *gross wages* or *gross pay*).

Take-home pay is less than the gross pay because of *deductions* such as income tax and National Insurance.

The amount left after deductions, the take-home pay, is called the *net weekly wage* (or *net pay*).

Net pay = gross pay − deductions

Exercise 4 **M**

A Copy and complete the table:

	Gross pay	Deductions	Net pay
1.	£115	£30	
2.	£150	£37	
3.	£140	£39	
4.	£95	£19.50	
5.	£190	£52.50	
6.	£137	£33.50	
7.	£122.50	£25	
8.	£170.50	£46	
9.	£103.50	£21.50	
10.	£235.50	£71.50	

B 1. Mr O'Reilly's gross wage was £151. How much did he take home if deductions came to £36?

2. Mr Page earned £114 per week gross. Find his net pay if deductions came to £28.

3. Mrs Quick earned £172 per week. Deductions totalled £47. Find her take-home pay.

4. If Mrs Rose earned £134 gross and deductions totalled £29.50, find her net pay.

5. Mrs Sharp earned £195.50 per week. From that, deductions of £53 were taken. Find her net pay.

6. Deductions of £47.50 were taken from Mr Tutt's gross pay of £181.50. Calculate his net pay.

7. Mrs Unsworth's gross pay was £129. If £26.50 was deducted, find her take-home pay.

8. Deductions from Mr Vernon's earnings came to £38. His gross weekly wage was £146.50. Find his net pay.

9. Out of Mr Wood's gross earnings of £163, £44.50 was deducted. Calculate his net pay.

10. Mrs Xavier earned £98.50 a week. Deductions totalled £19. Calculate her net pay.

11. Mrs Young earned £180 per week. If she paid £16 National Insurance and £34 income tax, calculate her net pay.

12. Mr Zeller earned £152 gross per week. His income tax was £32.50 while his National Insurance payment was £13.50. Find his take-home pay.

Exercise 5 M

A Copy and complete the table:

	Gross pay	Deductions	Net pay
1.	£123	£26.40	
2.	£172.60	£47	
3.	£108.90	£21.50	
4.	£143.50	£39.70	
5.	£193.20	£53.70	
6.	£138.90	£34.45	
7.	£156.25	£38.50	
8.	£184.75	£49.95	
9.	£106.15	£21.75	
10.	£142.65	£35.72	

B **1.** Mrs Allen earned £163.20 a week gross. Deductions totalled £39.50. Calculate her net pay.

2. Mr Booth's weekly earnings totalled £197.60. Deductions came to £54.70. Find his net pay.

3. Mr Cross earned £109.25. Deductions came to £22.40. Calculate his net pay.

4. Mrs Darwen's gross weekly earnings totalled £146.30. If £41.85 was deducted find her take-home pay.

5. Mr Eccles earned £151.95 per week. £35.63 was deducted. Find his take-home pay.

6. £27.65 was deducted from Mr Foster's gross wage of £125.41. Calculate his net pay.

7. £50.18 was deducted from Mrs Gregson's gross wage of £185.72. Calculate her take-home pay.

8. Mrs Harris earned £143.35 per week. Deductions totalled £44.45. Find her net pay.

Exercise 6 M

A Copy and complete the table:

	Gross pay	Deductions	Net pay
1.	£120	£24	
2.	£145.50	£41	
3.		£19	£82
4.		£32	£99
5.	£192		£138
6.		£29	£84
7.	£106		£84
8.	£175	£48	
9.		£37.50	£117
10.	£136		£104.50

B **1.** Mrs Isaacs earned £184 per week. Deductions totalled £48. Calculate her net pay.

2. Mr Jackson's take-home pay was £119. Deductions totalled £42. Find his gross pay.

3. Deductions totalling £26 were taken from Mr Kellett's gross pay. He was left with £99. Calculate his gross pay.

4. Mrs Land took home £88. If deductions totalled £29, calculate her gross pay.

5. Mr Mayor's gross pay was £176. His net pay was £127. Find the total deductions.

6. Mrs Noble's take-home pay was £118.50. Her gross pay was £155. Calculate the deductions.

7. Mrs Orr took home £137.40. Her gross pay was £191. Find the deductions.

8. Mr Palmer took home £122.90 after deductions totalling £45.40. Calculate his gross pay.

Exercise 7

1. Mrs Quirk worked for 35 hours at £4 an hour. Deductions came to £37. Calculate:
(*a*) her gross weekly wage, (*b*) her net weekly wage.

2. Mr Ritson earned £5 an hour. He worked 32 h in a week. If deductions totalled £42, find:
(*a*) his gross weekly wage, (*b*) his net weekly wage.

3. Mr Stuart worked 33 h at a rate of £6 an hour. Deductions came to £55.75. Calculate:
(*a*) his gross weekly wage, (*b*) his net weekly wage.

4. Toni earned £2.50 an hour. She worked 30 h. Deductions came to £16. Calculate:
(*a*) her gross weekly wage, (*b*) her take-home pay.

5. Vic worked for 25 h at £7 an hour. Deductions totalled £46.32. Calculate:
(*a*) his gross weekly wage, (*b*) his net weekly earnings.

Spending Money

Exercise 8

A Find the change out of 50 p when you spend:

1. 41 p **2.** 36 p **3.** 33 p **4.** 29 p **5.** 18 p

B Find the change out of £1 when you spend:

1. 75 p	**4.** 61 p	**7.** 44 p	**10.** 23 p
2. 87 p	**5.** 52 p	**8.** 47 p	**11.** 15 p
3. 69 p	**6.** 58 p	**9.** 34 p	**12.** 19 p

C Find the change out of £5 when you spend:

1. £4.25	**4.** £2.59	**7.** £2.68	**10.** £3.37
2. £3.80	**5.** £1.74	**8.** £2.43	**11.** £2.14
3. £3.42	**6.** £3.06	**9.** £1.98	**12.** £1.21

D Find the change out of £10 when you spend:

1. £9.30	**4.** £5.28	**7.** £4.13	**10.** £1.81
2. £7.95	**5.** £0.72	**8.** £5.77	**11.** £3.66
3. £8.55	**6.** £7.04	**9.** £6.45	**12.** £2.79

E Find the change out of £20 when you spend:

1. £18.10	**4.** £12.35	**7.** £14.22	**10.** £3.56
2. £15.40	**5.** £8.89	**8.** £9.07	**11.** £11.94
3. £17.85	**6.** £16.71	**9.** £7.63	**12.** £10.38

Exercise 9

Find the total value of the coins listed in each question:

A **1.** One 50 p coin
One 20 p coin
Two 10 p coins

2. One 50 p coin
Three 10 p coins
One 5 p coin

3. One 50 p coin
Two 20 p coins
Two 10 p coins
One 5 p coin

4. Two 50 p coins
One 20 p coin
Three 10 p coins
One 5 p coin

5. Two 50 p coins
One 20 p coin
Three 5 p coins
Two 2 p coins

6. Three 50 p coins
Two 20 p coins
One 5 p coin
Two 2 p coins

7. Three 50 p coins
Three 20 p coins
Two 10 p coins
Four 2 p coins

8. Five 50 p coins
Three 20 p coins
One 5 p coin
Two 2 p coins

B **1.** One 50 p coin, two 20 p coins, two 10 p coins and two 2 p coins.

2. Two 50 p coins, three 10 p coins, one 5 p coin and two 2 p coins.

3. Three 50 p coins, one 20 p coin, two 10 p coins, one 5 p coin and one 2 p coin.

4. Six 50 p coins, four 20 p coins, three 10 p coins and one 5 p coin.

5. Three £1 coins, one 50 p coin, three 20 p coins, one 10 p coin, five 5 p coins and two 2 p coins.

Exercise 10

Throughout this exercise, use only the coins £1, 50 p, 20 p, 10 p, 5 p, 2 p and 1 p.

In each question an amount of money is given. Show how you would pay that amount using the smallest possible number of coins.

e.g. £2.87

Use £1, £1, 50 p, 20 p, 10 p, 5 p and 2 p

Note that seven coins are needed. The coins £1, £1, 20 p, 20 p, 20 p, 20 p, 5 p, 2 p give the same total but eight coins are used. (The correct answer uses seven.)

1. £1.45
2. £1.76
3. £1.84
4. £2.12
5. £2.39
6. £2.70
7. £2.98
8. £3.04
9. £3.61
10. £4.26
11. £4.43
12. £4.77

Exercise 11

1. Debra bought some apples for 57 p. What change should she have received from £1?

2. Mr McBride bought an article for £2.74, then sold it for £3.50. Find the profit made.

3. Mrs Ronson bought an article for £28.46, then sold it for £37.89. Find the profit made.

4. Toby bought a notepad for 34 p and a pen for 39 p.
(*a*) How much did he spend?
(*b*) What change did he receive from £1?

5. Angie bought 4 batteries at 49 p each. How much did she have to pay?

6. (*a*) How many 12 p stamps can I buy for £1.20?
(*b*) How many 12 p stamps can I buy for £1.80?
(*c*) How many 15 p stamps can I buy for £3?
(*d*) How many 17 p stamps can I buy for £3.57?

7. Lee paid £18.99 for a pair of jeans. How much change did he get out of a twenty pound note?

8. Mr and Mrs Walsh paid £289 for a washing machine and £160 for a tumble dryer. Find the total amount paid for both items.

9. Sandra bought 4 mugs at £1.75 each. The £1 coin holder shows her money. How much more did she need to pay for the mugs?

10. Dawn, Helen and Jill went out for a meal. The bill came to £23.85. If they each paid the same amount, what was the equal share?

11. Charles, Roger and Hilary bought their mum a purse for £4.99, a bunch of flowers for £1.60 and a card for 28 p. A sheet of wrapping-paper cost 18 p. They each paid the same amount. How much was the equal share?

12. Two brothers saw a second-hand snooker table advertised for £48. Their mum gave them £10 towards buying it. The brothers shared the rest of the cost equally. How much did each brother need to pay?

Exercise 12 Shopping Bills **M**

Copy and complete the following:

			£	p
1. (*a*)	5 lb potatoes	@ 20 p a pound		
	3 lb carrots	@ 28 p a pound		
	2 lb apples	@ 34 p a pound		
		Total		

(*b*) Find the change out of a £5 note.

			£	p
2. (*a*)	3 kg potatoes	@ 44 p a kilogram		
	2 kg carrots	@ 62 p a kilogram		
	1 kg apples	@ 75 p a kilogram		
		Total		

(*b*) Find the change out of a £5 note.

			£	p
3. (*a*)	4 tins of beans	@ 27 p each		
	3 tins of soup	@ 26 p each		
	2 packets of cereal	@ 83 p each		
	2 jars of jam	@ 46 p each		
		Total		

(*b*) Find the change out of a £5 note.

			£	p
4. (*a*)	2 boxes of chocolates	@ £2.40 each		
	3 boxes of jelly babies	@ 39 p each		
	2 packets of liquorice allsorts	@ 72 p each		
		Total		

(*b*) Find the change out of a £10 note.

5.

Quantity	Description of goods	Unit price	£	p
3	books	£2.50		
4	books	£3.95		
2	books	£4.35		
		Total		

6.

Quantity	Description of goods	Unit price	£	p
1	toilet bag	£5.99		
1	bath gift set	£4.95		
2	toilet soap packs	£2.99		
2	tins of talcum powder	£1.95		
		Total		

7.

Quantity	Description of goods	Ref. No.	Unit price	£	p
2	Tall double wardrobe	TW1001	£89.99		
1	Interior fitment	IW100	£39.99		
2	3-drawer chests	XC1003	£32.99		
			Total		

Exercise 13 Discount

For each advertisement calculate the missing sum of money:

1.

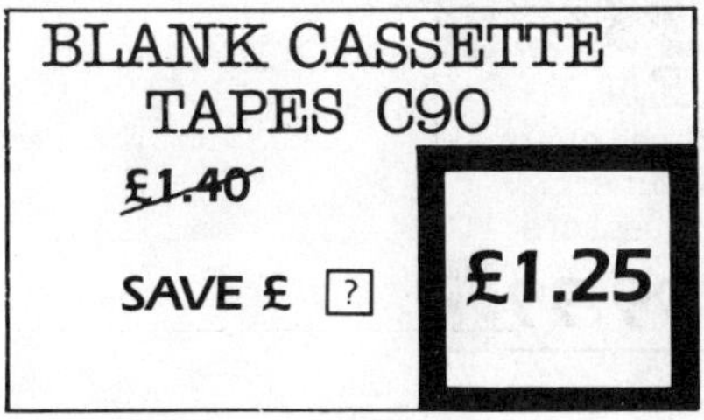

2.

3.

4.

5. Find the sale price.

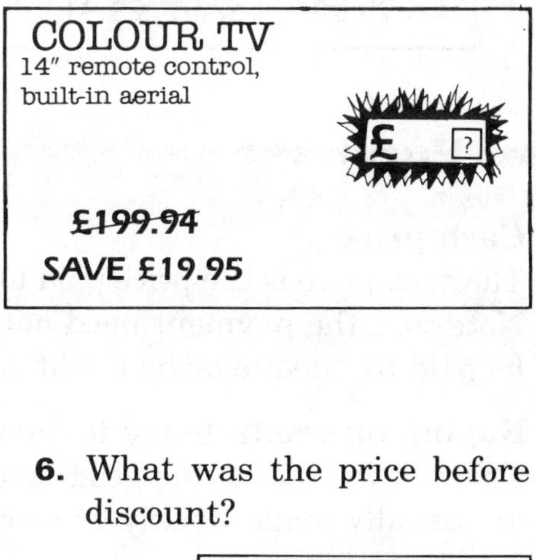

6. What was the price before discount?

7.

8. Find the sale price.

VACUUM CLEANER
including tools
£99.90 **SAVE £9.91** **£ ?**

Hire-Purchase

Cash price
The cash price is the price paid for goods at the time of buying. Note that the payment need not be cash. The cash price may be paid by cheque or by credit card.

Buying on credit (easy terms)
This is a way of buying goods over a period of time (payments are usually made weekly or monthly).

Hire-purchase (HP)
HP is one way of buying on credit. Goods do not belong to you (and are only on hire) until the last payment has been made.

Credit sales
This is another way of buying on credit. The goods belong to you as soon as the first payment is made.

Deposit
The amount that is paid straight away when agreement is made to buy on credit is called the deposit.

Balance (amount left)
The balance is the amount that is left at a particular time.

Instalments
These are the payments that are made at regular intervals when buying on credit.

Interest charges (interest or credit charge)
The extra charge made for buying on credit.

HP price − deposit = balance

Exercise 14 M

A Copy and complete the table:

	HP price	Deposit	Balance
1.	£60	£15	
2.	£145	£49	
3.	£99	£36.50	
4.	£189.99	£67.49	
5.	£249.90	£84.70	

	HP price	Deposit	Balance
6.	£208		£136
7.		£52	£104
8.		£85	£187.50
9.	£149.99		£97.50
10.		£97.24	£225.75

B **1.** A lawn mower costs £109.99. Find the balance payable after a deposit of £29.99 has been paid.

2. A greenhouse costs £179.99 on HP After a deposit of £37.49, find the balance.

3. A food processor costs £74.50 on HP If a deposit of £7.45 is paid, find the balance.

4. A fishing-rod costs £49.99 on HP If a deposit of £12.49 is paid, find the balance.

5. A bicycle costs £129.99. If a deposit of £45.51 is paid, find the balance.

HP price − cash price = credit charge

Exercise 15

A Copy and complete the table:

	HP price	Cash price	Credit charge
1.	£60	£54.50	
2.	£145	£137.50	
3.	£144.99	£129.99	
4.	£84.99	£79.49	
5.	£169.49	£157.99	

	HP price	Cash price	Credit charge
6.		£78	£15
7.	£66.75		£7.25
8.		£154.50	£18.50
9.		£209.70	£19.40
10.	£169.50		£17.99

B 1. If the HP price of an exercise bike is £61.50 while the cash price is £49, find the credit charge.

2. The cash price of a frame tent is £199.99. If the HP price is £258.50, find the credit charge.

3. The HP price of £415 for dining room furniture includes a credit charge of £85.50. Find the cash price.

4. A television costs £259.50 cash. There is an extra credit charge of £75.25 when buying on HP. Find the HP price.

5. A sewing machine costs £144.95 cash. It costs £42.76 extra on HP. Find the HP price.

When money is borrowed from a finance company, the company charges interest (the credit charge).

Loan + interest = HP price

Exercise 16 **M**

Here is a table giving a finance company's interest charges. Copy and complete it.

	Loan	Interest	HP price		Loan	Interest	HP price
e.g. 1	£5	£1.50	£6.50	**9.**	£80	£24	?
e.g. 2	£10	£3	£13	**10.**	?	£27	£117
1.	£20	£6	?	**11.**	£100	£30	?
2.	£25	£7.50	?	**12.**	£150	?	£195
3.	£30	£9	?	**13.**	£200	£60	?
4.	£40	?	£52	**14.**	£250	£75	?
5.	£50	£15	?	**15.**	£300	?	£390
6.	£60	?	£78	**16.**	?	£120	£520
7.	?	£21	£91	**17.**	£500	£150	?
8.	£75	?	£97.50	**18.**	?	£180	£780

Exercise 17

1. A tennis racket can be bought from a mail order catalogue for 20 payments of £1.20. How much does the tennis racket cost?

2. A pair of roller skates can be bought for 20 weekly payments of £1.29. Find the cost of the skates.

3. A keyboard can be bought for 12 monthly payments of £16.65. Find the total cost of the keyboard.

4. A computer costs £228. If it is paid in 12 equal instalments, find each instalment.

5. A lead guitar costs £130. If it is paid in equal instalments over 20 weeks, find each instalment.

6. A TV set is paid for by a deposit of £49 followed by 12 monthly instalments of £21. Find the total cost.

7. A tumble dryer was advertised for a deposit of £39.95 followed by 20 equal payments of £5.50. Find the total cost of the tumble dryer.

8. A sewing machine was paid for by a deposit of £32.95 as well as 12 equal payments of £9.75. Find the total cost of the sewing machine.

5 Units of Measurement

Time

Exercise 1

1. How many days are there in 3 weeks?

2. How many days are there in 4 weeks?

3. How many days are there in 2 weeks and 4 days?

4. There are 24 h in 1 day. How many hours are there in 5 days?

5. How many minutes are there in 3 h?

6. How many minutes are there in 4 h 20 min?

7. How many seconds are there in 2 min?

8. How many seconds are there in 1 min 25 s?

Exercise 2

1. If it is evening, this clock shows the time 18.25 using the 24-hour clock notation and 6.25 p.m. using the 12-hour clock notation.

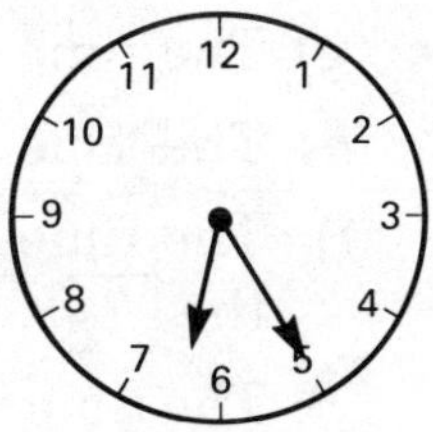

Write the time shown on this second clock if it is showing the time in the evening.

(*a*) Use the 12-hour clock notation.

(*b*) Use the 24-hour clock notation.

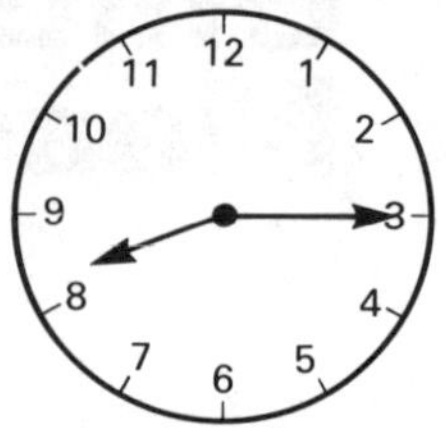

2. Write the time shown on the given clock if it is showing the time in the morning.

(*a*) Use the 12-hour clock notation.

(*b*) Use the 24-hour clock notation.

3. Write these 12-hour clock times using the 24-hour clock:

(*a*) 3.00 p.m. (*b*) 4.15 a.m. (*c*) 7.25 p.m.

4. Write these 24-hour clock times using the 12-hour clock:

(*a*) 18.00 (*b*) 21.30 (*c*) 08.55

5. Copy and complete these sentences:

(*a*) The time is 07.20, that is, [?] past seven in the morning.

(*b*) The time is 15.30, that is, half past [?] in the afternoon.

(*c*) The time is 5.10 p.m., that is, [?] past five in the evening.

(*d*) The time is 6.15 a.m., that is, [?] past six in the [?].

(*e*) The time is 21.40, that is, twenty to [?] in the [?].

(*f*) The time is 14.50, that is, [?] to [?] in the afternoon.

(*g*) The time is 23.35, that is, [?] to twelve in the [?].

(*h*) The time is 8.55 p.m., that is, [?] to nine in the [?].

(*i*) The time is 04.12, that is, [?] minutes past four in the [?].

(*j*) The time is 18.51, that is, [?] minutes to [?] in the [?].

6. (*a*) It is morning and the clock shown is ten minutes fast. What is the correct time?

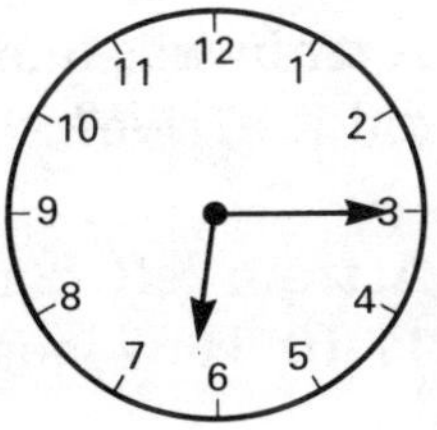

(*b*) A clock showing the time as 14.10 is twenty minutes fast. What is the correct time?

(*c*) A clock showing the time as 07.55 is a quarter of an hour slow. What is the correct time?

(*d*) A clock showing 20.25 is half an hour slow. What is the correct time?

Exercise 3

1. Write how many minutes there are to the next hour if the time is now:

(*a*) 08.40	(*e*) 15.25	(*i*) 19.35
(*b*) 7.45 p.m.	(*f*) 6.20 a.m.	(*j*) 21.43
(*c*) 2.15 a.m.	(*g*) 11.10	(*k*) 18.17
(*d*) 17.55	(*h*) 10.05 p.m.	(*l*) 9.32 p.m.

2. A timer needs to be set for 1 h 50 min. How many minutes is that?

3. A TV programme started at 19.30 and finished at 21.15. How many minutes did it last?

4. A mathematics lesson started at 09.20 and finished at 10.30. How long was the lesson?

5. A bus journey took 20 min. If the bus set off at 16.15, at what time would the journey end?

6. A train journey lasts $1\frac{1}{2}$ hours. If the train set off at 09.10, at what time would it reach the end of the journey?

7. A bus arrived at a bus stop at 11.15 a.m. It was 10 min late. At what time should it have arrived?

8. A train was due at 17.15. It was 20 min late. At what time did it arrive?

9. A train left London at 14.30 and arrived in Swansea at 17.10. How long did the journey take?

10. Eve should have arrived at her friend's house at 19.20. She arrived at 19.34. How many minutes late was she?

11. A film was shown from 14.10 to 15.50. How long did the film last?

12. A battery needs to be charged for 6 h. If I start charging it at 5 o'clock in the evening, at what time will it be ready?

13. A battery needs to be charged for 14 h. If I start charging it at 21.40 at what time will it be ready?

14. Mr Finch needs to be at work by 09.00. His journey takes 45 min. At what time must he set off?

15. Mrs Ince set off for work at 07.15. Her walk to the bus took 10 min and her bus journey took 25 min. She then had a 5 min walk to work. At what time did she arrive at work?

Exercise 4

1. Which is the sixth month of the year?

2. How many days are there in September?

3. If 6 April was a Monday, what day was 11 April of the same year?

4. The first Tuesday in August was 3 August. Write the dates of all the Tuesdays in August of that year.

5. 17 May is a Wednesday. What is the date of the following Wednesday?

6. If today's date is Wednesday 18 July, what date will a week on Friday be?

7. 14 September 1986 can be written as 14.9.86. Write these dates using figures:
(*a*) 13 March 1985 (*b*) 25 August 1988

8. Write these dates using the name of the month:
(*a*) 14.7.76 (*b*) 8.2.89

9. Bob was born on 5 September 1971. How old was he on 5 September 1986?

10. Jackie was thirteen years old on 7.4.84. When was she born?

11. How many days are there from 6 January to 11 January?

12. How many days are there from 24 May to 6 June?

13. The Greaves family went on holiday on 22 July and returned on 5 August. For how many nights were they away?

14. The Johnsons went on holiday on Friday 14 March for 10 nights.
(*a*) On what date did they return?
(*b*) On which day did they return?

Exercise 5 A Date Calculator

The following calculator helps you to find the day of the week on which a certain date falls. It can be used to give the day for any date between 1940 and 2019.

The calculator is made from three discs that are fastened together with a paper fastener. Make the discs out of card or stiff paper.

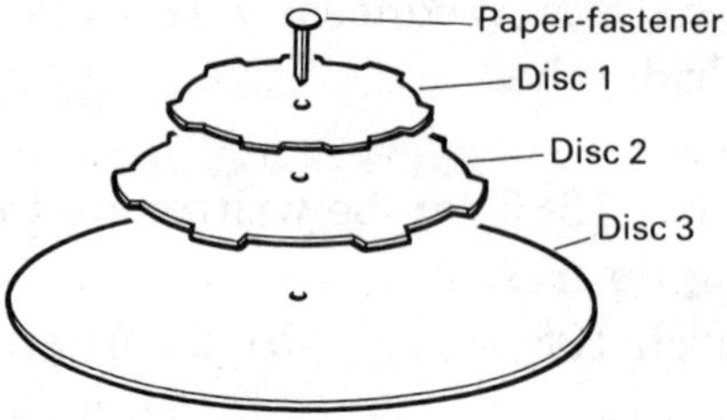

Disc I should have a radius of 30 mm, disc II should have a radius of 40 mm and the radius of disc III should be 75 mm.

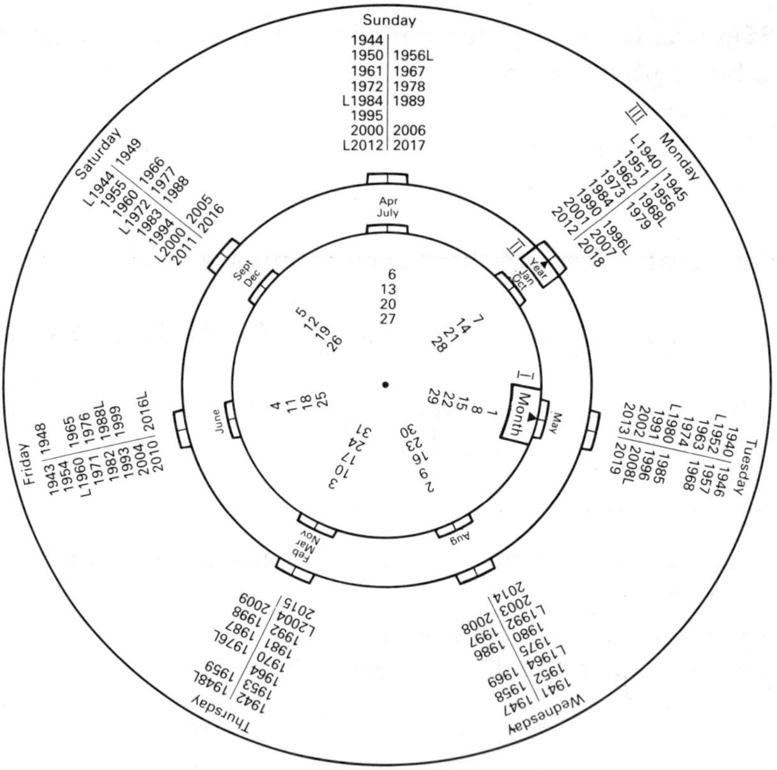

To find what day of the week 18 May 1979 was:

1. Turn disc II until  points to 1979 on disc III.

2. Keep discs II and III still.

3. Turn disc I until 

points to May on disc II.

4. Find 18 on disc I and read the day of the week that is in line with that date but on disc III.

Note: If the date required is in January or February of a leap year then use the year marked with the L. Other months in the leap year are found in the normal way. For example, 1984 was a leap year. It is marked on the calculator twice. L 1984 is only for January or February dates while 1984 should be used for all other months.

Length

Exercise 6

The line below is 10 cm long:

A Estimate the lengths of the following:

1.

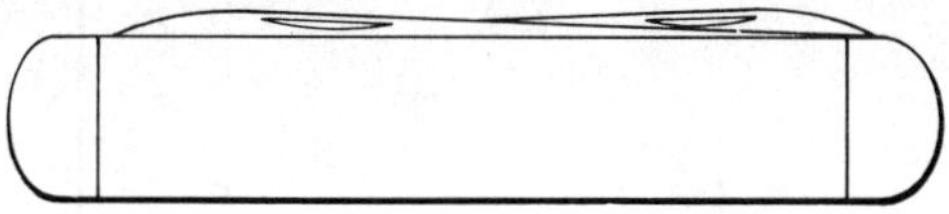

2.

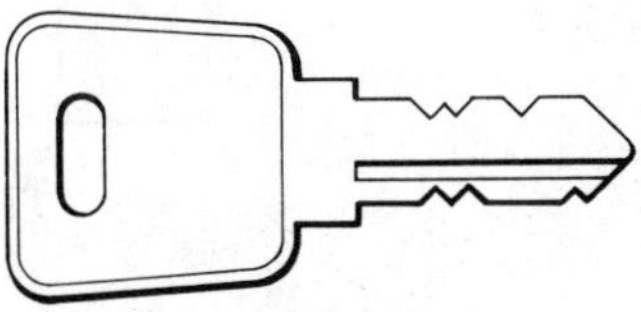

3.

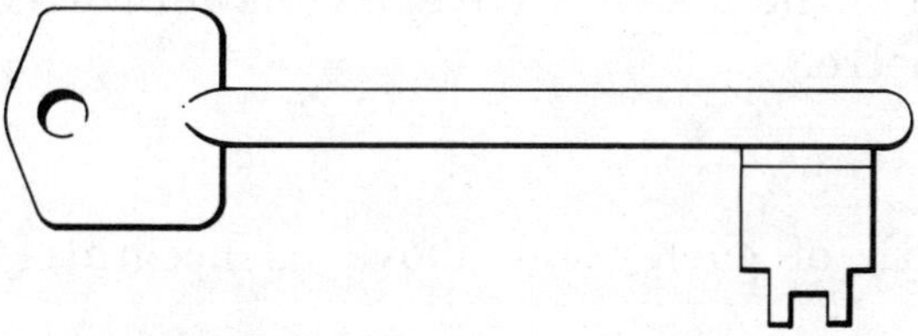

4.

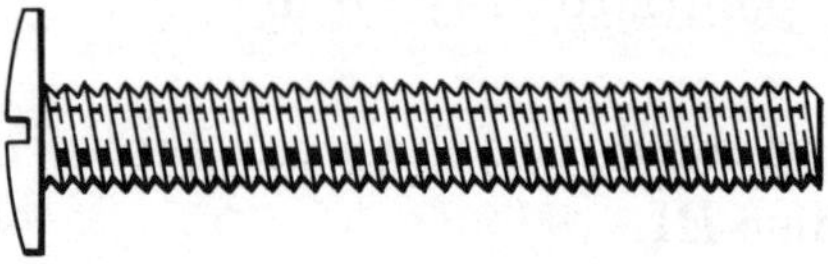

8.

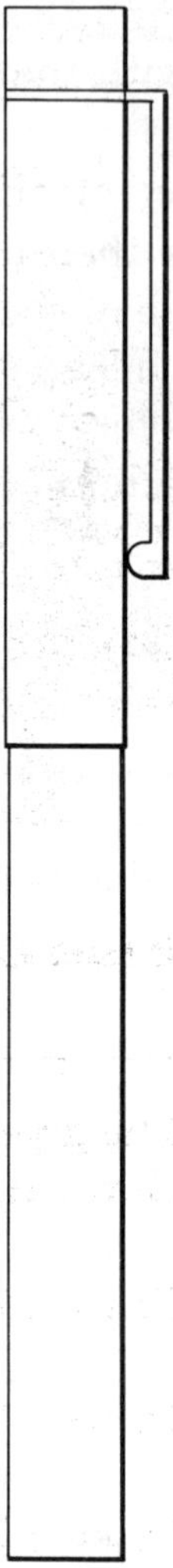

5.

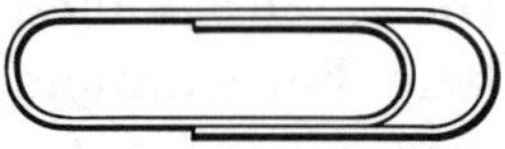

6.

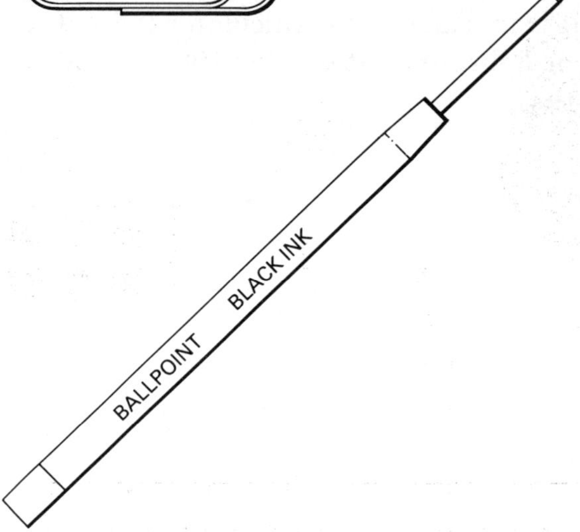

7.

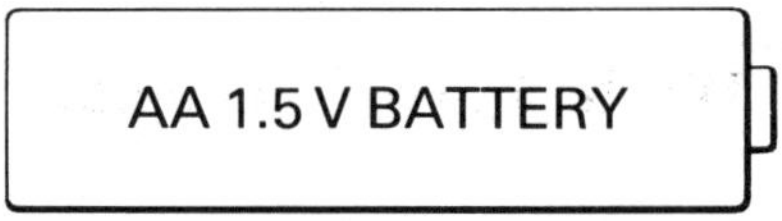

B Now measure each of the above. Give each answer correct to the nearest centimetre.

C Measure the length of each item above as accurately as possible.

Exercise 7

In the picture below, the man is of average height.
Estimate the height of:

1. the house **2.** the telegraph pole **3.** the tree

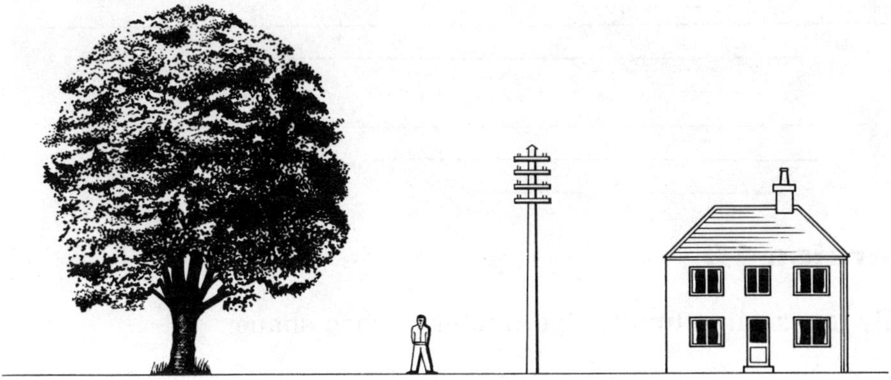

Exercise 8

1. 

The given line has been divided into centimetres.
How long is the line?

2. Draw a line 7 cm long. Divide it into centimetres.

3. Draw a line 5 cm long. Divide it into $\frac{1}{2}$ cm parts.

4. Draw a line 12 cm long. Divide it into 3 cm parts.

5. Draw a line 8 cm long. Divide it into 4 equal parts.

6. Draw a line 12.5 cm long. Divide it into 5 equal parts.

7. Draw a rectangle 100 mm by 40 mm.
(*a*) Divide the rectangle into 5 equal columns.
(*b*) Divide the rectangle into rows that are 5 mm wide.
(*c*) How many small rectangles have you made?

Exercise 9

Measure the given lines. Draw them in order of size, shortest first. Write the length, in centimetres, on each line:

Exercise 10

By measuring, find the perimeter of each shape:

1.

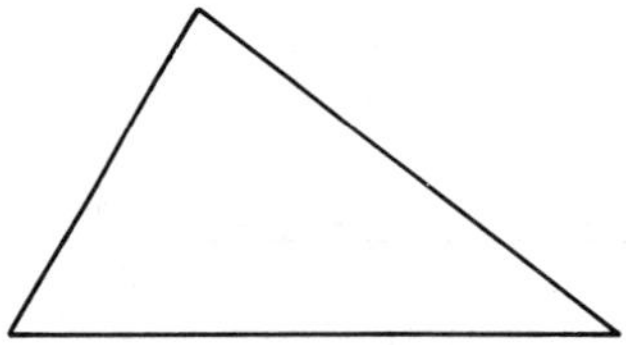

2.

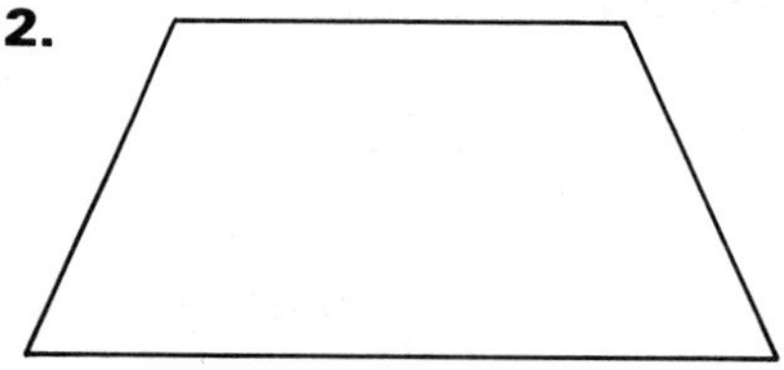

3.

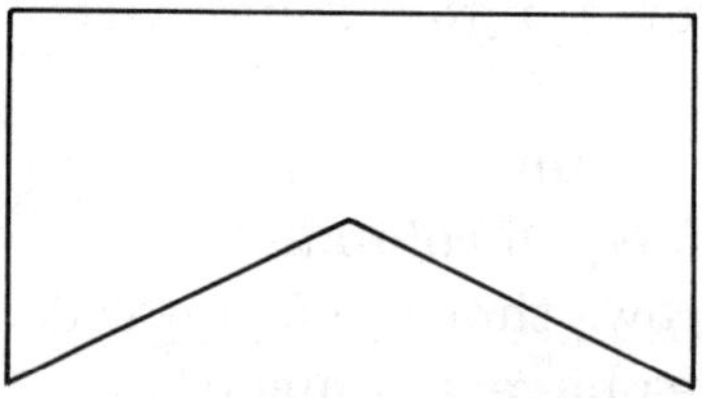

4.

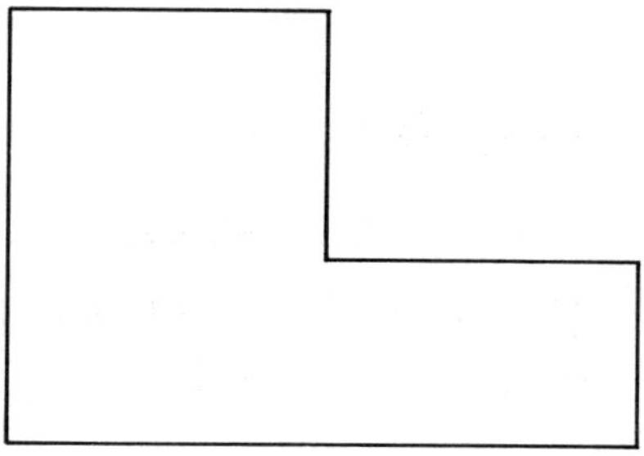

5.

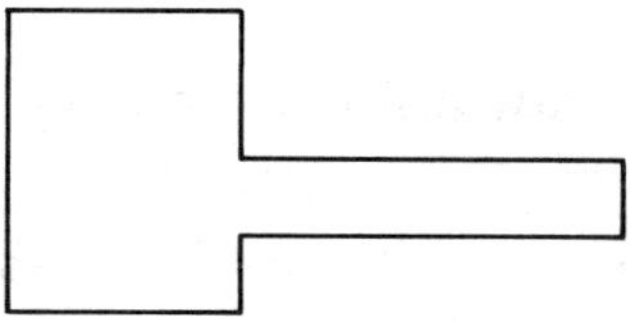

6.

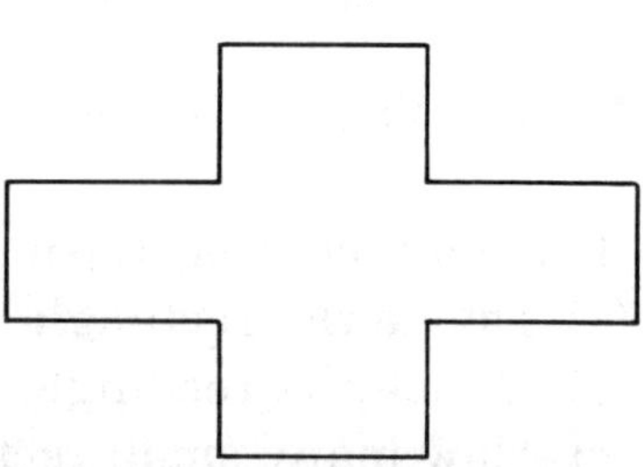

7.

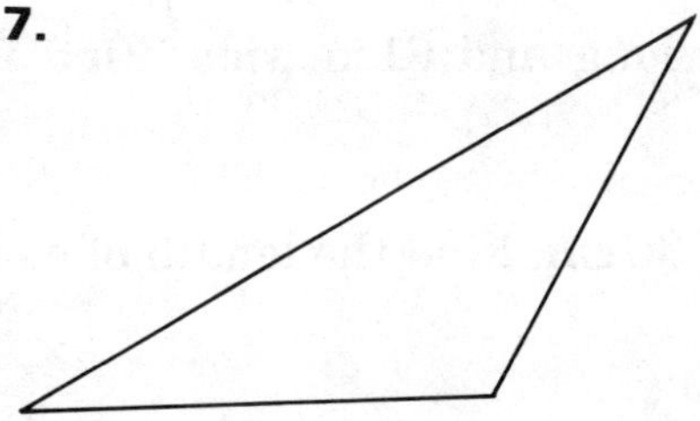

8.

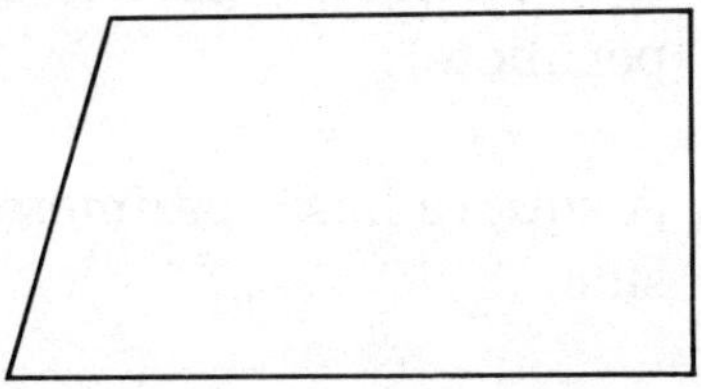

Exercise 11

Find the perimeter of each shape:

1.

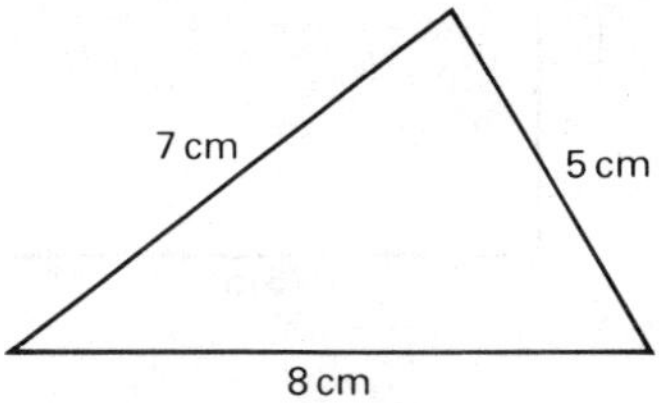

2.

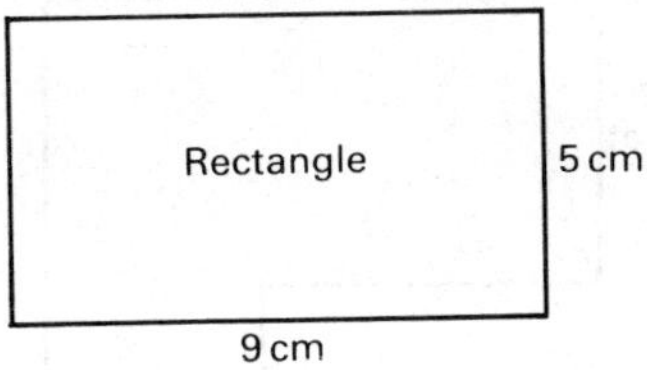

3.

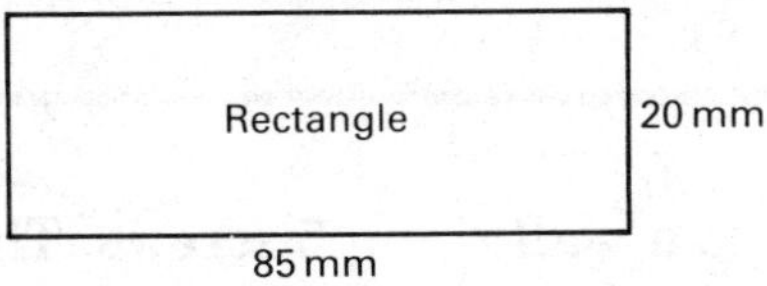

4.

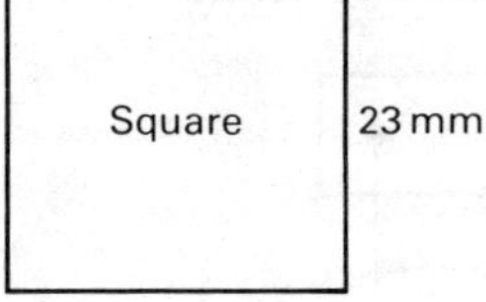

5.

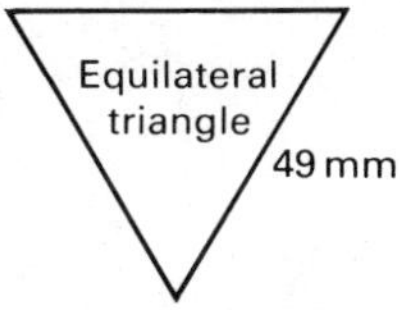

6.

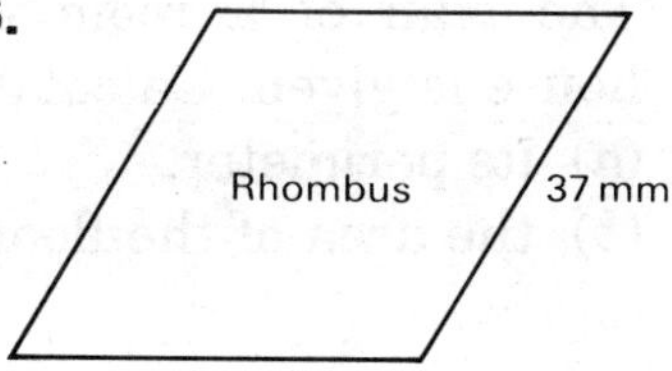

7.

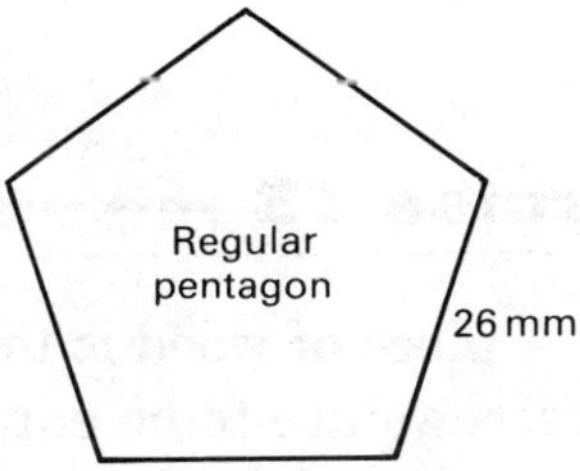

8.

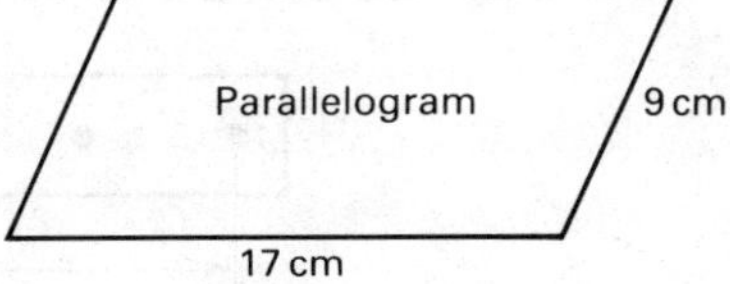

Exercise 12

1. A rectangular field is 82 m long and 61 m wide. Find its perimeter.

2. A square has a perimeter of 36 cm. Find the length of each side.

3. A regular octagon has a perimeter of 24 cm. Find the length of each side.

4. The perimeter of a rectangular room is 18 m. If it is 5 m long, find its width.

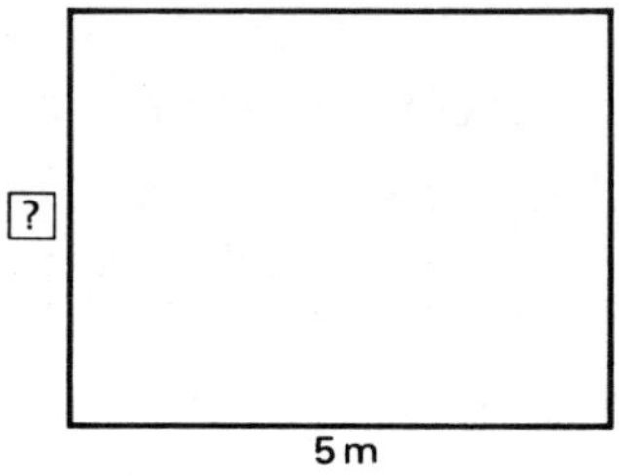

5. The plan of a room in a house is given. Calculate:
(*a*) its perimeter,
(*b*) the area of the floor.

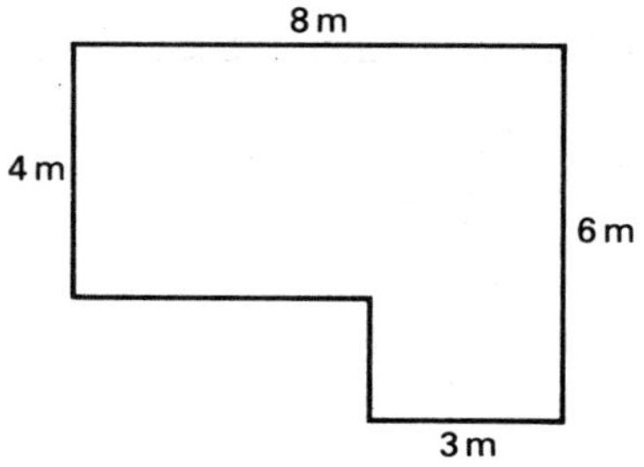

Exercise 13

1. A piece of wood is to be fixed to a wall using 5 screws. The screws are to be equally spaced. The distance between the centres of the two end screws is 40 cm. What is the distance between the centre of one screw and the next?

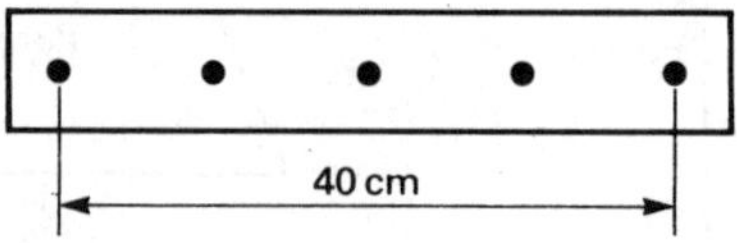

2. A cardigan has 4 buttonholes, all equally spaced. The distance between the top and bottom buttonholes is 36 cm. What is the distance between the buttonholes?

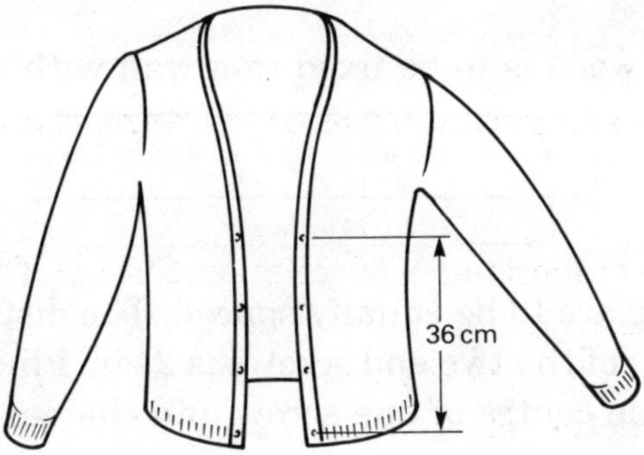

3. A gas fire of length 80 cm is fixed against a straight wall.
The wall is 4 m long. The fire is fixed in the middle of the wall. What length of space is there at each side of the fire?

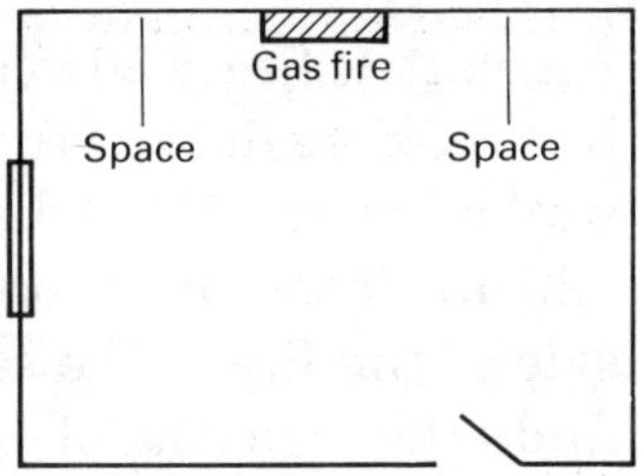

4. (*a*) To knit a sweater for a chest size of 86 cm, a pattern gives a tension of 10 sts to 5 cm (that is, 10 stitches to 5 cm). If the back of the sweater needs 100 stitches, how wide is it?
(*b*) The same pattern states that 12 rows should be knitted in a length of 5 cm. How many rows should be knitted in a length of 30 cm?

Exercise 14

1. A kerbstone is 3 ft long.
(*a*) What length of pavement will 20 kerbstones make?
(*b*) Change the answer in part (*a*) to yards given that 3 feet = 1 yard.

2. How many 3 ft kerbstones are needed for a distance of:
(*a*) 90 ft? (*b*) 120 ft? (*c*) 50 yd?

3. A piece of wood is to be fixed to a wall with 4 screws.

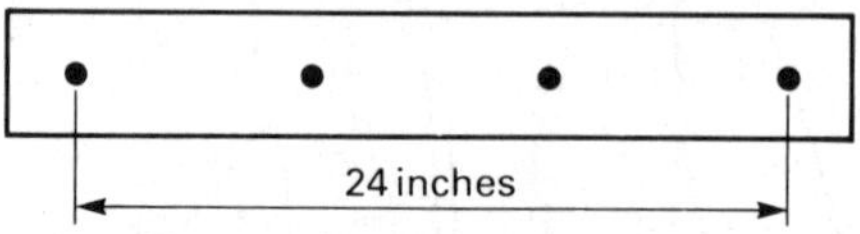

The screws are to be equally spaced. The distance between the centres of the two end screws is 24 in. Find the distance between the centre of one screw and the next.

4. An electric fire is 30 in long. It is fixed against a straight wall of length 10 ft (that is, 120 in). If the space at each side of the fire is the same, find the length of each space.

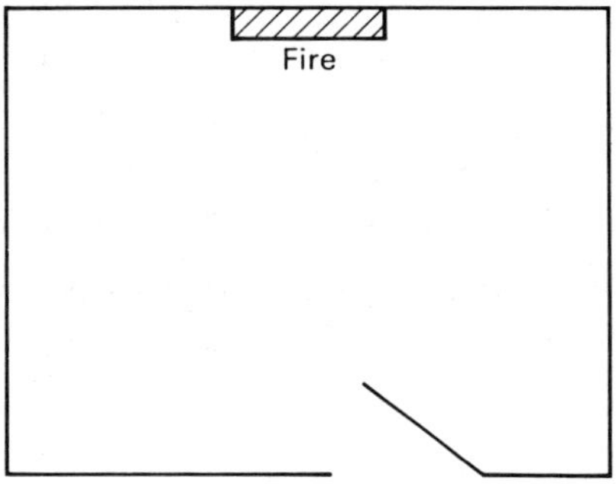

Mass

Exercise 15

Choose the best answer:

1. A bar of soap has a mass of about:
A. 14 mg B. 140 mg C. 140 g D. 1.4 kg

2. A 53-seater coach weighs about:
A. 700 kg B. 70 000 g C. 7000 mg D. 7 t

3. A small tin of beans, including the tin, has a mass of about:
A. 2.9 kg B. 290 g C. 290 mg D. 2900 mg

4. A bicycle weighs about:
A. 5 kg B. 15 kg C. 65 kg D. 450 g

Exercise 16 M

1. A bag of sugar has a mass of 1 kg. If 650 g are used, how many grams are left?

2. A box contains the masses:

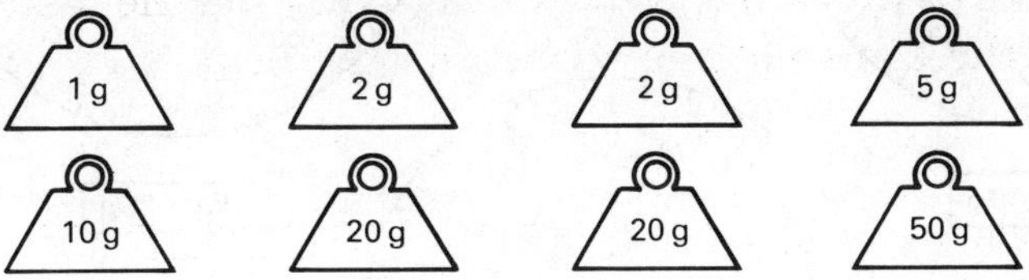

Which masses would you use to weigh the following?
(*a*) 26 g (*b*) 87 g (*c*) 73 g (*d*) 99 g

3. A jar of jam has a mass of 680 g. If the jam weighs 454 g, how heavy is the jar?

4. (*a*) Find the mass of 3 eggs if each egg weighs 56 g.
(*b*) The ingredients for a coffee and walnut cake are:

225 g butter
225 g caster sugar
3 eggs
225 g self-raising flour
30 mℓ coffee essence
75 g chopped walnuts
some coffee buttercream

Assuming each egg weighs 56 g and ignoring the coffee essence and buttercream find the total mass of the ingredients.

5. If a common brick has a mass of 2.6 kg, find the total mass of 6 bricks.

6. Copy the dials, taken from some scales. Choosing the correct dial in each case, show 2 lb 10 oz, 2.5 kg and 180 g. If a mass can be shown on more than one dial choose the more accurate one.

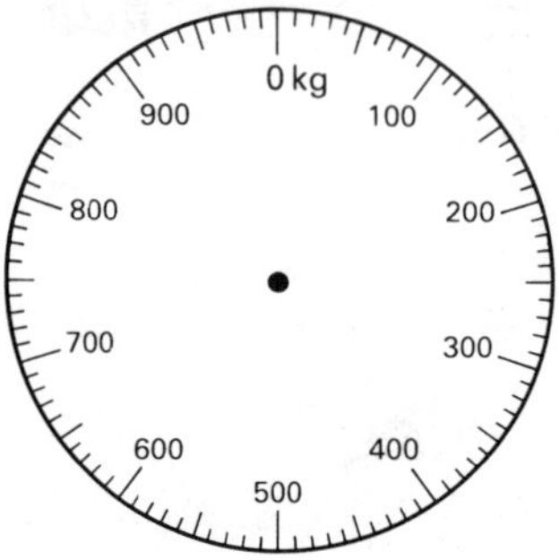

1 kg × 10 g

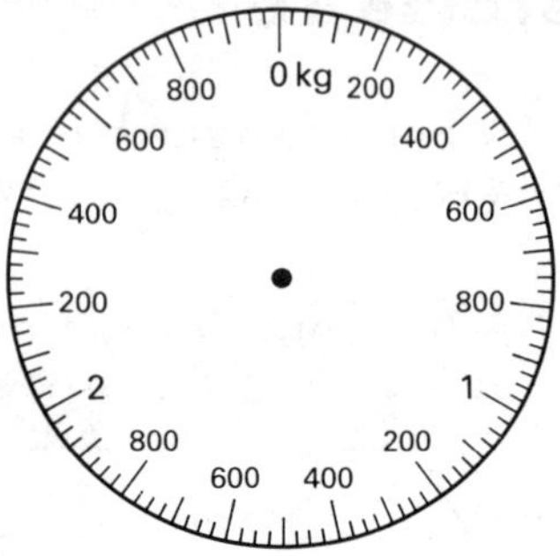

3 kg × 25 g

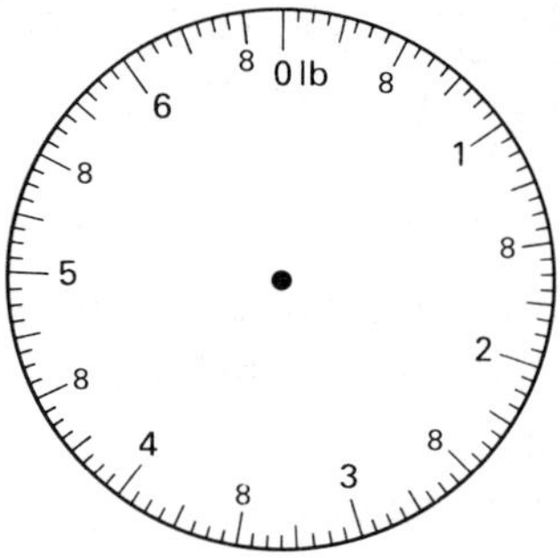

6 lb 8 oz × 1 oz

Capacity

Exercise 17

Choose the best answer:

1. A washing-up bowl holds about:
A. 950 mℓ B. 3 ℓ C. 6 ℓ D. 20 ℓ

2. A wine glass holds about:
A. 15 mℓ B. 150 mℓ C. 75 mℓ D. 250 mℓ

3. A tablespoon holds about:
A. 15 mℓ B. 150 mℓ C. 5 mℓ D. 50 mℓ

4. A bucket holds about:
A. 20 ℓ B. 9 ℓ C. 200 ml D. 4 ℓ

5. A milk saucepan holds about:
A. 1.2 ℓ B. 120 mℓ C. 12 ℓ D. 12 mℓ

Exercise 18

1. There is 450 mℓ of milk in a jug. If 200 mℓ is poured into a glass, how much is left in the jug?

2. A milk carton holds 500 mℓ of milk. 320 mℓ is poured into a glass. How much is left in the carton?

3. A car's petrol tank holds 54 ℓ. After a journey there are 35 ℓ left in the tank. How many litres have been used?

4. A medicine bottle holds 150 mℓ. How many 5 mℓ doses can be taken from the bottle?

5. A tall glass of capacity 300 mℓ is used to fill a container with water. The capacity of the container is 2.4 ℓ (which is 2400 mℓ). Find the number of full glasses needed to fill the container completely.

6. A cow after calving gives the following amounts of milk on five different days: 29 ℓ, 31 ℓ, 29 ℓ, 30 ℓ, 30 ℓ. Find the total number of litres.

7. A full mug holds 240 mℓ. How many millilitres will four of the mugs hold?

8. A bottle of tomato feed contains 200 mℓ. The cap holds 5 mℓ. The instructions state: "Use 10 mℓ per gallon of water."
(*a*) How many millilitres should be used with 6 gal of water?

(*b*) How many capfuls are needed with 5 gal of water?
(*c*) How many capfuls can be filled from one full bottle?
(*d*) How many gallons of water are needed to dilute one full bottle?
(*e*) If a watering-can contains $1\frac{1}{2}$ gal of water, how many capfuls should be added?

Change of Units

	To change		What to do
cm to mm	——	——	× 10
mm to cm	——	——	÷ 10
m to cm	——	——	× 100
cm to m	——	——	÷ 100
m to mm	g to mg	ℓ to mℓ	× 1000
mm to m	mg to g	mℓ to ℓ	÷ 1000
km to m	kg to g	——	× 1000
m to km	g to kg	——	÷ 1000

Exercise 19

1. Change to millimetres:
(*a*) 6 cm (*b*) 9.2 cm (*c*) 30 cm (*d*) 54 cm

2. Change to metres:
(*a*) 200 cm (*b*) 650 cm (*c*) 4000 cm (*d*) 146 cm

3. Change to metres:
(*a*) 7000 mm (*b*) 3500 mm (*c*) 1800 mm (*d*) 6250 mm

4. Change to metres:
   (*a*) 9 km (*b*) 4.7 km (*c*) 2.25 km (*d*) 0.8 km

5. Change to milligrams:
   (*a*) 4 g (*b*) 9.7 g (*c*) 52 g (*d*) 7.3 g

6. Change to kilograms:
   (*a*) 6000 g (*b*) 3200 g (*c*) 8470 g (*d*) 20 000 g

7. Change to grams:
   (*a*) 5 kg (*b*) 30 kg (*c*) 14 kg (*d*) 0.7 kg

8. Change to litres:
   (*a*) 4000 mℓ (*b*) 30 000 mℓ (*c*) 1800 mℓ (*d*) 7320 mℓ

Exercise 20

Copy and complete:

1. 8.7 cm = [?] mm
2. 48 mm = [?] cm
3. 8 m = [?] cm
4. 8 m = [?] mm
5. 9000 g = [?] kg
6. 2000 mg = [?] g
7. 3 kg = [?] g
8. 7 ℓ = [?] mℓ
9. 4.5 ℓ = [?] mℓ
10. 20 ℓ = [?] mℓ
11. 250 mm = [?] cm
12. 7000 mg = [?] g
13. 4.5 m = [?] cm
14. 2.63 m = [?] mm
15. 3.9 m = [?] cm
16. 8.14 kg = [?] g
17. 4290 mg = [?] g
18. 2.46 ℓ = [?] mℓ
19. 8160 mℓ = [?] ℓ
20. 6.51 g = [?] mg

Exercise 21

1. Which is heavier, $1\frac{1}{2}$ kg or 1400 g?

2. A packet of tea weighs 125 g. What is the total mass of 10 packets?

3. A piece of wood is 1.4 m long. A length measuring 65 cm is cut off. What length is left?

4. Out of 2.5 kg of flour, 800 g is used. How much is left?

5. Out of 1.5 ℓ of milk, Sally drinks 300 mℓ. How many litres are left?

6. Wendy used 6 pieces of wood, each 75 cm long. What was the total length used, in metres?

7. Callum used 26 pieces of wood, each 65 cm long. What was the total length used, in metres?

8. Owen cycled 4.7 km to work. At lunchtime he cycled to a cafe, then returned to work after lunch, a distance of 600 m each way. In cycling home he made a detour of 850 m. How far, in kilometres, did Owen cycle that day?

9. (*a*) How many 150 mℓ glasses of fruit juice can be poured from 1 ℓ?
(*b*) How many millilitres are left?

10. A tin of talcum powder has a mass of 160 g. Find, in kilograms, the total mass of:
(*a*) 8 tins of talc, (*b*) 15 tins of talc.

6 Use of Tables and Charts

Exercise 1

Great Rivers in North America

River	Length (km)
Missouri (USA)	4342
Mackenzie (Canada)	4022
Mississippi (USA)	3760
St Lawrence (Canada – USA)	3760
Yukon (Canada – USA)	3166
Rio Grande (USA – Mexico)	2880
Arkansas (USA)	2320
Colorado (USA – Mexico)	2240
Ohio (USA)	2090
Red (USA)	2080
Saskatchewan (Canada)	1928
Columbia (USA)	1920
Peace (Canada)	1686
Snake (USA)	1661

1. How long are the following rivers?
(*a*) the St Lawrence,
(*b*) the Rio Grande,
(*c*) the Arkansas.

2. How many kilometres longer than the Colorado is the Missouri?

3. By how many kilometres is the Mackenzie longer than the Snake?

4. (a) Which is the fifth longest river given in the table?
(b) How long is it?

Exercise 2

Average Mass and Height of Young People

Girls			Boys	
Mass (kg)	Height (cm)	Age (years)	Mass (kg)	Height (cm)
35.7	145	11	35.2	144
39.7	152	12	38.3	147
45	157	13	42.2	155
49.2	160	14	48.8	163
51.5	161	15	54.5	168
53.1	162	16	58.8	172
54	163	17	61.8	173
54.4	163	18	63.1	175

1. What is the average height of a boy aged 18 years?

2. What is the average mass of a girl aged 16 years?

3. What age group of girls has an average height of 152 cm?

4. Give the average height of 11-year-old boys in metres.

5. Find the difference in the masses of 17-year-old boys and girls.

6. How much heavier is an 18-year-old girl when compared with a 12-year-old girl?

7. How much heavier is an 18-year-old boy when compared with a 12-year-old boy?

Exercise 3

Tides for October

Date	Morning		Evening	
	Time	Height (m)	Time	Height (m)
1 Thur	00.40	9.3	12.58	9.2
2 Fri	01.08	9.1	13.27	9.0
3 Sat	01.38	8.7	13.57	8.6
4 Sun	02.09	8.3	14.32	8.2
5 Mon	02.49	7.9	15.17	7.8
6 Tue	03.41	7.4	16.16	7.5
7 Wed	04.53	7.1	17.32	7.3
8 Thur	06.19	7.1	18.55	7.5
9 Fri	07.42	7.5	20.05	8.0
10 Sat	08.43	8.1	20.58	8.6
11 Sun	09.28	8.8	21.43	9.3
12 Mon	10.09	9.3	22.24	9.8
13 Tues	10.49	9.7	23.05	10.1
14 Wed	11.31	10.0	23.49	10.3

Date	Morning		Evening	
	Time	Height (m)	Time	Height (m)
15 Thur	–	–	12.14	10.1
16 Fri	00.32	10.2	12.56	10.0
17 Sat	01.16	10.0	13.39	9.6
18 Sun	02.02	9.5	14.26	9.2
19 Mon	02.55	8.9	15.20	8.6
20 Tue	03.57	8.2	16.27	8.1
21 Wed	05.15	7.8	17.49	7.8
22 Thur	06.44	7.7	19.17	7.9
23 Fri	08.03	8.1	20.29	8.4
24 Sat	09.01	8.5	21.21	8.8
25 Sun	09.45	8.9	22.01	9.1
26 Mon	10.22	9.2	22.36	9.3
27 Tue	10.56	9.3	23.09	9.3
28 Wed	11.27	9.4	23.41	9.3
29 Thur	11.59	9.3	–	–
30 Fri	00.12	9.2	12.28	9.2
31 Sat	00.41	9.0	12.57	9.1

1. (*a*) Find the height of the morning tide on 22 October.
(*b*) At what time was that tide?

2. (*a*) Find the height of the evening tide on 11 October.
(*b*) What was the time of that tide?

3. On which date was there no morning tide?

4. On which date was there no evening tide?

5. Give the date and time of the highest tide.

6. (*a*) Give the times of both tides on 28 October.
(*b*) How many hours and minutes are there between those two tides?

7. (*a*) Give the times of both tides on 7 October.
(*b*) How many hours and minutes are there between those two tides?

8. Give the difference in the heights of the morning tides on 5 October and 18 October.

9. Give the difference in the heights of the evening tides on 6 October and 26 October.

10. Ignoring the dates 15 October and 29 October, on how many days was the evening tide higher than the morning tide?

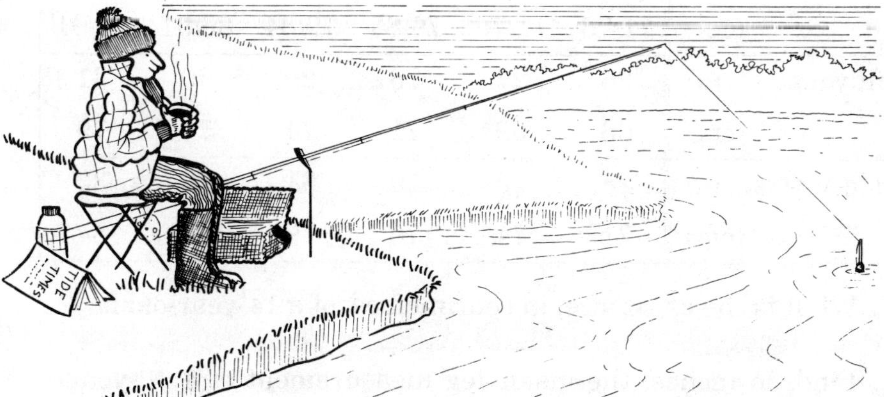

Exercise 4

The following table shows the average sizes of boys from the age of 4 years to 16 years:

Boyswear

Approx. age (years)		4	5	6	7	8	9	10
Collar size	(in)	11	11	$11\frac{1}{2}$	$11\frac{1}{2}$	12	12	13
	(cm)	28	28	29	29	30/31	30/31	33
To fit chest	(in)	22/23	22/23	24/25	24/25	26/27	26/27	28/30
	(cm)	56/58	56/58	61/63	61/63	66/69	66/69	71/76
To fit waist	(in)	22	$22\frac{1}{2}$	23	$23\frac{1}{2}$	24	$24\frac{1}{2}$	25
	(cm)	56	57	58	60	61	62	63
Inside leg trs	(in)	18	19	22	23	25	26	27
	(cm)	46	48	56	58	63	66	69

Approx. age (years)		11	12	13	14	15	16
Collar size	(in)	13	13	$13\frac{1}{2}$	$13\frac{1}{2}$	$14\frac{1}{2}$	$14\frac{1}{2}$
	(cm)	33	33	34/35	34/35	37	37
To fit chest	(in)	28/30	28/30	31/33	31/33	34/36	34/36
	(cm)	71/76	71/76	79/84	79/84	86/91	86/91
To fit waist	(in)	26	27	28	29	30	31
	(cm)	66	69	71	74	76	79
Inside leg trs	(in)	28	29	30	31	32	33
	(cm)	71	74	76	79	81	84

1. What is the chest size, in centimetres, of a 14-year-old boy?

2. Find, in inches, the inside-leg measurement of a 10-year-old boy.

3. Find the collar size of 15-year-old boys. Give your answer in:
(*a*) inches, (*b*) centimetres.

4. What is the waist size, in centimetres, of 13-year-old boys?

5. Which age group has a waist size of 60 cm?

6. Which age group has an inside leg measurement of 84 cm?

7. Which age groups have a chest size of 71/76 cm?

8. Find the difference, in centimetres, of the waist sizes of 15-year-old and 5-year-old boys.

Exercise 5

Monthly Savings Accumulation Table

Monthly savings	Value of investment, assuming the continuance of the current rate of interest, at the end of a period of:		
	1 year	3 years	5 years
£	£	£	£
5	63.04	207.59	380.80
10	126.09	415.19	761.60
20	252.17	830.38	1523.20
50	630.44	2075.95	3808.01
100	1260.87	4151.90	7616.02
200	2521.75	6303.80	15 232.05
500	6304.37	20 759.49	38 080.12

1. If you save £50 a month, what is the value of your investment after 3 years?

2. If you save £200 a month, what is the value of your investment after 5 years?

3. If you save £5 a month, what is the value of your investment after 1 year?

4. How much needs to be invested for 5 years to reach a value of £1523.20?

5. If you save £50 a month for 1 year:
   (*a*) Work out how much you would save altogether.
   (*b*) From the table, find the value of your investment.
   (*c*) How much interest would you make?

Exercise 6

Distances are given to the nearest mile.

Cardiff	Dover	Edinburgh	Liverpool	Newcastle	York	London
233						
395	493					
202	297	225				
320	367	109	173			
245	293	199	101	52		
153	78	405	211	252	209	

Answer the following questions using the distance chart above.

1. How far is it from Cardiff to London?

2. How far is it from Edinburgh to York?

3. How far is it from Newcastle to Dover?

4. Mr Kane travels from Liverpool to London, then travels to Dover. How far does he travel altogether?

5. How much further is it from Newcastle to Cardiff than from Dover to York?

6. Mrs O'Connor travels from Edinburgh to Newcastle, then to Liverpool. From Liverpool she returns to Edinburgh. How many miles did she travel altogether?

7. Find the distance from York to Cardiff. Change that distance to kilometres by multiplying by 1.6.

8. Find the distance from Edinburgh to London, in kilometres. (Multiply the number of miles by 1.6.)

Exercise 7 M

Two different types of distance chart are given below. In the first, the distances are given in kilometres. Copy the second chart and fill in the blank spaces by finding the distances, using the first chart.

Distances are given in kilometres.

	London	Southampton	Sheffield	Nottingham	Inverness	Carlisle
Aberystwyth	340	317	269	257	793	377
Carlisle	494	539	259	301	417	—
Inverness	908	948	655	695	—	—
Nottingham	206	275	72	—	—	—
Sheffield	269	327	—	—	—	—
Southampton	126	—	—	—	—	—

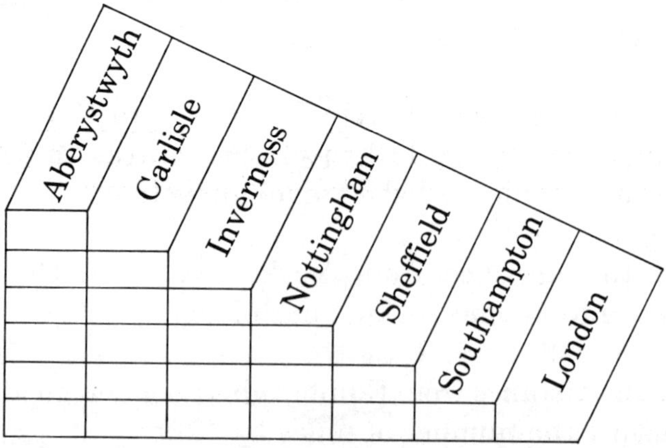

Exercise 8 Using a Timetable

Here is a timetable. It shows the times of some trains between Peterborough and Birmingham.

Peterborough → Birmingham New Street

		◇ ☕			◇ ☕	◇
Peterborough ....d	08 15	09 34	. .	16 05	18 29	20 35
Melton Mowbray ....d	09 04	. .	. .	16 55	. .	21 24
Leicester ....a	09 25	10 39	. . . .	17 16	19 35	21 43
d	09 26	10 44	13 45	17 27	19 38	21 46
Nuneaton ....d	10 00	11 15	14 16	18 00	20 08	22 18
Birmingham New St. ...a	10 36	11 50	14 53	18 38	20 44	22 51

Birmingham New Street → Peterborough

	◇ ☕	◇ ☕			◇ ☕	
Birmingham New St. ...d	08 10	10 20	12 21	13 15	16 10	20 20
Nuneaton ....d	08 45	10 56	12 53	13 51	16 49	20 54
Leicester ....a	09 17	11 23	13 24	14 22	17 16	21 24
d	09 19	11 25	13 27	. .	17 20	21 27
Melton Mowbray ....d	. .	11 48	13 51	. .	17 43	21 47
Peterborough ....a	10 23	12 40	14 42	. .	18 34	22 38

a Arrival time
d Departure time
◇ Seats may be reserved on trains to which this symbol applies in timetable
☕ Buffet service of drinks, hot and cold snacks for whole or part of journey

Use the timetable on page 81.

1. Find the time that the 16.05 train from Peterborough arrives at Birmingham.

2. Find the time that the 21.24 train from Melton Mowbray arrives at Birmingham.

3. Find the time that the 10.20 train from Birmingham arrives at Leicester.

4. Mr T. Rains arrived at Peterborough at 12.40. At what time did he leave Nuneaton?

5. A train, travelling from Melton Mowbray, arrives at Nuneaton at 6 o'clock in the evening. What time did it set off from Melton Mowbray?

6. Does the 08.15 train from Peterborough arrive at Leicester just before or just after half past nine in the morning?

7. A train arrives at Leicester just after quarter past five in the evening. At what time did it leave Birmingham?

8. What is the latest train from Birmingham that arrives at Melton Mowbray shortly before 2 o'clock in the afternoon?

9. Which trains from Peterborough do not stop at Melton Mowbray?

10. The 19.38 train from Leicester arrived at Birmingham New Street 5 min late. What was its time of arrival?

11. The 08.45 train from Nuneaton arrived at Peterborough 2 minutes early. What was its time of arrival?

12. On which trains travelling from Peterborough to Birmingham is it possible to get a drink?

13. What does the symbol ◇ mean?

14. How long does the 10.44 train from Leicester take to get to Birmingham?

15. How long does the 13.27 train from Leicester take to get to Peterborough if it arrives in Peterborough 10 min late?

Exercise 9 A Conversion Graph

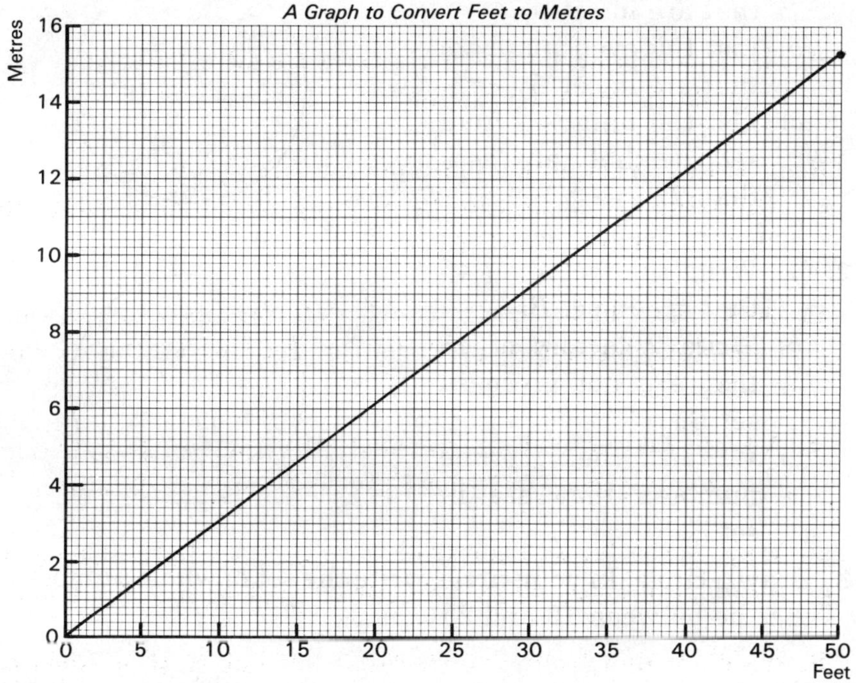

Use the graph to help with these questions:

A Change to metres (to the nearest tenth of a metre):

1. 23 ft	**3.** 13 ft	**5.** 33 ft	**7.** 41 ft	**9.** 49 ft
2. 36 ft	**4.** 46 ft	**6.** 10 ft	**8.** 18 ft	**10.** 28 ft

B Change to feet (to the nearest 0.5 ft):

1. 11 m **2.** 14 m **3.** 9 m **4.** 7 m **5.** 12.5 m

Revision Exercises I to VI

Revision Exercise I

1. (*a*) List the whole numbers that are in set *A*.
(*b*) List the even numbers that are in set *A*.
(*c*) List the numbers from set *A* that are less than 7.

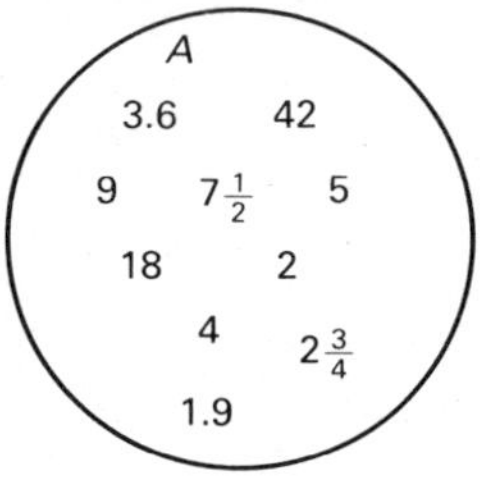

2. (*a*) Write the factors of 45 that are in set *B*.
(*b*) Write 2 factors of 45 that are not given in set *B*.
(*c*) Write the prime numbers that are in set *B*.

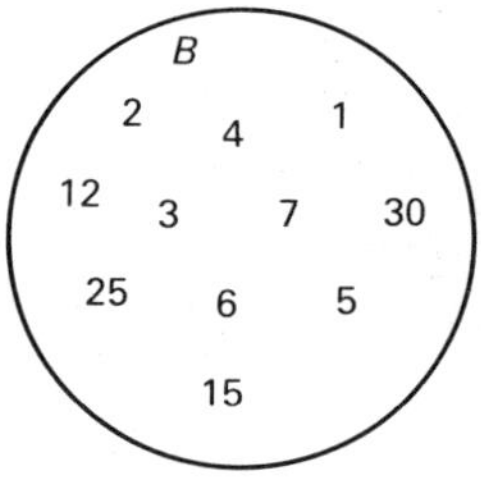

3. Draw a circle. In your circle, write the set of whole numbers that lie between 19 and 25.

4. Draw a circle. In your circle, write the multiples of 7 that are less than 40.

5. Write the following numbers in order of size, smallest to largest, then find the next two numbers in the sequence:

16, 30, 9, 2, 23

6. (*a*) Write the square number that lies between 60 and 70.
(*b*) Draw the dot pattern that shows the triangular number 21.

Revision Exercise II

1. Is 384 closer to 300 or to 400?

2. Is 7490 closer to 7000 or to 8000?

3. Round 48 to the nearest ten.

4. (*a*) Round 653 to the nearest ten.
(*b*) Round 653 to the nearest hundred.

5. (*a*) Round 1279 to the nearest ten.
(*b*) Round 1279 to the nearest hundred.
(*c*) Round 1279 to the nearest thousand.

6. Give the length of the following line correct to the nearest centimetre:

7. (*a*) Write 5.6 cm correct to the nearest centimetre.
(*b*) Write 8.2 km correct to the nearest kilometre.
(*c*) Write 1.81 km correct to the nearest kilometre.
(*d*) Write 4.3 correct to the nearest whole number.
(*e*) Write 67.499 correct to the nearest whole number.

8. (*a*) Round £2650 correct to the nearest thousand pounds.
(*b*) Round 583 m correct to the nearest hundred metres.

9. A house cost £32 420. How much was that, correct to the nearest thousand pounds?

10. A bar of chocolate had a mass of 460 g. Write its mass to the nearest hundred grams.

11. Write correct to 1 significant figure:
(*a*) 87 (*b*) 42.5 (*c*) 19.25 (*d*) 683

12. Estimate answers to the following.
Work with 1 significant figure.
(*a*) 37×64 (*b*) 41×380 (*c*) 766×321

13. A tray holds 12 eggs. If there are 63 full trays, are there more than or less than 600 eggs?

14. A tin has a mass of 624 g. Estimate the mass of 27 of these tins, giving your answer in kilograms.

15. Which calculation has the greater value,
73×42 or 70×40?

Revision Exercise III

1. Mr G. O. North drove 249 miles in a day. Write the distance using words.

2. There are one thousand seven hundred and sixty yards in a mile. Write that number in figures.

3. Write these numbers in order of size, putting the largest first:

769, 679, 796, 976, 970, 709, 670

4. Copy the following and fill in the missing operation, $+$, $-$, $\times$ or $\div$, to make each answer correct:
(*a*) $24 \boxed{?} 4 = 6$ (*b*) $8 \boxed{?} 3 = 32 \boxed{?} 8$

5. Find the value of $712 + 899$.

6. Work out $8024 - 3746$.

7. Find the remainder when:
(*a*) 6324 is divided by 10,
(*b*) 2063 is divided by 5,
(*c*) 4821 is divided by 9.

8. Carry out these calculations:

(*a*)
$$\begin{array}{r} 6183 \\ +\ 1798 \\ \hline \end{array}$$

(*b*)
$$\begin{array}{r} 7238 \\ -\ 2475 \\ \hline \end{array}$$

(*c*)
$$\begin{array}{r} 546 \\ \times\ \ \ 3 \\ \hline \end{array}$$

(*d*) $5\overline{)3205}$

9. Set out in columns then work out:

(*a*) $6452 + 2968$ (*d*) $2912 \div 7$
(*b*) $7605 - 2697$ (*e*) 61×18
(*c*) 5×1307 (*f*) 74×32

10. 7 people are each given 24 sheets of paper. How many sheets is that altogether?

11. If $82 + 37 = 119$, find the value of:

(*a*) $37 + 82$ (*b*) $119 - 37$ (*c*) $119 - 82$

12. If $31 \times 8 = 248$, find the value of:

(*a*) 8×31 (*b*) $248 \div 8$

13. Find the value of:

(*a*) $19 + 3 \times 4$ (*b*) $18 - 6 \div 2$

14. Find the value of:

(*a*) $15 + 6 \div 3$ (*b*) $(15 + 6) \div 3$

15. Find the value of:

(*a*) $(4 + 6) \times (7 - 5)$ (*b*) $(9 - 2) - (6 - 2)$

16. Draw dot patterns to show:

(*a*) the square number 36,
(*b*) the triangular number 21,
(*c*) the rectangular number 10.

17. Here is a list of temperatures:

$14\,°C,\ ^-7\,°C,\ ^-2\,°C,\ 6\,°C,\ 1\,°C,\ ^-10\,°C,\ ^-6\,°C$

(*a*) Which is the third lowest temperature in the list?

(*b*) Which temperatures in the list lie between $^-8\,°C$ and $^-4\,°C$?

(*c*) If the temperatures in the list were written in order from lowest to highest, which of the temperatures would be in the middle?

18. (*a*) What temperature, in degrees Celsius, is shown on the thermometer?

(*b*) If the temperature shown rises by 8 °C, what would the new temperature be?

(*c*) If the temperature shown went down by 10 °C, what would the new temperature be?

(*d*) Through how many degrees Celsius would the temperature shown need to rise to reach 17 °C?

(*e*) Through how many degrees Celsius would the temperature shown need to drop to reach $^-4\,°C$?

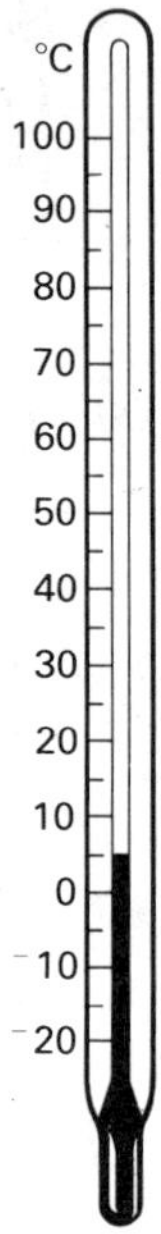

19. At 2 o'clock in the morning the temperature was $^-6\,°C$. By 11 o'clock in the morning it had risen by 15 °C. What was the temperature at 11 o'clock?

20. If the temperature fell from 1 °C to $^-14\,°C$, what was the change in temperature?

Revision Exercise IV

1. Mrs Collins earns £3 an hour. How much does she earn in 12 h?

2. Mandy's hourly rate was £2.30. How much would she earn in 34 h?

3. Using the ready reckoner on page 33 find the weekly earnings of someone who works 37 h at £3.50 an hour.

4. Mr Isles earned £121 per week gross. If deductions came to £22.50, find his net pay.

5. £31.78 was deducted from Mrs McCreary's gross earnings of £130.25. Calculate her take-home pay.

6. Miss Evans earned £152.30 gross. Her take-home pay was £112.80. Calculate the deductions.

7. Wyn earns £2.40 an hour. During a certain week he works 32 h and his deductions total £15.90. Calculate his net pay.

8. Find the change from 50 p when you buy a stamp at 18 p.

9. Find the change from £1 when you spend 82 p.

10. If you spend £3.54, what change do you get out of £5?

11. Find the change out of £10 when you spend £7.91.

12. Find the total value of the listed coins:

(*a*) One 50 p coin
Two 20 p coins
Three 10 p coins
One 5 p coin

(*b*) Three 50 p coins
One 20 p coin
Four 10 p coins
Three 5 p coins
Two 2 p coins

13. Find the total value of two £1 coins, one 50 p coin, three 20 p coins, three 10 p coins, five 5 p coins and two 2 p coins.

14. Using only the coins £1, 50 p, 20 p, 10 p, 5 p, 2 p and 1 p, show how you would pay £2.87 using the smallest possible number of coins.

15. Two sisters bought a second-hand television set for £65. Their mum gave them £15 towards buying it. The sisters shared the remaining cost equally. How much did each sister pay?

16. (*a*) Copy and complete the bill:

	£	p
2 jars of honey @ 86 p each		
3 jars of strawberry jam @ 46 p each		
1 jar of marmalade @ 68 p each		
Total		

(*b*) Find the change out of a £10 note.

17. Copy and complete the bill:

Quantity	Description of Goods	Unit price	£	p
1	pair of jeans	£17.99		
2	sweatshirts	£13.99		
3	pairs of socks	£1.85		
		Total		

18. Pop-up toaster:
Normal Price £21.99
Sale Price £18.50
Save £ ?

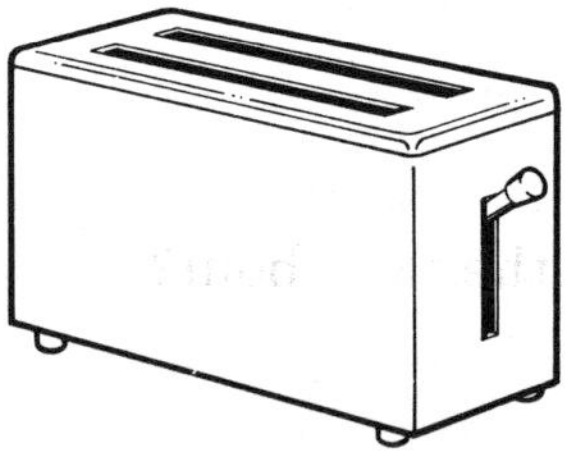

How much do you save in buying the toaster at the sale price?

19. A garden shed costs £345.49. Find the balance payable after a deposit of £113.05.

20. If the HP price of an article is £75 and the cash price £62.50, find the credit charge.

21. A finance company charges £36 interest on a loan of £120. Find the HP price.

22. A cassette recorder can be bought from a mail order catalogue for 20 weekly payments of £1.40. What is the total cost of the cassette recorder?

Revision Exercise V

1. Write the time shown on the given clock if it is showing the time in the morning.
(*a*) Use the 12-hour clock notation.
(*b*) Use the 24-hour clock notation.

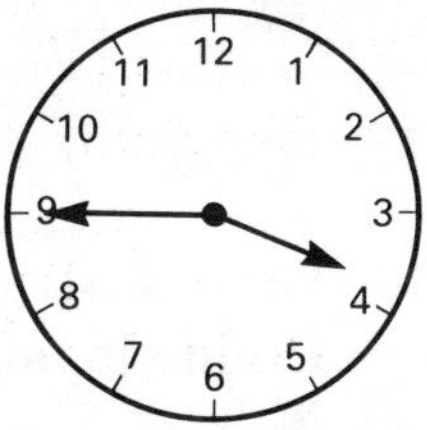

2. Write the time shown on the given clock if it is showing the time in the evening.
(*a*) Use the 12-hour clock notation.
(*b*) Use the 24-hour clock notation.

3. Write these 12-hour clock times using the 24-hour clock:
(*a*) 2.30 a.m. (*b*) 6.15 p.m. (*c*) 1.35 p.m.

4. If the time is now 13.35, how many minutes are there to the next hour?

5. A TV programme started at 20.30 and finished at 21.45. How long was the programme?

6. How many days are there in April?

7. If 12 June was a Thursday, what day was 25 June of the same year?

8. 5 January was the first Friday in January. Write the dates of all the Fridays in January.

9. Ashok was 14 years old on 21 August 1986. What was his date of birth?

10. Measure the length of the key.

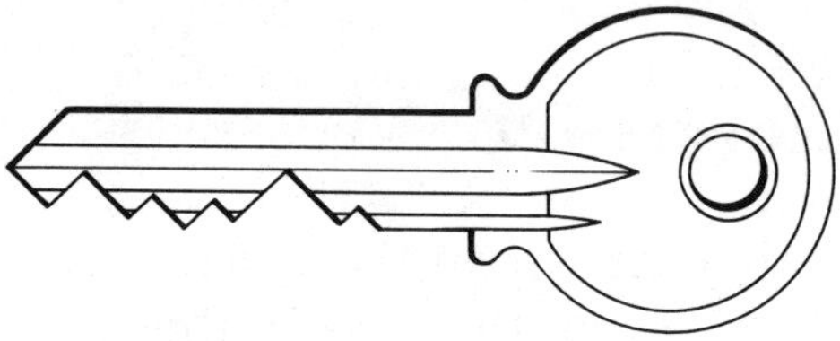

11. Draw a straight line 9 cm in length.
Divide it into 6 equal parts.

12. By measuring, find the perimeter of the quadrilateral.

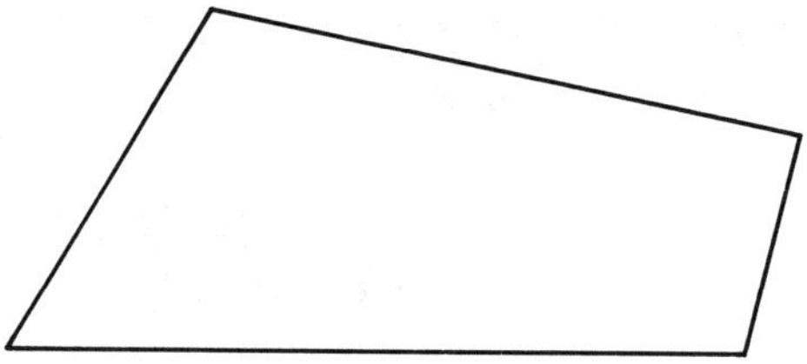

13. Find the perimeter of the triangle.

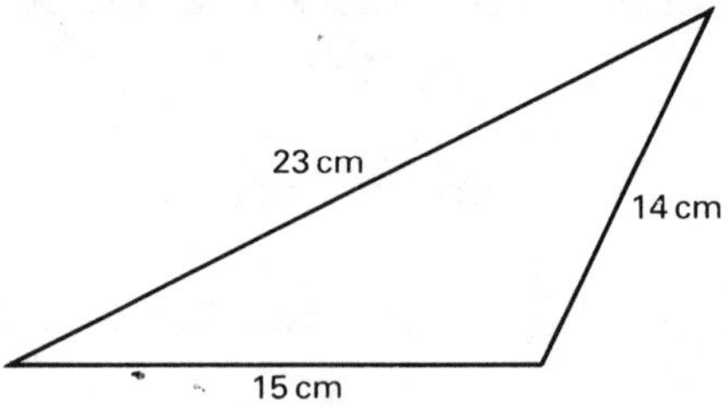

14. The perimeter of a rectangular room is 20 m. If the room is 4 m wide, how long is it?

15. Len wants to make a set of bookshelves so that each shelf will hold about 25 books. If each book is on average about 2 cm thick:

(*a*) How long should each shelf be?

(*b*) How much wood is needed for four shelves?

(*c*) If the two uprights each measure 900 mm, what is the total length of wood needed to make bookshelves with four shelves?

16. A painting measures 32 in by 21 in. To hang it, a cord is tied to two eyelets as shown. The eyelets are 20 inches apart. How far is each eyelet from the upright edges of the frame?

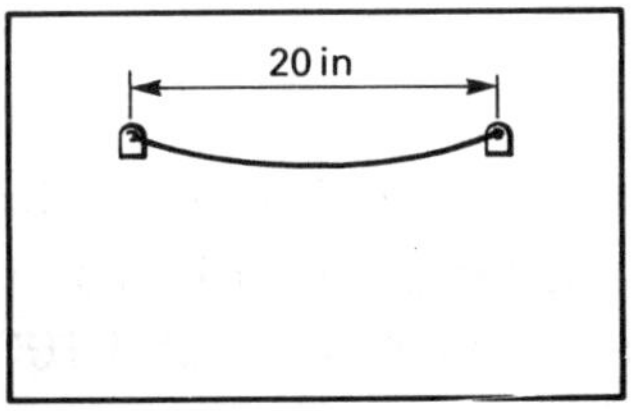

17. A tin of beans, including the tin, weighs 288 g. If the beans weigh 225 g, find the mass of the tin.

18. Out of 400 mℓ of juice, 180 mℓ was drunk. How many millilitres are left?

19. Copy and complete:

(*a*) 2400 g = [?] kg (*b*) 4.25 m = [?] mm

20. A piece of string 4.35 m long was cut into five equal pieces. Find the length of each piece in centimetres.

Revision Exercise VI

1. Answer these using the table in Exercise 1 on p.71.
(*a*) Find the length of the Columbia.
(*b*) How many kilometres longer than the Saskatchewan is the Ohio?

2. Using the table in Exercise 2 on p.72, find:
(*a*) the average height of a 14-year-old girl,
(*b*) the difference in mass in grams of 13-year-old girls and boys.

3. Using the table in Exercise 3 on pp.73–4, find:
(*a*) the height, and
(*b*) the time of the morning tide on 23 October.

4. Find the waist measurement of a 9-year-old boy (in centimetres and inches) using the table in Exercise 4 on p.76.

5. Using the table in Exercise 5 on p.77, find:
(*a*) the value of an investment of £10 a month over 5 years,
(*b*) the value of an investment of £100 a month over 3 years.

6. Using the distance chart on p.79, find the distance from:
(*a*) Liverpool to Dover,
(*b*) Newcastle to York.

7. Using the distance chart on p.80, find:
(*a*) the distance from Sheffield to Carlisle,
(*b*) how much further it is from Aberystwyth to Inverness than from London to Nottingham.

8. Using the timetable on p.81, find:
(*a*) the time of arrival at Melton Mowbray of the train that left Nuneaton at 12.53,

(*b*) the time of arrival at Leicester of a train that left Peterborough at 20.35 and was 10 min late in arriving.

9. Using the conversion graph on p.83, find:
(*a*) the number of metres in 21 ft,
(*b*) the number of feet in 10 m.

7 Shapes and Solids

Solids

A cube is shown.
One *face* is shaded.

This is an *edge*.

The corners are called *vertices*.
Each corner is called a *vertex*.

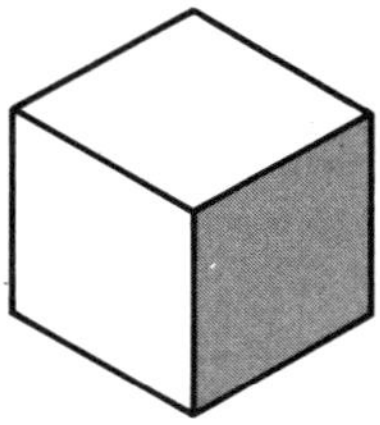

Here is a different sort of drawing of a cube. Hidden edges have been drawn using broken lines.

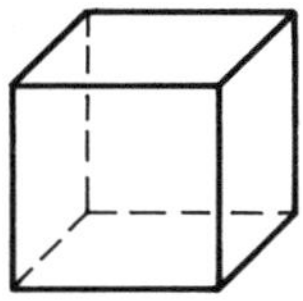

Exercise 1

1. (*a*) How many edges has a cube?
(*b*) How many faces has a cube?
(*c*) How many vertices has a cube?

A paperweight has the shape of a *tetrahedron* (*triangular-based pyramid*).
(*a*) How many edges has it?
(*b*) How many faces has it?
(*c*) How many vertices has it?

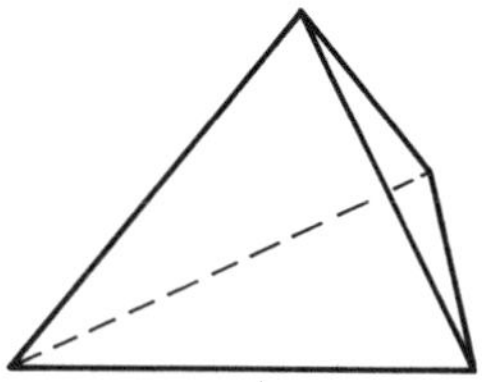

3. (*a*) If QRST is a square, what sort of pyramid is shown?

(*b*) How many edges has it?

(*c*) How many faces has it?

(*d*) How many vertices has it?

(*e*) Which edges are the same length as PQ?

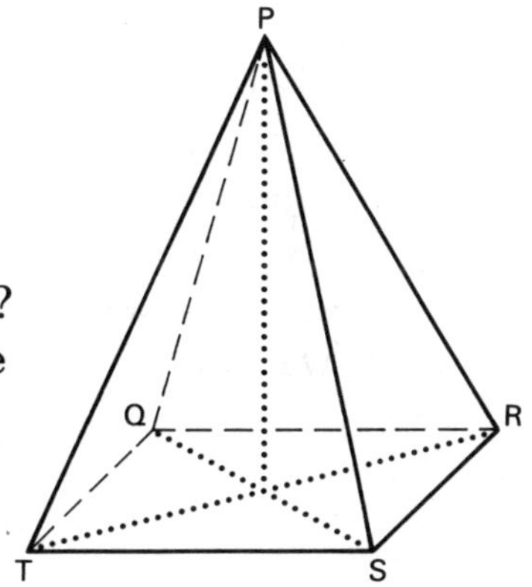

Exercise 2 M

1. A cuboid has been drawn on dotty paper. Copy it.

(*a*) How many edges does the cuboid have?

(*b*) How many faces does the cuboid have?

(*c*) How many vertices does the cuboid have?

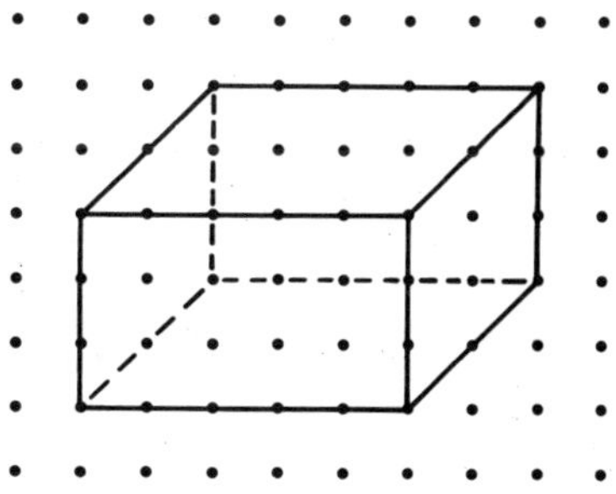

2. Part of a cuboid has been drawn on dotty paper. Copy and complete it.

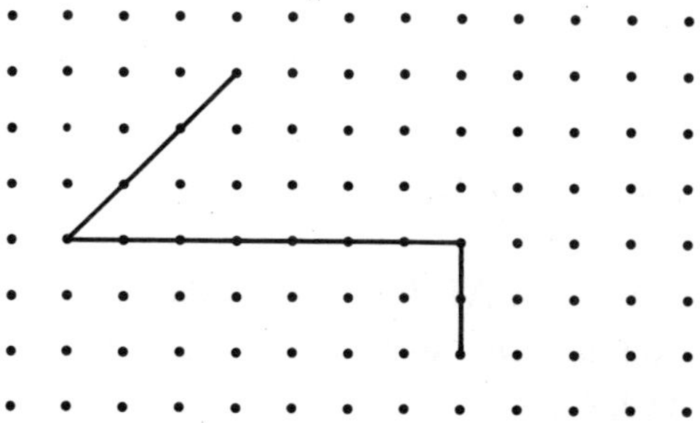

3. Here is another cuboid. This one has been drawn using dotty triangles. Copy it. Be sure to use your dotty paper the right way up.

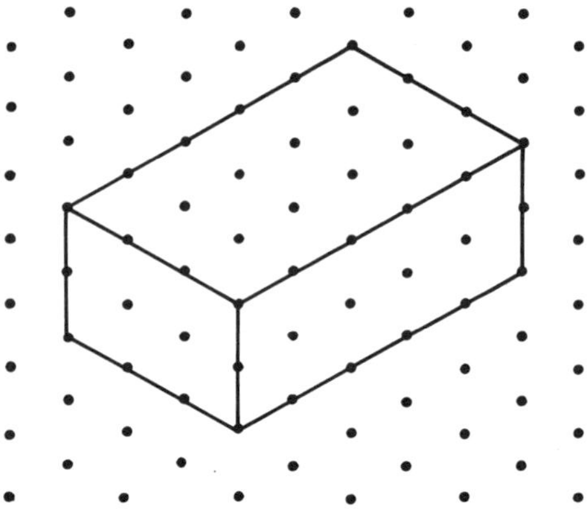

4. Here is part of a cuboid. Copy and complete it.

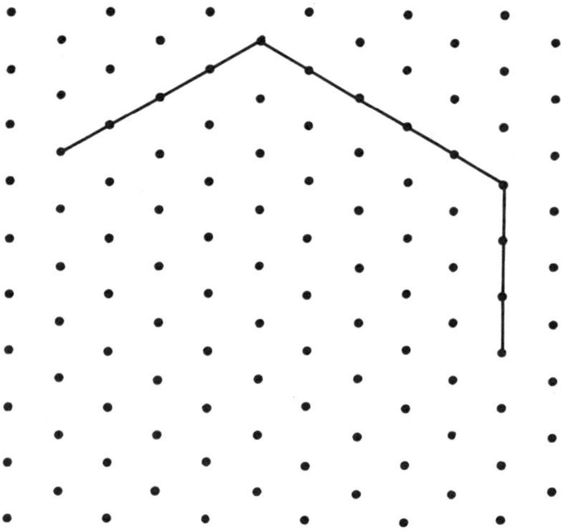

Exercise 3

1. In the given cuboid,
   (*a*) which edges are the same length as AB?
   (*b*) which edges are the same length as AD?
   (*c*) which edges are parallel to edge AD?
   (*d*) which face is parallel to face ABFE?

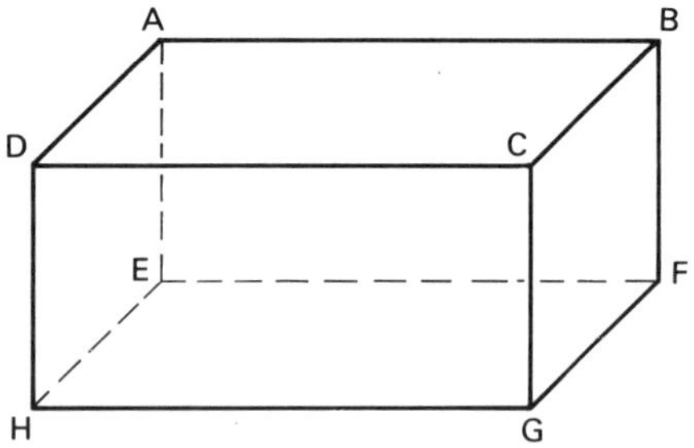

2. The drawings show two solids that were made from straws (pipe cleaners were used at the vertices to join the straws together).

(*a*)

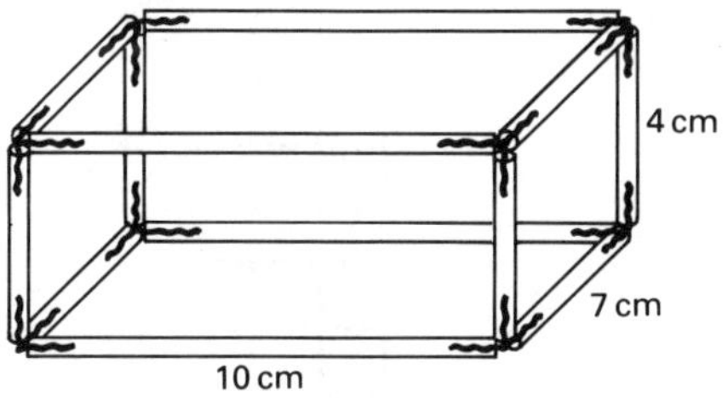

(*b*)

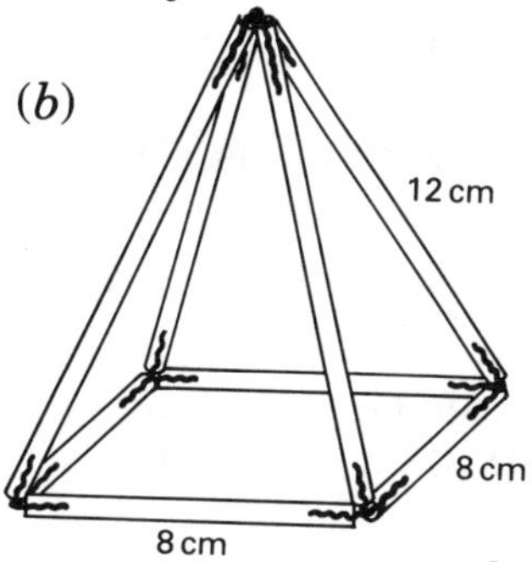

What is the total length of the straws used to make each solid?

Exercise 4

What is the smallest number of cubes that could be used to make each of these shapes?

1.

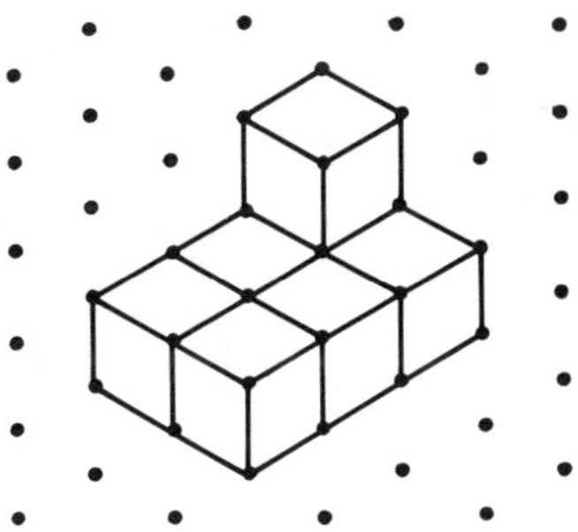

3.

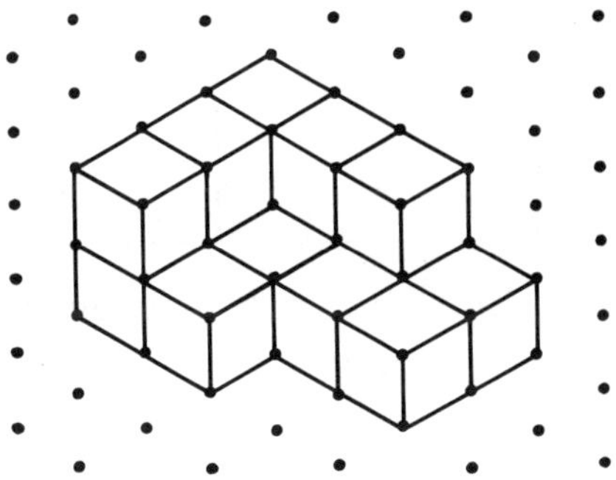

2.

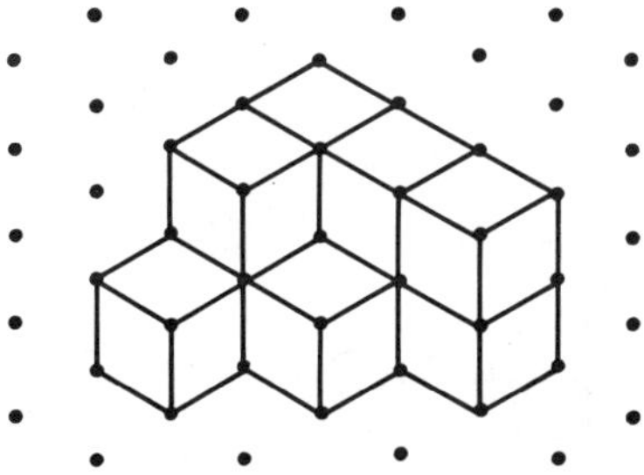

4.

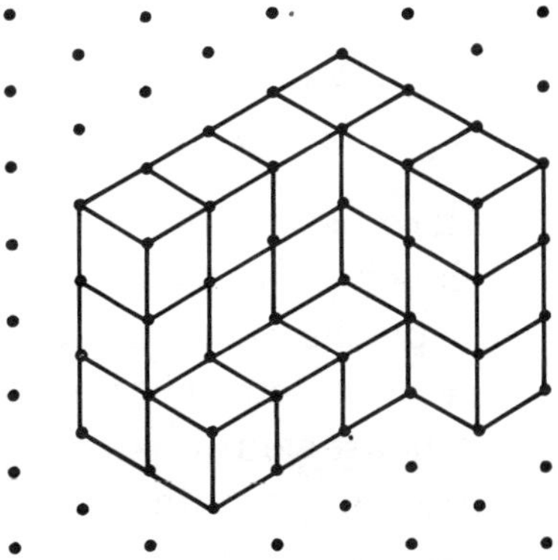

Nets of Polyhedra

Exercise 5

This shape is a net of an open box made with 5 square faces. (It folds into a box.)

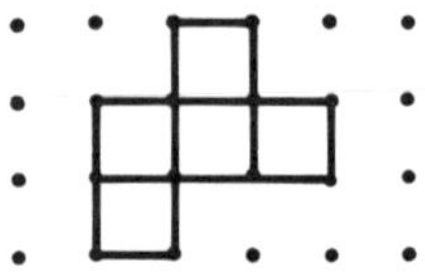

This shape is not a net. It does not work.

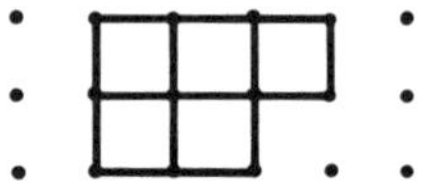

Try to find some different nets of an open box with square faces. Draw them on dotty paper or squared paper.

Exercise 6 **M**

1. Which of the following drawings are nets of a cube?

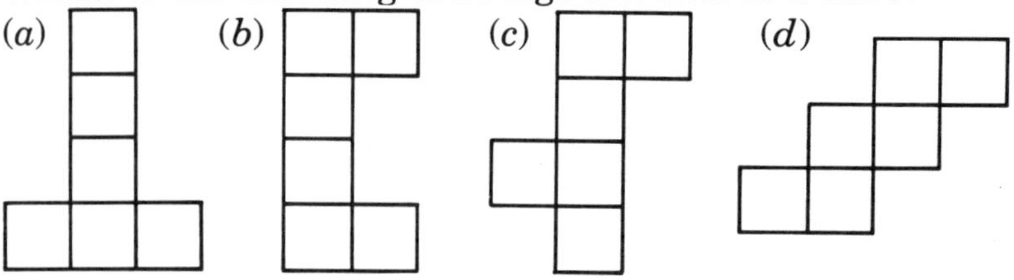

Make each of the above out of squared paper. Fold them to see if they work.

2. Try to find some more nets of a cube. (There are eleven.)

3. To make a cube, we need to put some tabs on some of the edges of some of the squares in the net.

Here is a net of a cube showing the tabs. Copy the net on to squared paper and cut it out. (Choose your own size.) Fold the net into a cube and stick down the tabs.

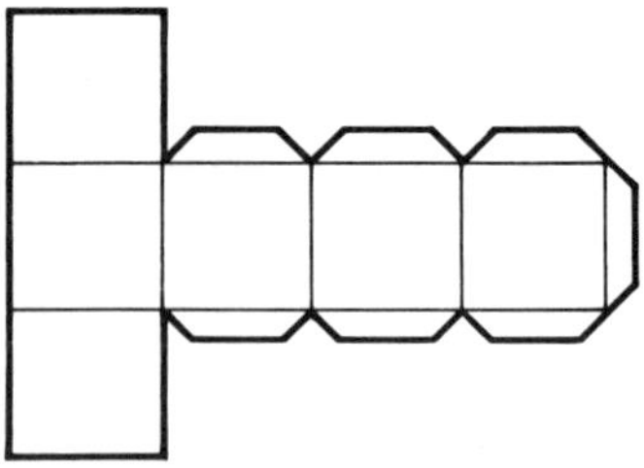

4. Here is a net of a cube. When the cube has been made from this net, which face would be opposite face E?

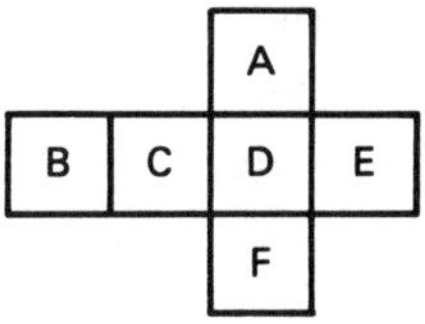

5. Here is a net of a cube. When the cube has been made from this net, which face would be opposite face E?

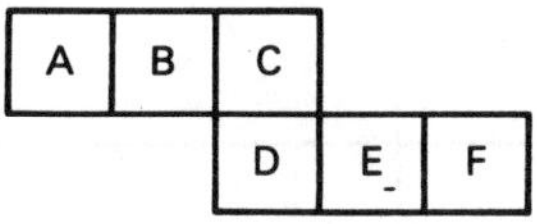

Exercise 7

1. Which of the following drawings below are nets of the cuboid shown?

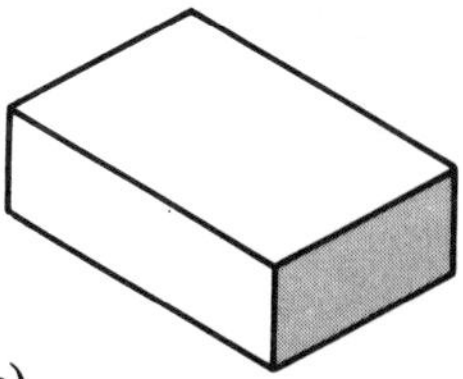

(*a*)

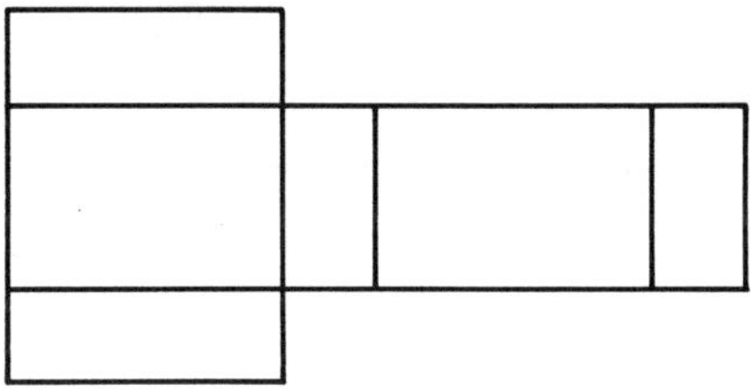

(*b*)

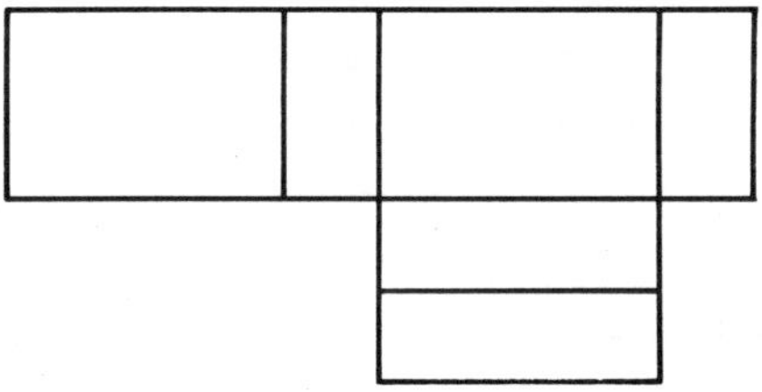

(*c*)

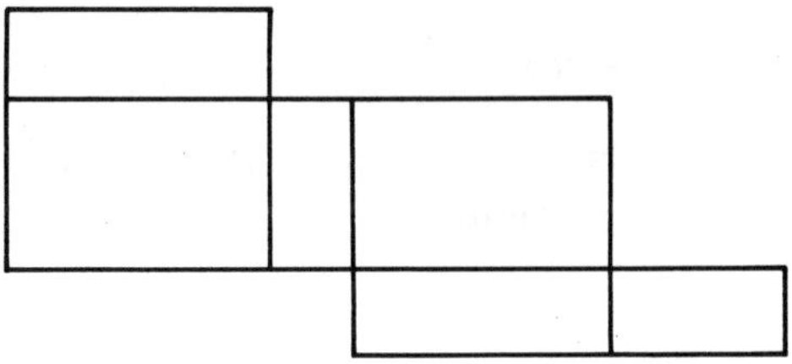

(*d*)

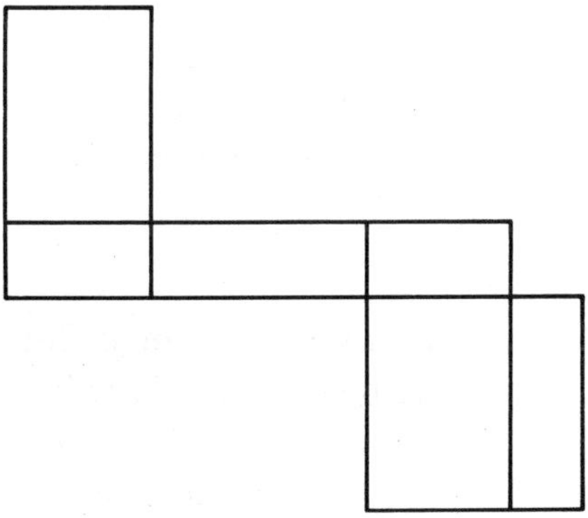

2. Which of the drawings below are nets of the cuboid shown?

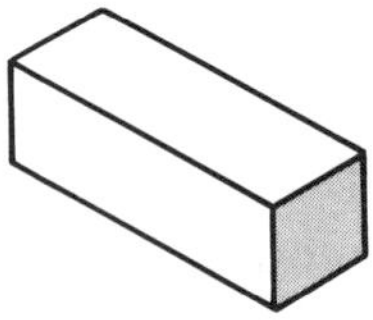

(*a*)

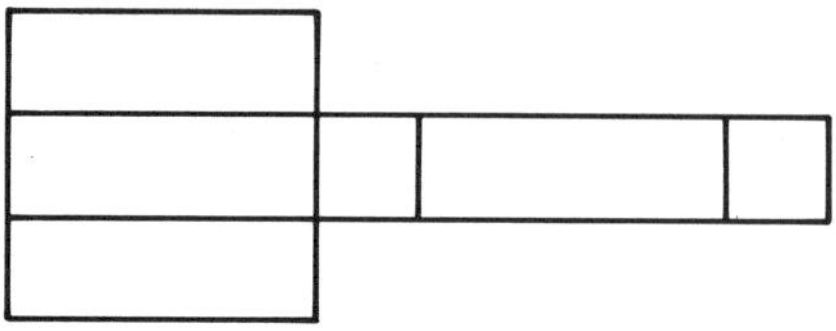

(*b*)

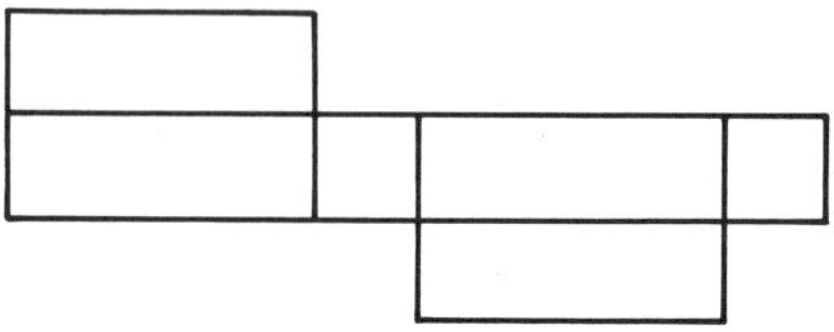

(*c*)

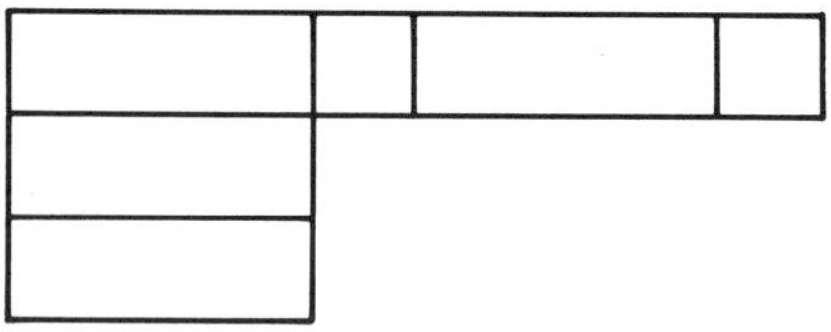

(*d*)

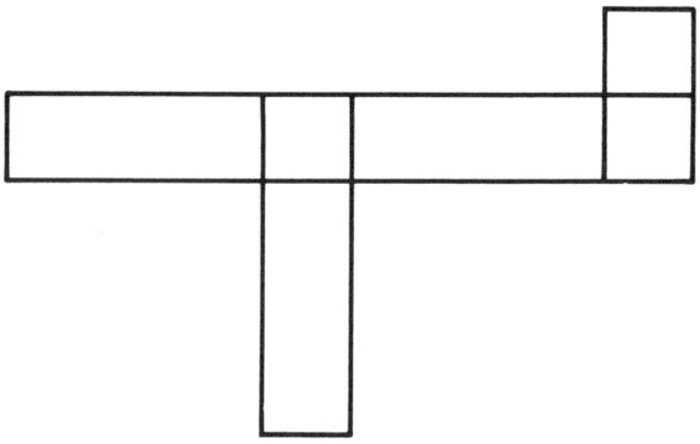

3. A rectangular box (a cuboid) is 8 cm long, 6 cm wide and 4 cm high.

Here is one possible net of this box drawn to one-quarter of its true size.

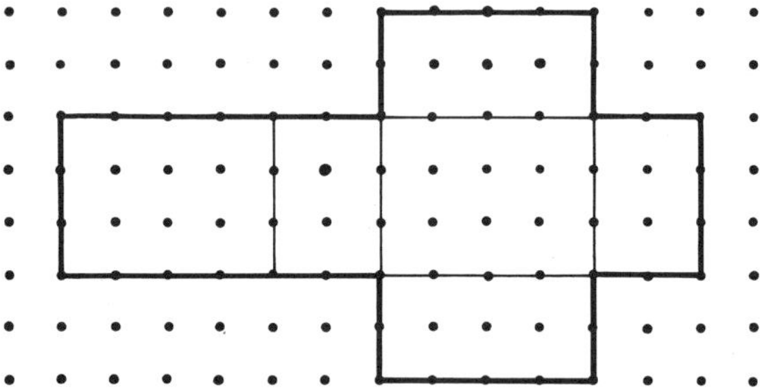

(*a*) Draw two other different nets of this box (both should be one-quarter of the true size).

(*b*) On squared (or dotty) paper draw a half-size net of a box that measures 10 cm by 8 cm by 2 cm.

Exercise 8

1. Here is a net of a triangular-based prism.
Copy this net, but make your copy larger.
(Make each rectangular face 8 cm long and 4 cm wide.)
Make your copy on paper (or thin card).
Put flaps on it. Cut out the net.
Fold it to make a triangular-based prism.

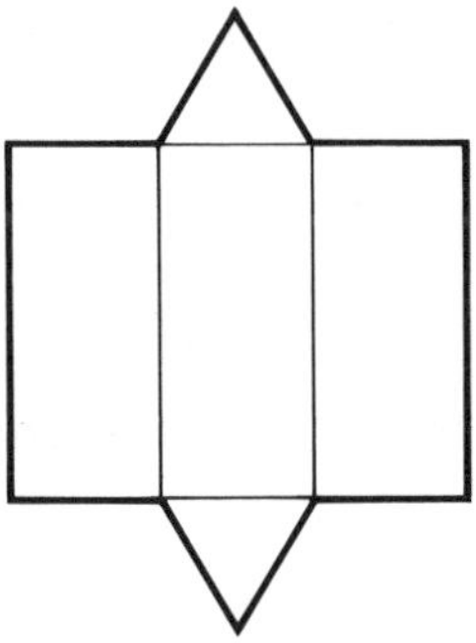

2. Draw a net of a square-based pyramid.
Use paper (or thin card). Put flaps on it.
Cut it out. Fold it to make a square-based pyramid.

3. Here is a net of a solid.
Name the solid that can be formed from the net.

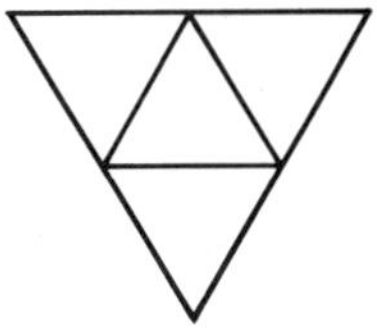

Exercise 9

Make a model of a building such as a church.

If you make a church, you will probably need to make cuboids, a triangular prism, and a pyramid.

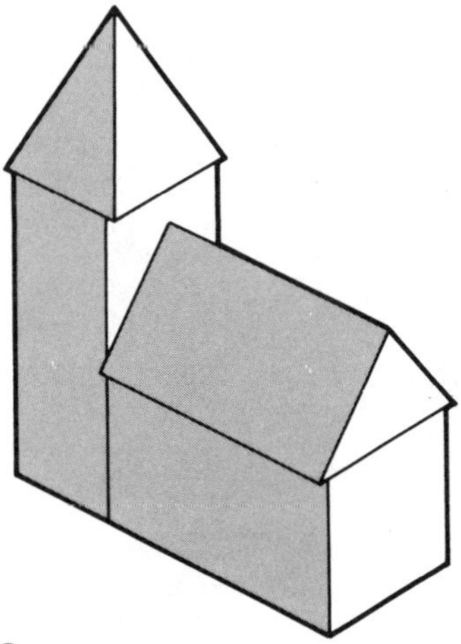

Exercise 10

Collect as many different-shaped cardboard containers as you can.

Cut along some of the edges so that they unfold to make a net.

Draw the nets in your exercise book.

Do not forget to give the names of the solids (e.g. net of a triangular prism).

Angles and Triangles

Exercise 11

For each angle write whether it is an acute angle, an obtuse angle, a reflex angle or a right angle.

1.

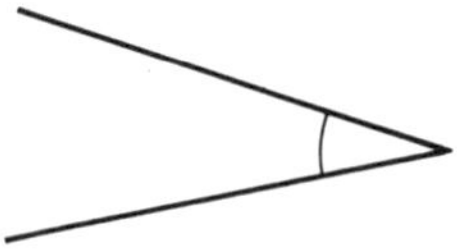

2.

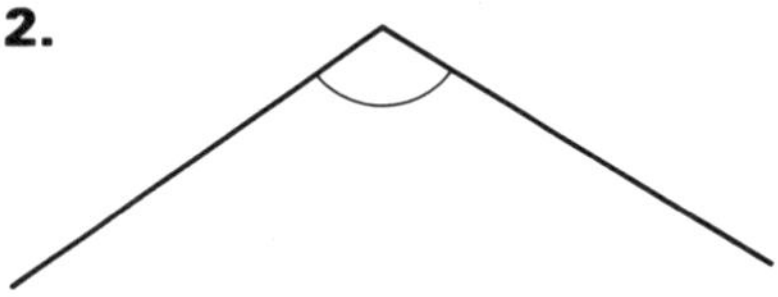

3.

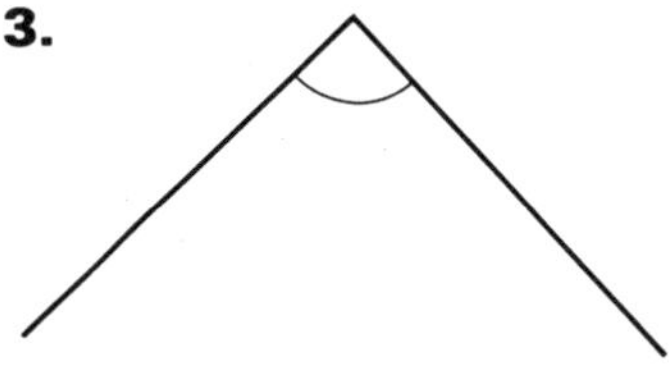

4.

5.

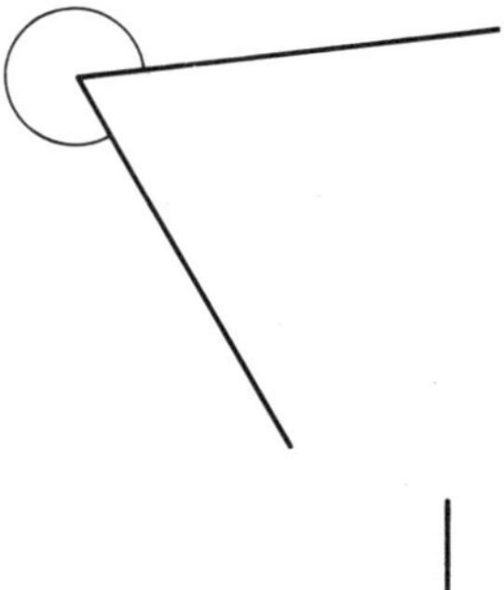

6.

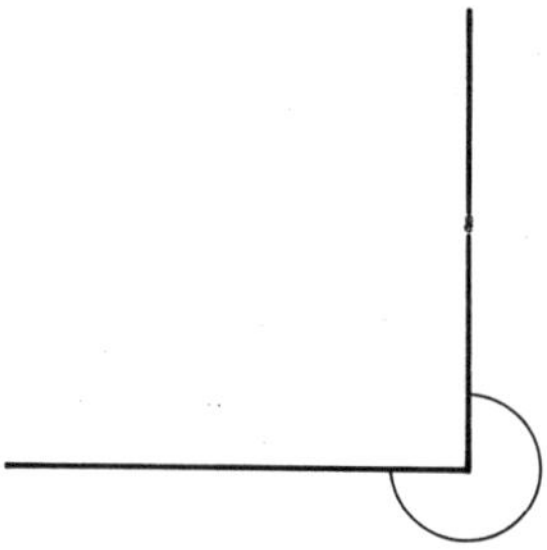

7.

8.

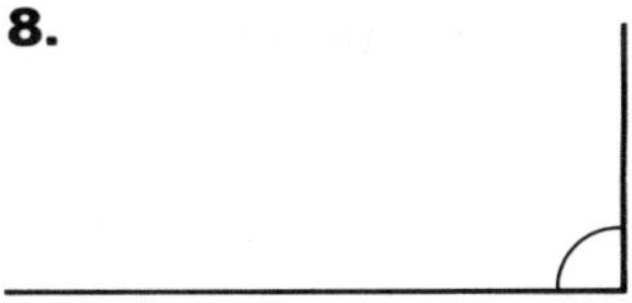

Exercise 12

1. In the given triangle, which is the largest angle, $\angle A$, $\angle B$ or $\angle C$?

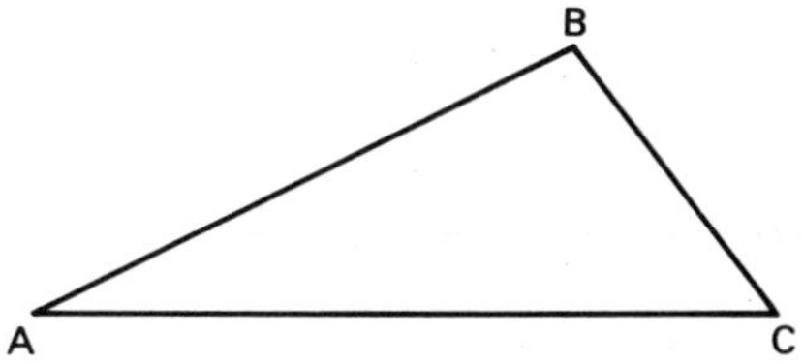

2. In the given triangle, which is the largest angle, $P\hat{Q}R$, $Q\hat{R}P$ or $R\hat{P}Q$?

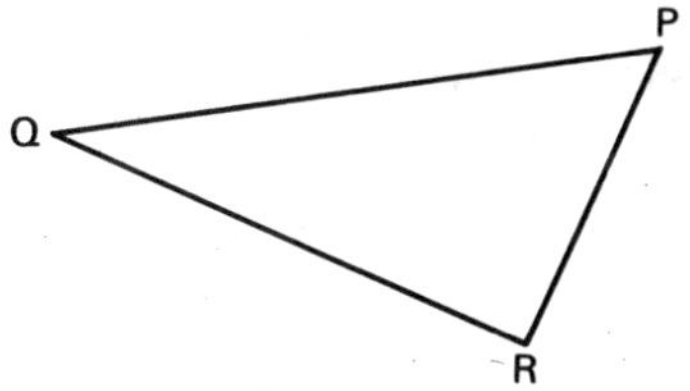

3. The diagram shows triangle ABC.

(*a*) Measure AB, BC and CA and write their lengths.

(*b*) Find the perimeter of the triangle.

(*c*) Measure the three angles of the triangle and write their sizes.

(*d*) Add together the three angles and write what you notice.

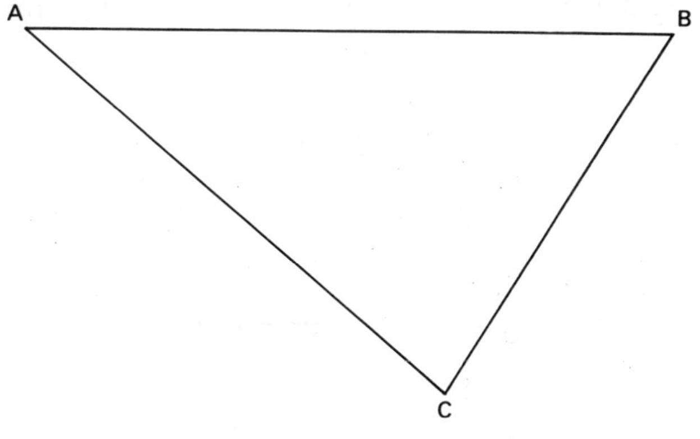

4. Using a protractor, measure $K\hat{M}L$.

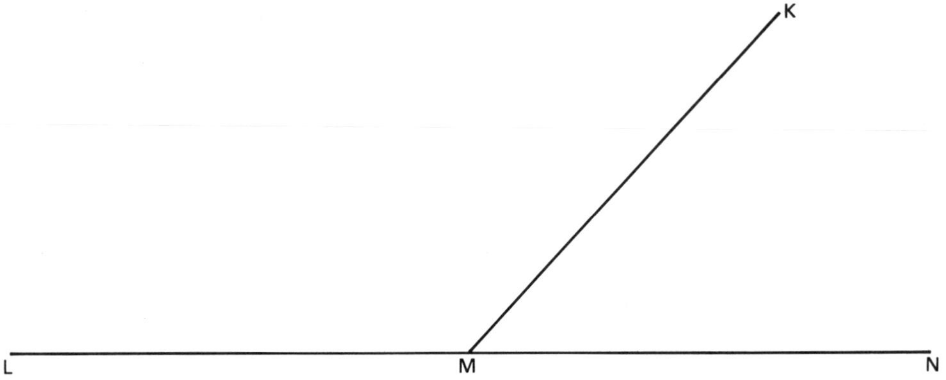

5. Using a protractor, find the size of the marked angle.

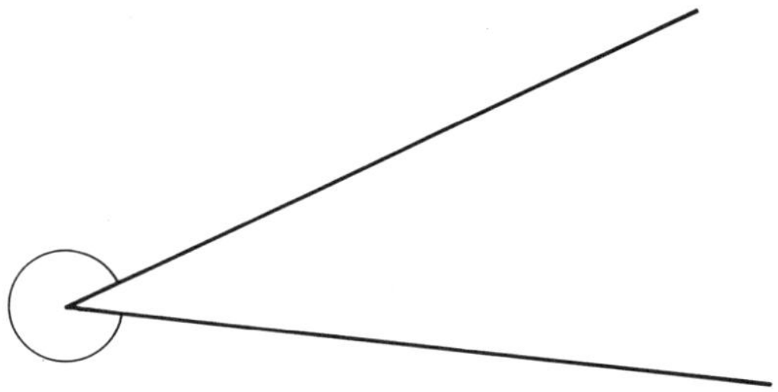

6. $\triangle$ PQR has been drawn accurately. The three angles measure 27°, 84° and 69°. By estimating, find which angle is which.

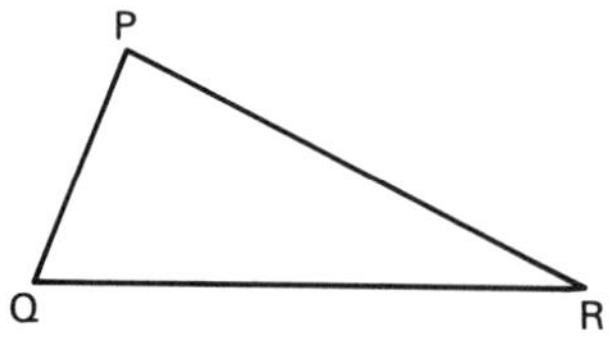

Exercise 13

Calculate the value of x in each of the following diagrams:

1.

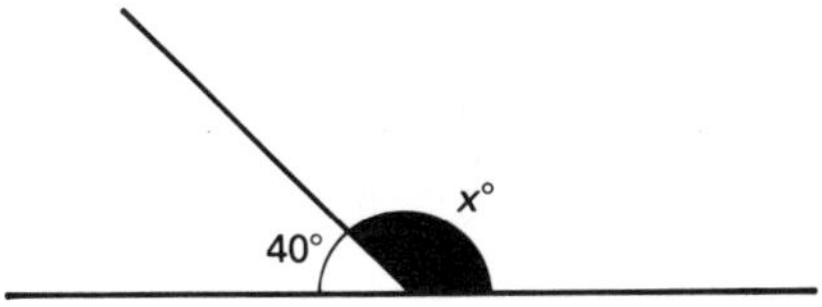

2.

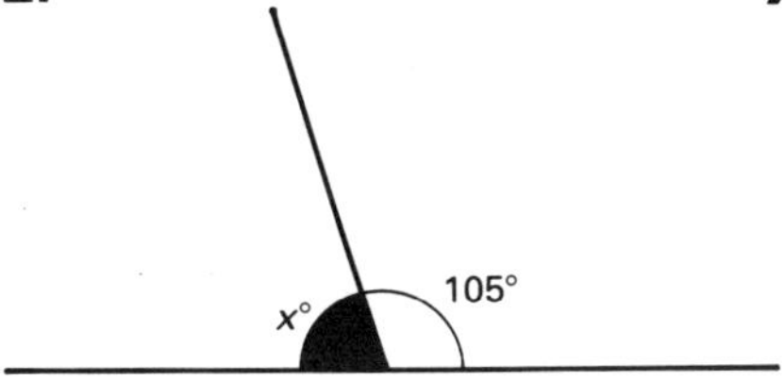

3.

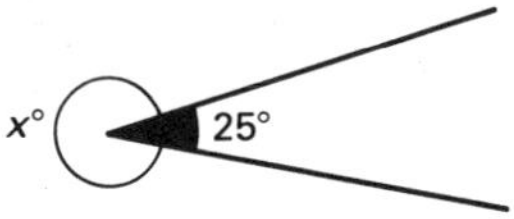

4.

5.

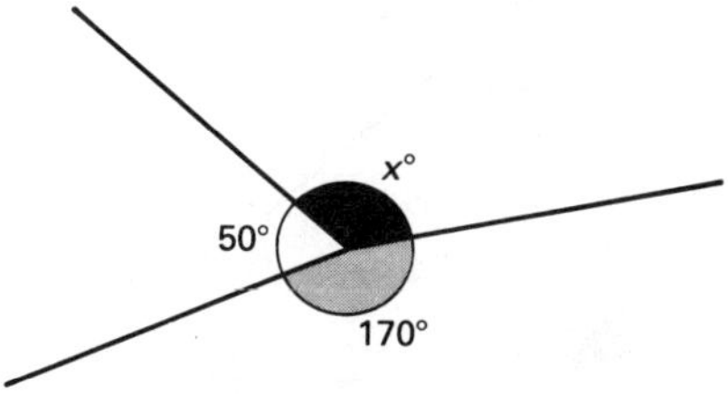

6.

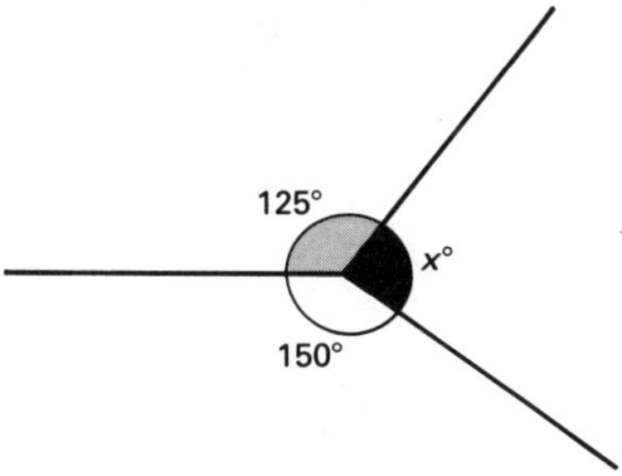

7.

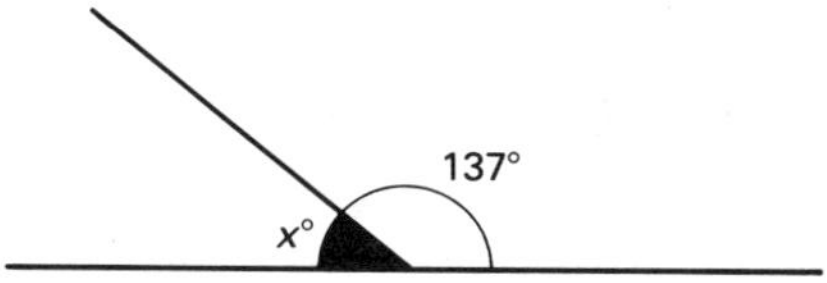

8.

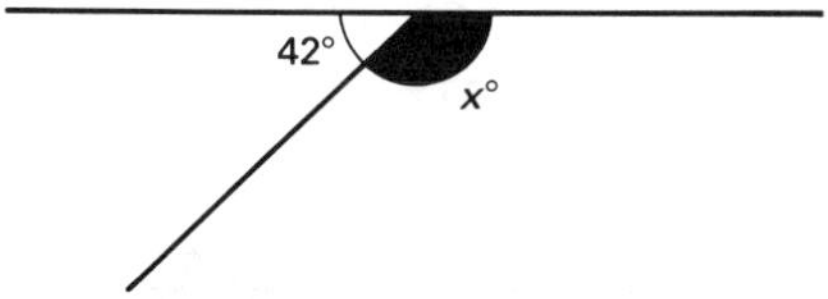

9.

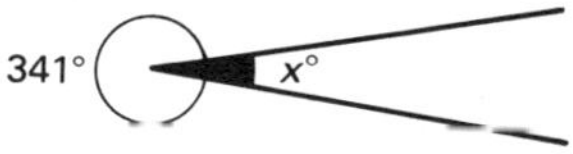

10.

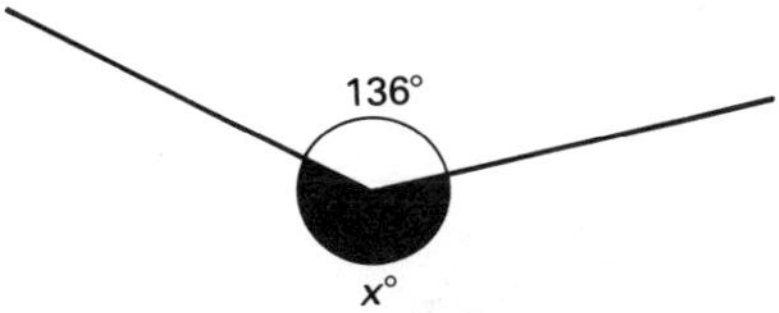

11.

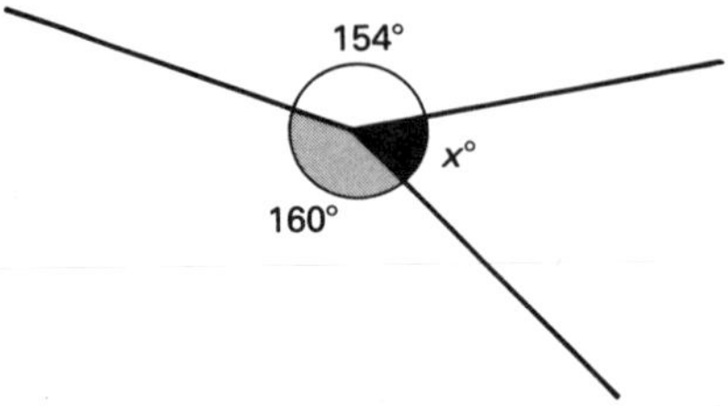

12.

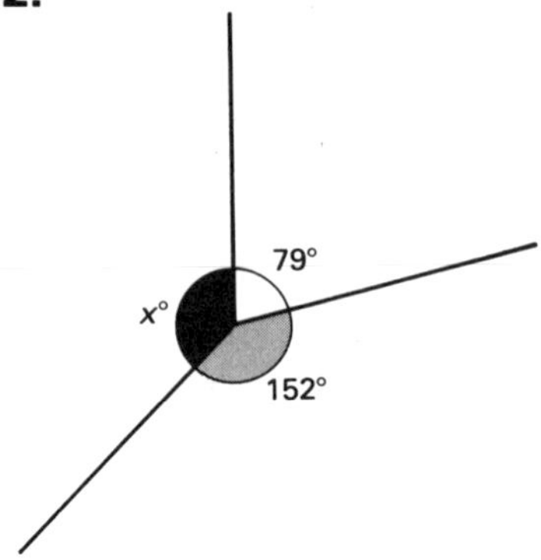

Exercise 14

Calculate the values of x:

1.

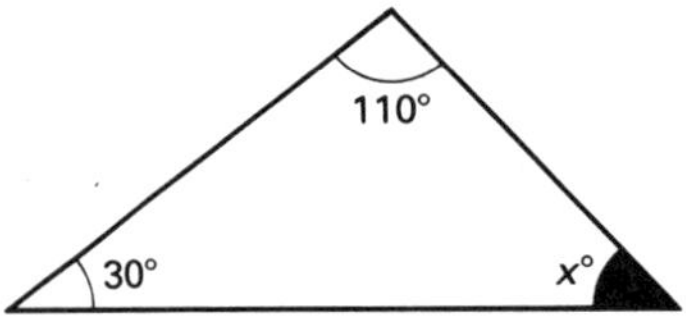

2.

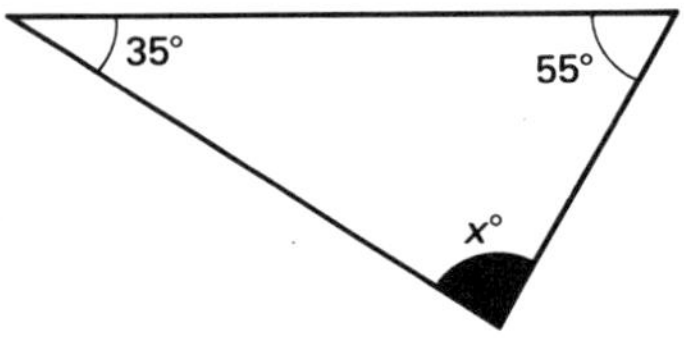

3.

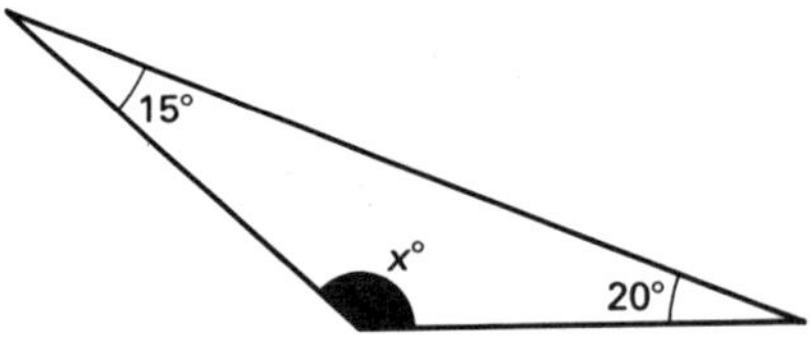

4.

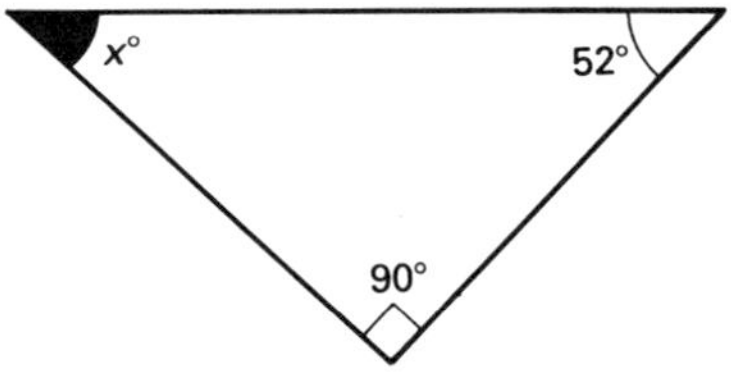

5.

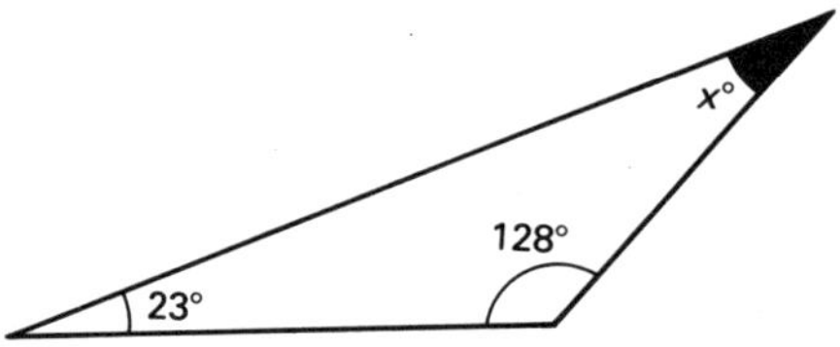

6.

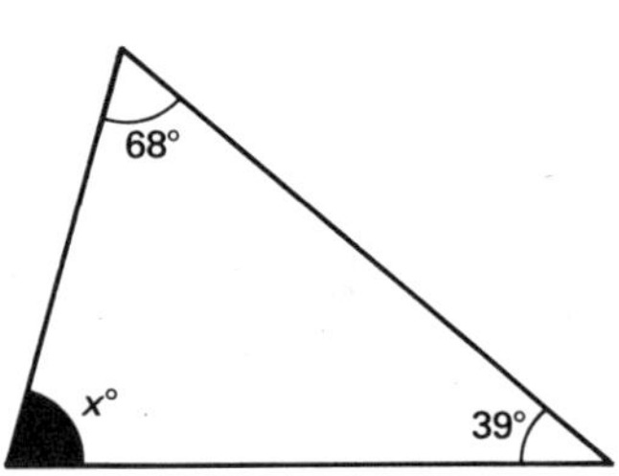

7.

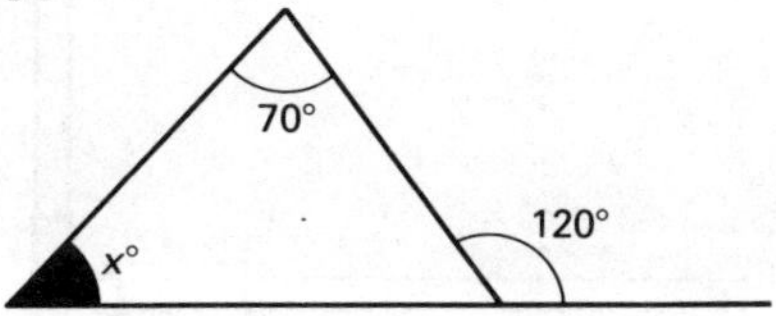

8.

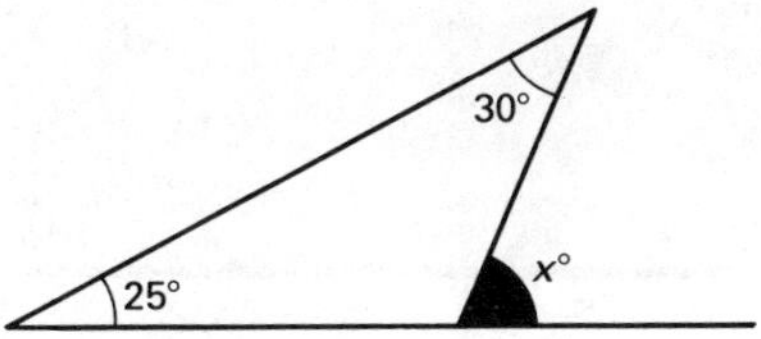

9.

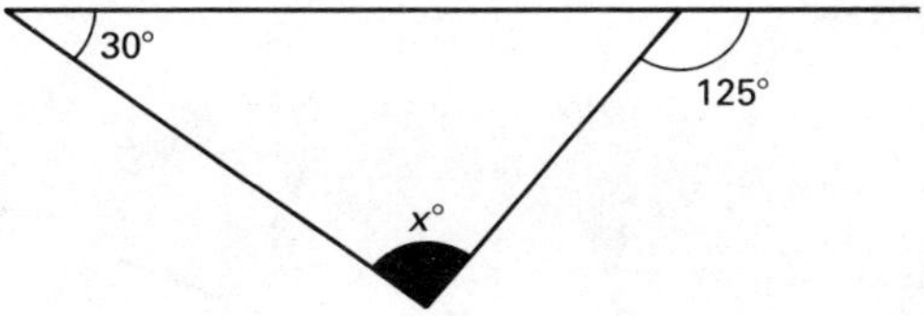

10.

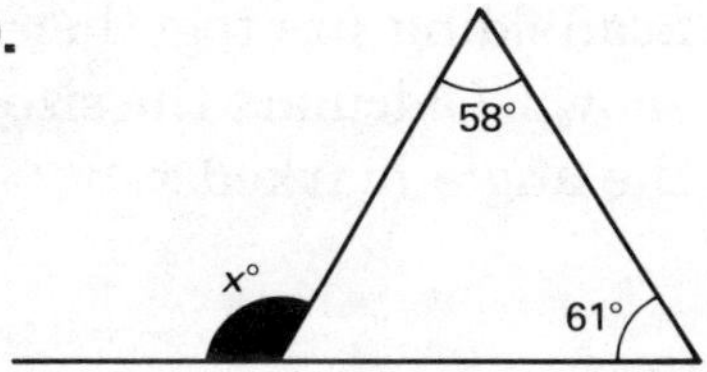

11.

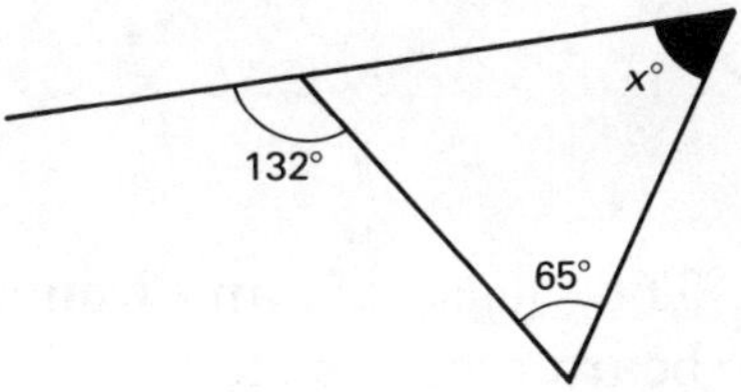

12.

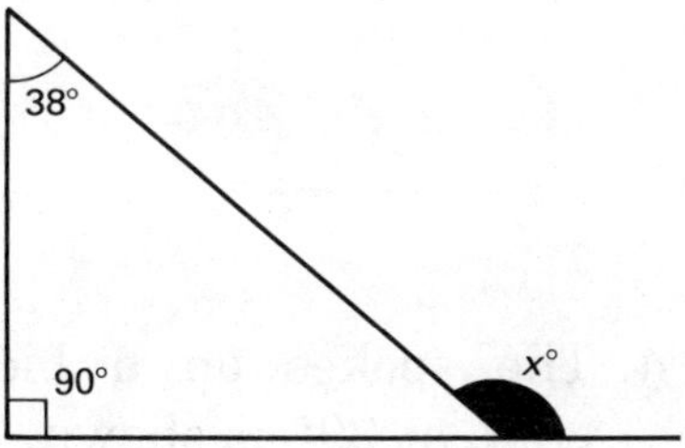

Exercise 15

1. In the given diagram of part of a house, what angle does the roof make with the wall?

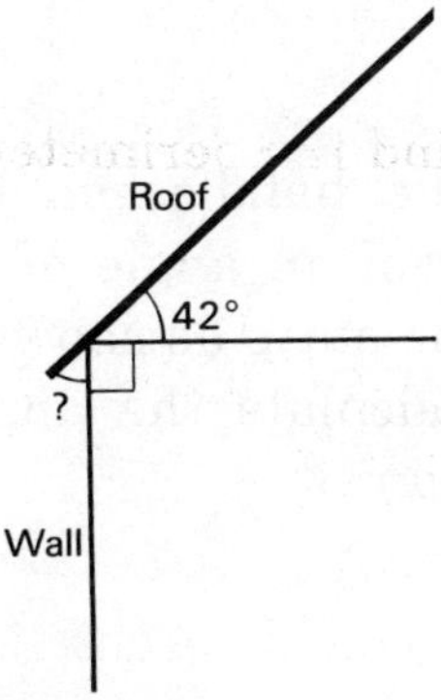

2. Scaffolding fits together as shown. Calculate the size of the angle marked x.

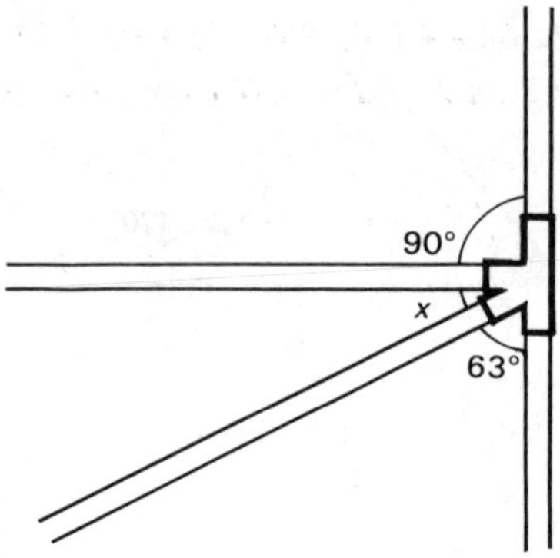

3. The legs of an ironing board cross as shown. Calculate the angles marked a, b and c.

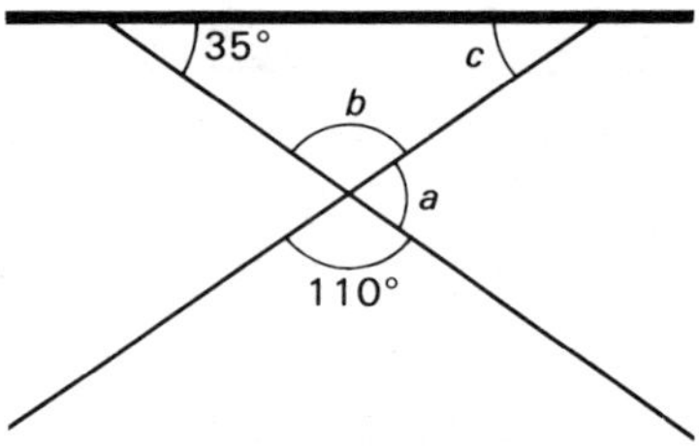

4. The spokes on a bicycle cross at 29° as shown. Calculate the angles marked d and e.

5. The handles on a mower form an angle of 83° with the stays as shown. Calculate the angle labelled f.

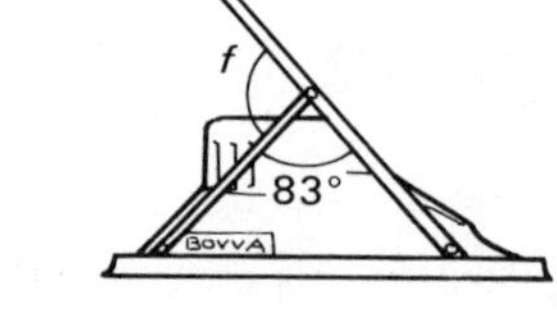

6. Part of the frame of an exercise bike is shown.

Angle $g = 57°$,
angle $h = 75°$.
Calculate angle i.

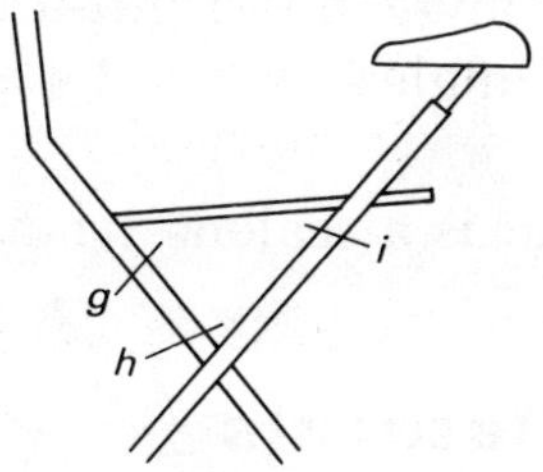

Triangles and Their Properties

Exercise 16

1. An equilateral triangle ABC is shown.
   (*a*) How long is each side?
   (*b*) What size are the angles?

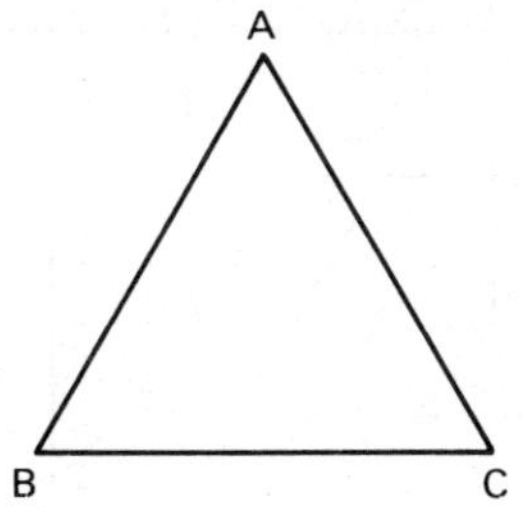

2. The sketch shows an isosceles triangle PQR in which PQ = PR and $\angle PQR = 68°$.
   Find (*a*) $\angle PRQ$, (*b*) $\angle QPR$.

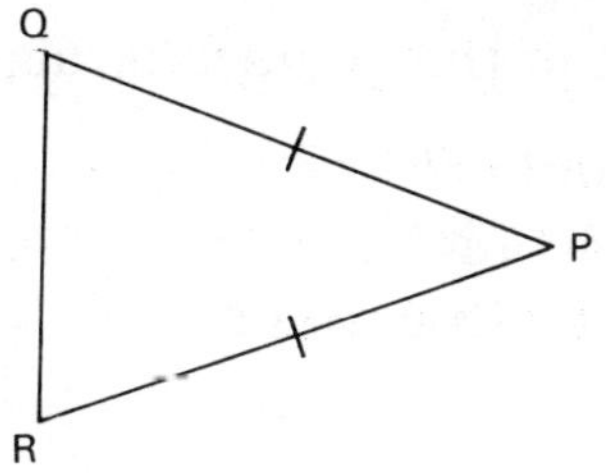

3. In isosceles triangle XYZ, $\angle XYZ = \angle YXZ = 72°$.
   (*a*) Sketch the isosceles triangle and label it.
   (*b*) Find the size of angle XZY.
   (*c*) Which side is equal in length to side XZ?

4. An angle of an isosceles triangle is 70°. Sketch two different isosceles triangles for which this can be true. Mark the three angles of each triangle and write the sizes of the angles on each diagram.

5. One angle of a right-angled triangle is 25°. Find the other two angles.

6. What is a scalene triangle?

Quadrilaterals

Exercise 17

Here are some quadrilaterals:

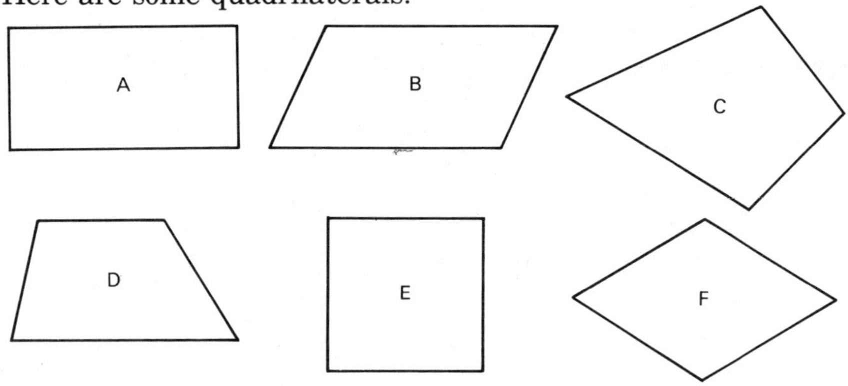

Which of the quadrilaterals is:

1. a square?
2. a rectangle?
3. a parallelogram?
4. a rhombus?
5. a kite?
6. a trapezium?

Exercise 18 M

For each of these questions you may measure first.

e.g. Copy and cut out the parallelogram. With one straight cut, make a trapezium.

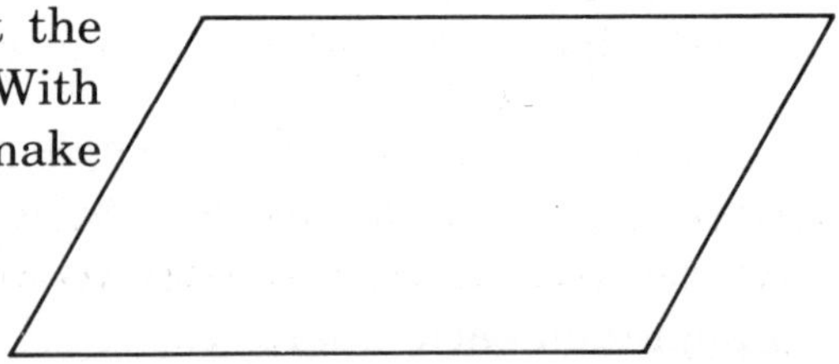

One possible answer is shown here. After cutting as shown, a trapezium is left.

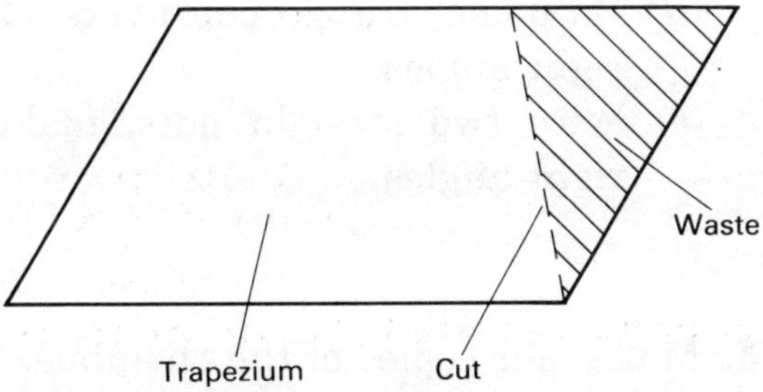

1. Make three copies of the parallelogram given above. Use a different copy for each part of the question.

(*a*) With one straight cut, make a rhombus.
(*b*) With two straight cuts, make a rectangle.
(*c*) With two straight cuts, make a square.

2. Make five copies of the rectangle. Use a different copy for each part of the question.

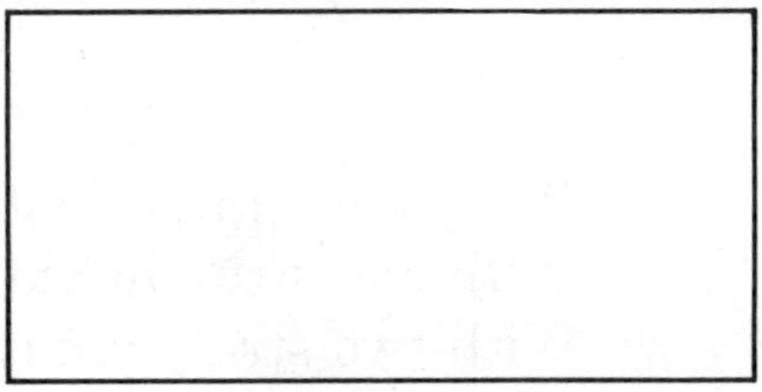

(*a*) With one straight cut, make a square.
(*b*) With two straight cuts, make a parallelogram.
(*c*) With two straight cuts, make a rhombus.
(*d*) With one straight cut, make a trapezium that contains right angles.
(*e*) With two straight cuts, make a trapezium without right angles.

3. Make five copies of the square. Use a different copy for each part of the question.

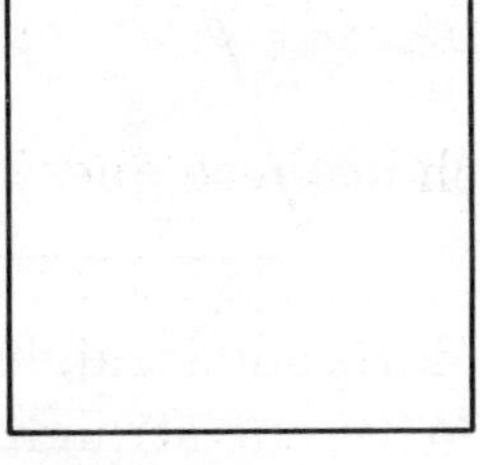

(*a*) With one straight cut, make a rectangle.
(*b*) With two straight cuts, make a parallelogram.
(*c*) With two straight cuts, make a kite.

(*d*) With one straight cut, make a trapezium that contains right angles.
(*e*) With two straight cuts, make a trapezium without right angles.

4. Make four copies of the rhombus. Use a different copy for each part of the question.

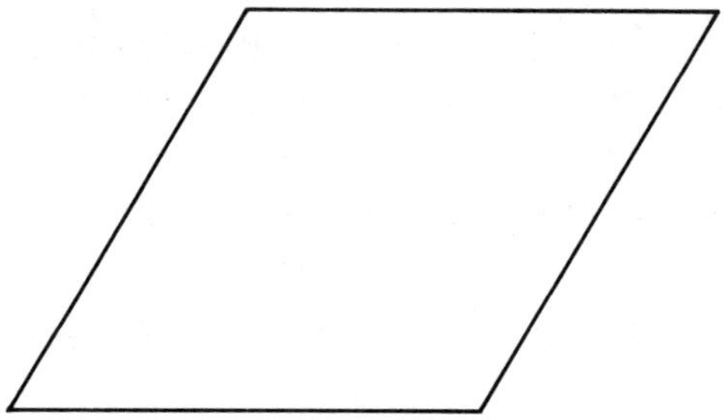

(*a*) With one straight cut, make a parallelogram.
(*b*) With one straight cut, make a trapezium.
(*c*) With two straight cuts, make a rectangle.
(*d*) With two straight cuts, make a kite.

5. Make five copies of the trapezium. Use a different copy for each part of the question.

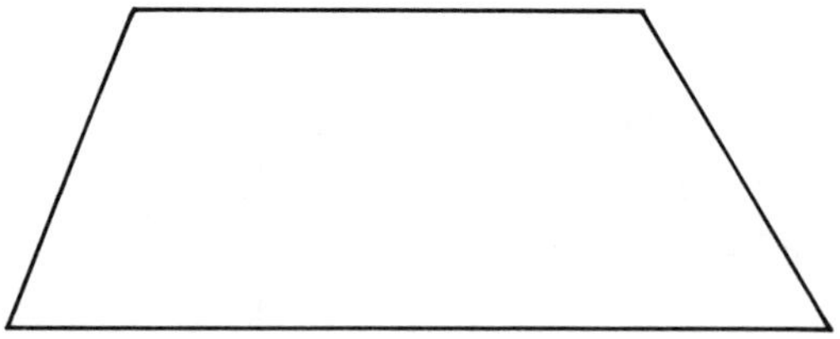

(*a*) With one straight cut, make a parallelogram.
(*b*) With one straight cut, make a rhombus.
(*c*) With two straight cuts, make a rectangle.
(*d*) With two straight cuts, make a square.
(*e*) With one straight cut, make a kite.

6. Make four copies of the kite. Use a different kite for each part of the question.

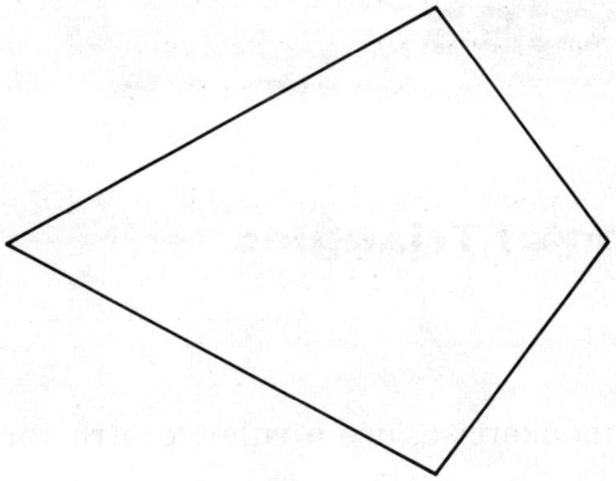

(*a*) With two straight cuts, make a rhombus.
(*b*) With two straight cuts, make a parallelogram.
(*c*) With two straight cuts, make a trapezium.
(*d*) With one straight cut, make a trapezium.

8 Constructions and Circles

Construction of Triangles

Exercise 1

Throughout this exercise, use a ruler, protractor and pair of compasses:

1. Copy the following triangles:

(*a*)

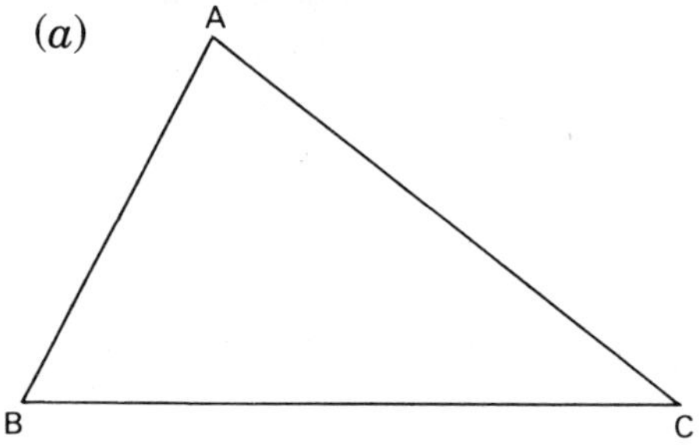

Measure angle ACB.

(*b*)

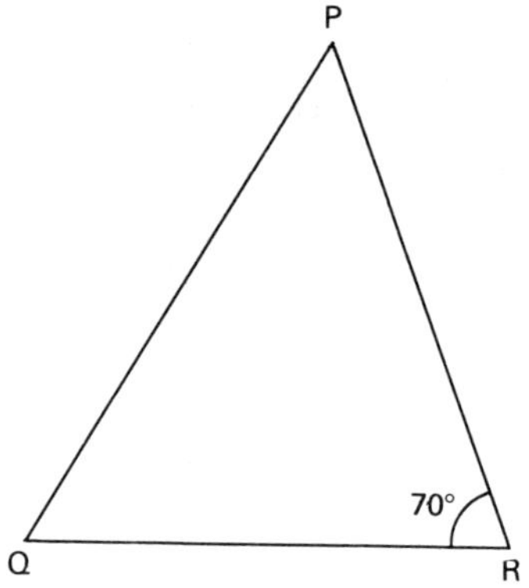

Measure PQ.

2. Construct the following triangles. (The given drawings are not to the proper size.)

(*a*)

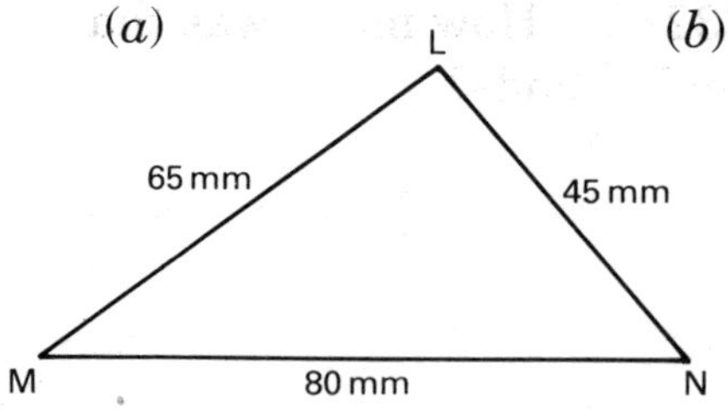

Measure ∠LMN.

(*b*)

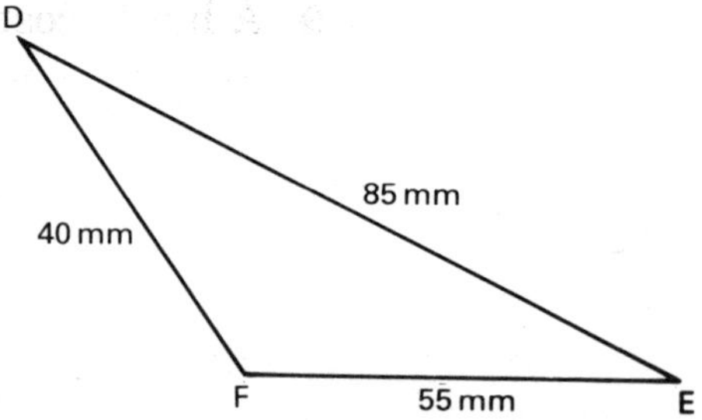

Measure XZ.

(c)

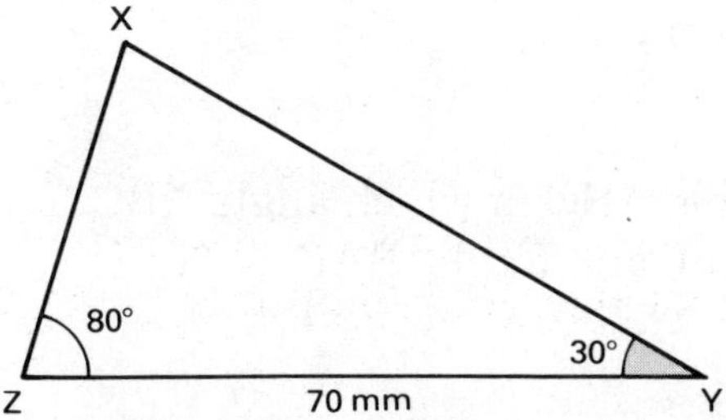

Measure ∠ DFE.

(d)

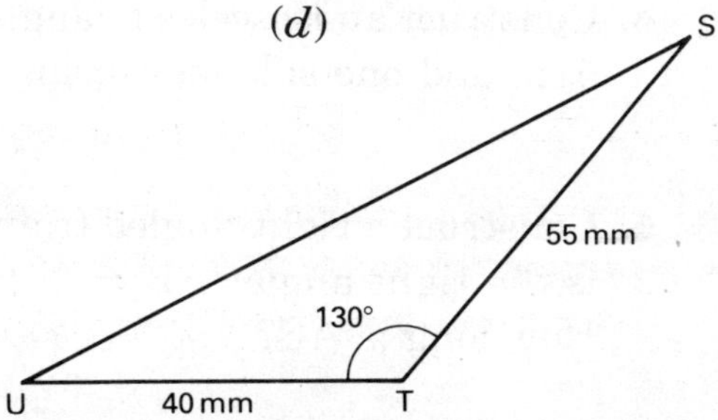

Measure US.

3. (*a*) Construct triangle JGM in which GM = 90 mm, GJ = 80 mm and MJ = 55 mm. Measure angle JGM.

(*b*) Construct triangle VWX in which VW = 10 cm, $V\hat{W}X = 35°$ and WX = 9 cm. Measure VX and give its length.

Exercise 2

1. Construct triangle RLM where RL = 6 cm, $R\hat{L}M = 60°$ and LM = 6 cm.

(*a*) Measure RM.

(*b*) What is the size of angle RML?

(*c*) What sort of triangle have you drawn?

2. Construct an equilateral triangle in which each side is 5 cm long.

3. Construct an isosceles triangle with a base 34 mm in length and having angles of 70° at each end of the base.

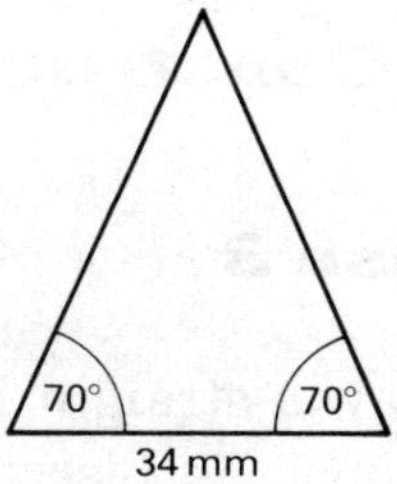

(*a*) How long are the other sides?

(*b*) Measure the third angle.

4. Construct an isosceles triangle having two sides measuring 5 cm and one side measuring 4 cm.

5. Construct a right-angled triangle ABC in which angle ABC is the right angle, AB = 35 mm and angle BAC = 60°. How long is AC?

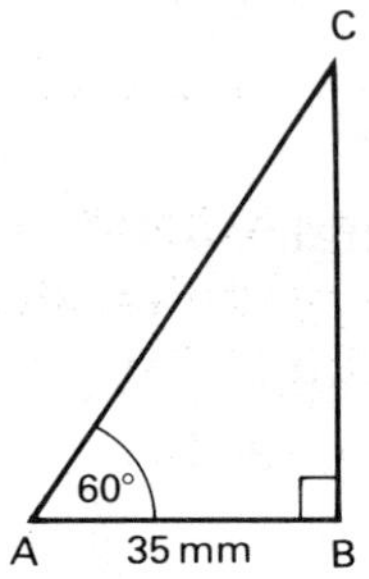

6. The *angle of elevation* of the top of a bungalow from a point 6 m away is 40° (as shown in the sketch).

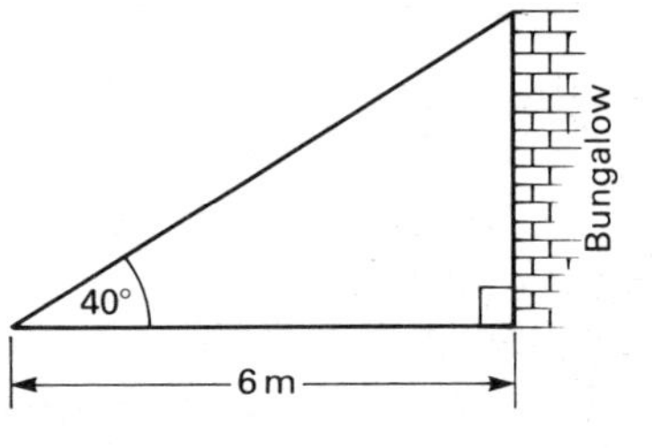

Using a scale drawing find the height of the bungalow. Use a scale of 1 cm to 1 m.

Other Constructions

Exercise 3

1. Draw a straight line AB, 67 mm in length. Using a pair of compasses, bisect line AB.

2. Draw any angle and label it PQR.
Using a pair of compasses, construct the bisector of $P\hat{Q}R$.

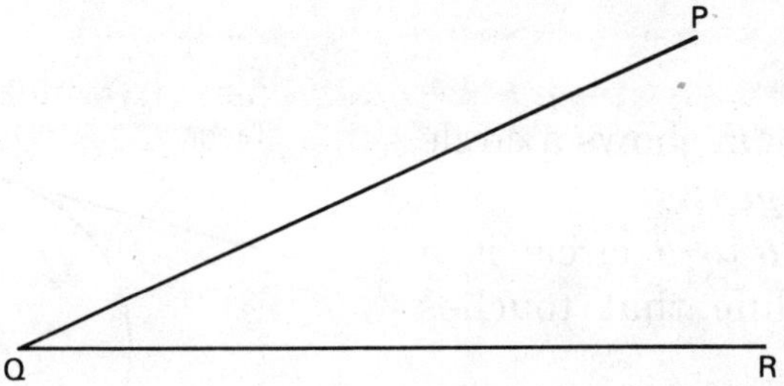

3. Draw any straight line AB and mark any point P above it.
Using a set square and ruler, draw a line parallel to AB that passes through point P.

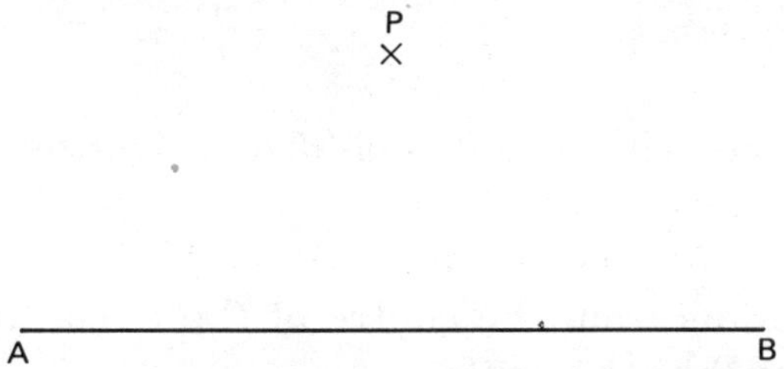

4. Draw a straight line, XY.
Make it 8 cm long.
Using a protractor, draw an angle XYZ equal to 42°.
Make YZ 9 cm long.

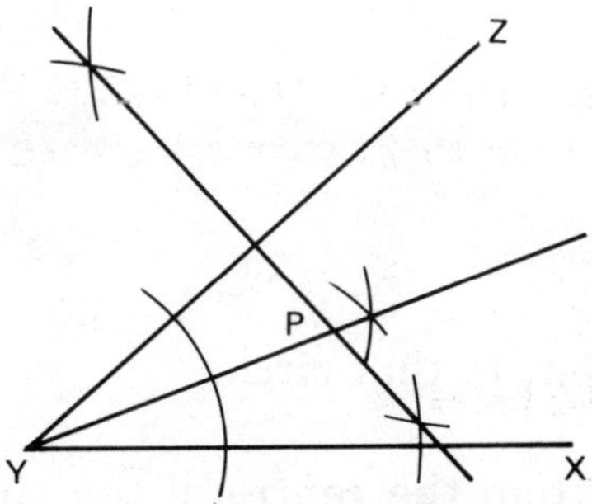

With a pair of compasses, bisect angle XYZ.
With a pair of compasses, bisect YZ.
Label the point where the two bisectors meet as P.
How long is YP?

The Circle

Exercise 4

1. The diagram shows a circle and a *tangent*.
   A *tangent* to a circle is a straight line that touches the circle.
   Draw a circle having a radius that is double the radius of the given circle.
   Draw a tangent to the circle.

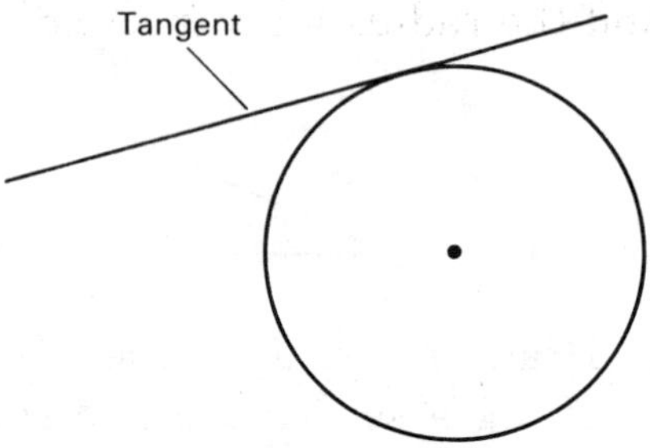

2. Draw a circle with a radius of 35 mm. On the circle, draw a chord.

3. Draw a radius from the centre of the circle to where the tangent touches the circle

4. Draw a circle with a radius of 28 mm. Show a sector on that circle.

5. Draw an arc of a circle.

Exercise 5

A 1. Draw any circle.

2. Draw any tangent to that circle.

3. Draw a radius from the centre of the circle to where the tangent touches the circle.

4. Measure the angle between the tangent and the radius. Write its size.

B Repeat part A for a different circle.

C Write what you notice about the angle between the tangent and the radius for any circle.

9 Squares, Cubes and Square Roots

Squares

Exercise 1

Find the value of:

1. 3^2	**4.** 2^2	**7.** 1^2	**10.** 8^2	**13.** 15^2
2. 5^2	**5.** 0^2	**8.** 7^2	**11.** 10^2	**14.** 25^2
3. 4^2	**6.** 6^2	**9.** 9^2	**12.** 12^2	**15.** 13^2

Exercise 2

25 is a square number.

Here is its dot pattern:

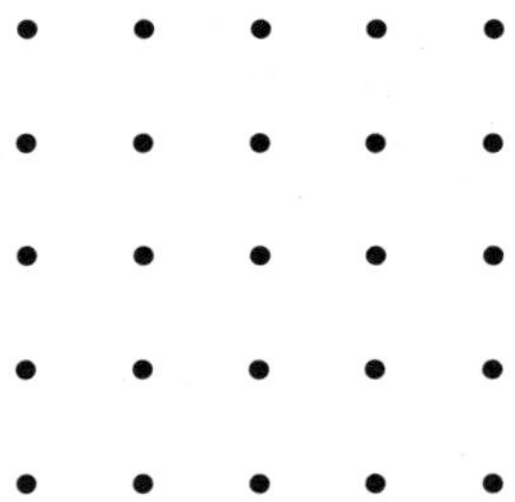

1. Write three more square numbers that are less than 50. Draw their dot patterns.

2. Write two square numbers that lie between 100 and 150.

3. Write a square number that lies between 200 and 300.

4. Write a square number that is bigger than 500.

Exercise 3

Estimate the following (work with 1 significant figure):

e.g. $3.8^2 \approx 4^2 = \underline{\underline{16}}$

1. 8.2^2
2. 9.1^2
3. 7.9^2
4. 2.7^2
5. 1.8^2
6. 6.6^2
7. 4.5^2
8. 6.4^2
9. 9.8^2
10. 5.7^2
11. 4.2^2
12. 9.6^2

Reminder on using a calculator for squaring numbers:
Here are three different ways of finding squares on a calculator.
Check which of the methods work on your calculator.

e.g. To find 5.8^2

Method 1 (use $5.8^2 = 5.8 \times 5.8$)

Key in: [AC] [5] [·] [8] [×] [5] [·] [8] [=]

Method 2

Key in: [AC] [5] [·] [8] [×] [=]

Method 3 (if the calculator has an x^2 key)

Key in: [AC] [5] [·] [8] [x^2]

Exercise 4

Use a calculator to find:

1. 5.2^2
2. 7.8^2
3. 3.1^2
4. 9.4^2
5. 2.9^2
6. 6.3^2
7. 45^2
8. 86^2
9. 74^2
10. 19^2
11. 3.65^2
12. 41.8^2
13. 10.7^2
14. 447^2
15. 623^2
16. 93.8^2
17. 218^2
18. 5.64^2
19. 82.5^2
20. 0.613^2
21. 706^2
22. 0.892^2
23. 98.7^2
24. 50.2^2

Cubes

Exercise 5

Find the value of:

1. 2^3	**4.** 10^3	**7.** 6^3	**10.** 9^3
2. 4^3	**5.** 3^3	**8.** 0^3	**11.** 8^3
3. 5^3	**6.** 1^3	**9.** 7^3	**12.** 11^3

Square Roots

Since $3^2 = 9$ we can also write $\sqrt{9} = 3$.

$\sqrt{\ }$ is called the square root symbol and $\sqrt{9} = 3$ is read as "the square root of 9 is 3".

The square root of a number (9 in this case) is the number, which, when multiplied by itself (for example, 3×3), gives the original number (which was 9).

Exercise 6

e.g.1 $\sqrt{9} = \underline{\underline{3}}$ (since $3 \times 3 = 9$ or $3^2 = 9$)

e.g.2 $\sqrt{64} = \underline{\underline{8}}$ (since $8 \times 8 = 64$ or $8^2 = 64$)

Find the square roots:

1. $\sqrt{4}$	**3.** $\sqrt{100}$	**5.** $\sqrt{16}$	**7.** $\sqrt{1}$
2. $\sqrt{25}$	**4.** $\sqrt{36}$	**6.** $\sqrt{49}$	**8.** $\sqrt{81}$

Exercise 7

Find the two closest whole numbers between which the following square roots lie.

e.g. $\sqrt{70}$ lies between $\underline{\underline{\text{8 and 9}}}$

(since $8^2 = 64$ and $9^2 = 81$ and 70 lies between 64 and 81).

1. $\sqrt{20}$	**5.** $\sqrt{8}$	**9.** $\sqrt{60}$	**13.** $\sqrt{42}$
2. $\sqrt{40}$	**6.** $\sqrt{30}$	**10.** $\sqrt{12}$	**14.** $\sqrt{98}$
3. $\sqrt{75}$	**7.** $\sqrt{18}$	**11.** $\sqrt{72}$	**15.** $\sqrt{23}$
4. $\sqrt{15}$	**8.** $\sqrt{90}$	**12.** $\sqrt{3}$	**16.** $\sqrt{50}$

Some calculators have a square root key on them.

It may look like this: $\boxed{\sqrt{\ }}$ or $\boxed{\sqrt{x}}$.

When the square root key is used, the square root of the number in the display is found.

Try this on a calculator:
Key in $\boxed{1}$ $\boxed{6}$ $\boxed{\sqrt{\ }}$. It should give 4 in the display.

Exercise 8

Using a calculator, find the square roots in Exercise 7.

For each answer shown in the display, write only the first three digits.

e.g.1 $\sqrt{70} = \underline{\underline{8.36}}$ *e.g.2* $\sqrt{2} = \underline{\underline{1.41}}$

Exercise 9

Using a calculator, find the following square roots. For each answer shown in the display, write only the first three digits.

1. $\sqrt{54}$	**7.** $\sqrt{3.5}$	**13.** $\sqrt{4.86}$	**19.** $\sqrt{1000}$
2. $\sqrt{150}$	**8.** $\sqrt{200}$	**14.** $\sqrt{1.92}$	**20.** $\sqrt{5000}$
3. $\sqrt{6}$	**9.** $\sqrt{1.5}$	**15.** $\sqrt{80.4}$	**21.** $\sqrt{2163}$
4. $\sqrt{12}$	**10.** $\sqrt{9.6}$	**16.** $\sqrt{71.9}$	**22.** $\sqrt{3084}$
5. $\sqrt{225}$	**11.** $\sqrt{328}$	**17.** $\sqrt{421}$	**23.** $\sqrt{1.08}$
6. $\sqrt{400}$	**12.** $\sqrt{823}$	**18.** $\sqrt{900}$	**24.** $\sqrt{72.25}$

10 Decimal and Vulgar Fractions

Reading Decimal Scales

Exercise 1

Write the decimal number to which each arrow points:

1.

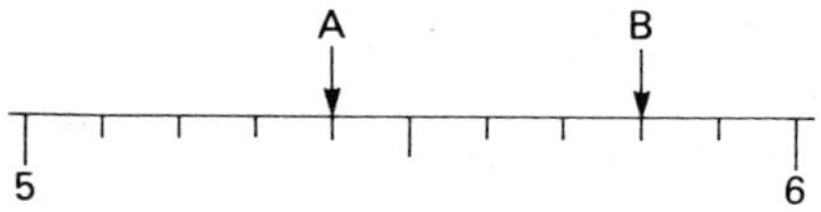

2.

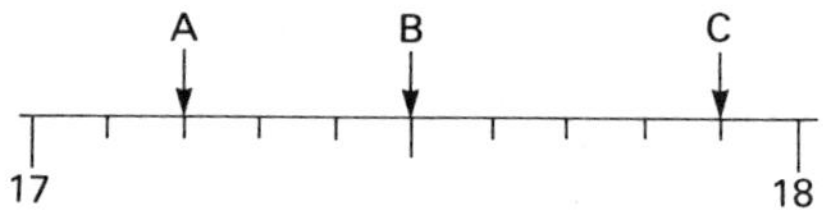

3.

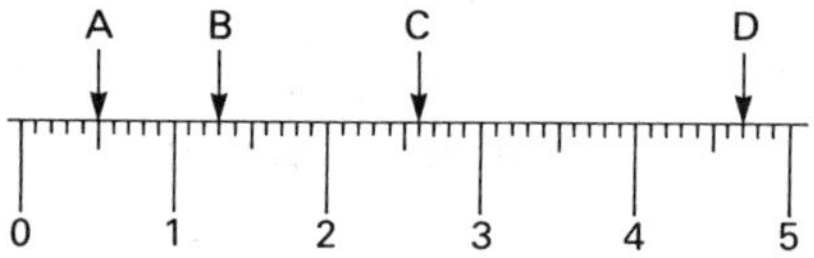

4.

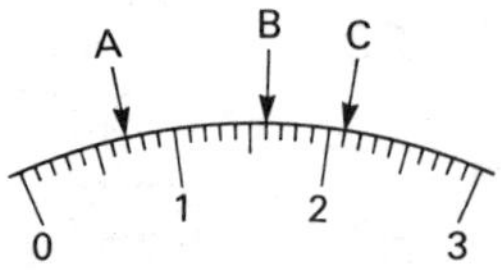

5.

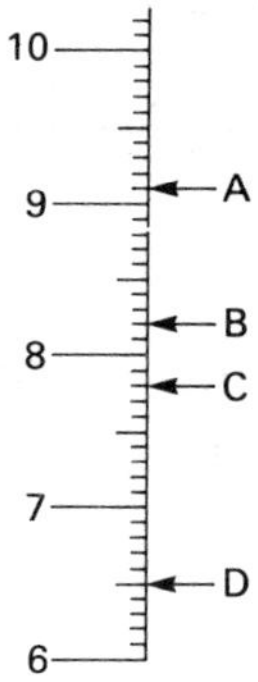

6.

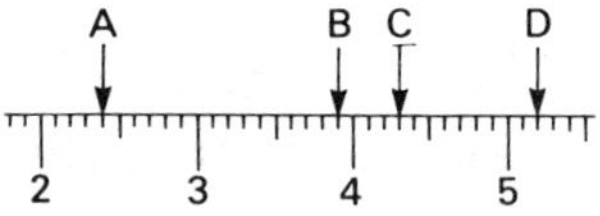

7.

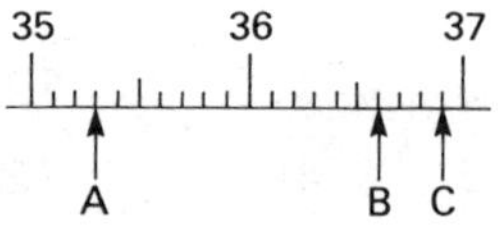

8.

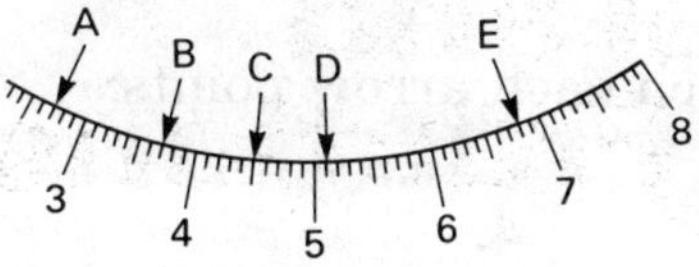

10.

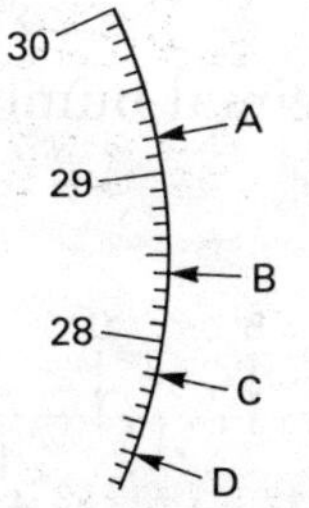

9.

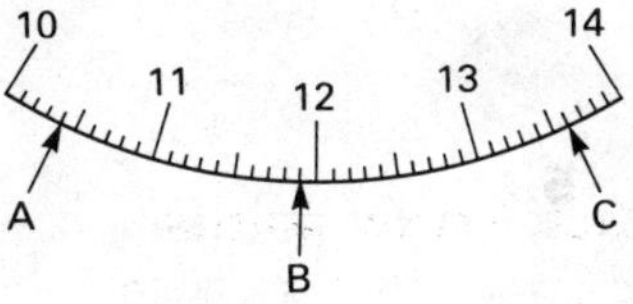

Exercise 2

Write the decimal number to which each arrow points:

1.

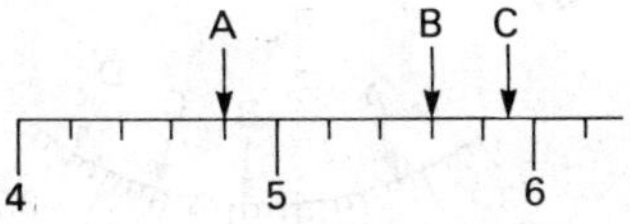

2.

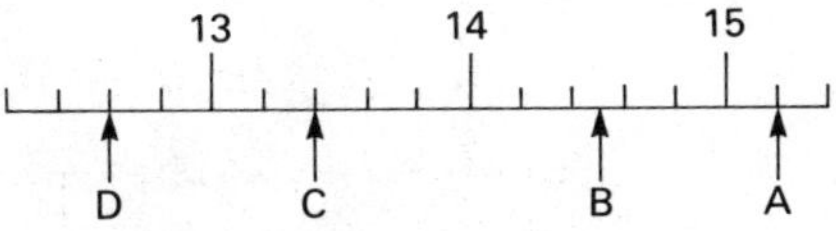

3.

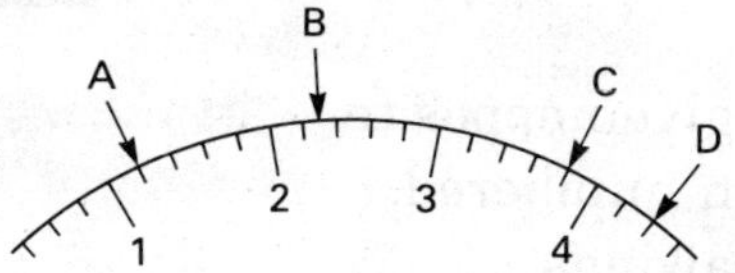

4.

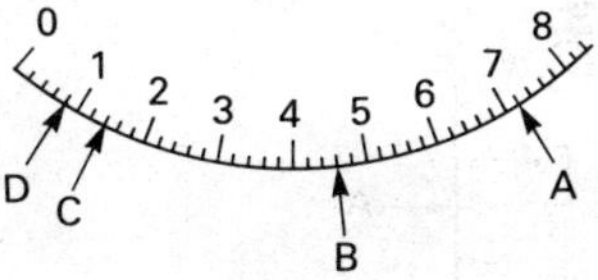

5.

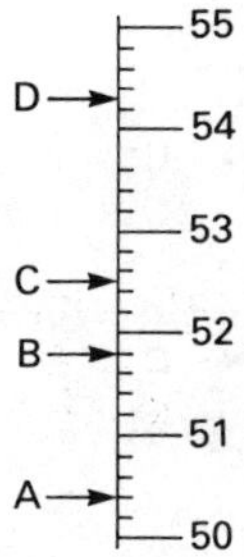

6.

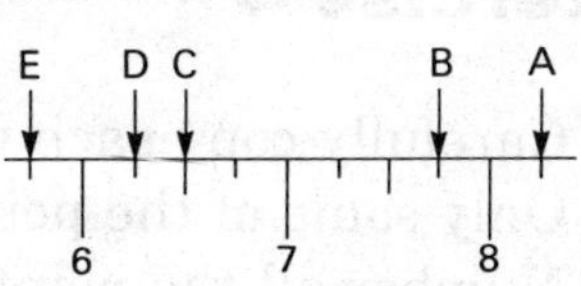

Exercise 3

Write the decimal number to which each arrow points:

1.

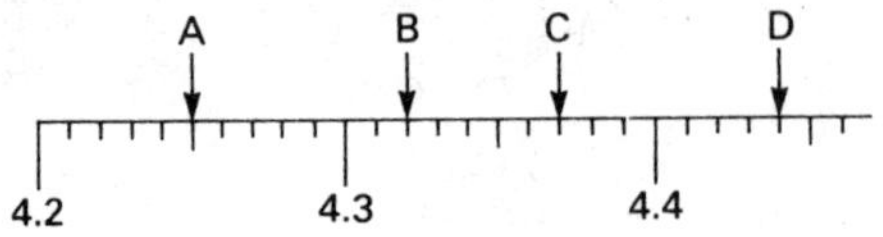

4.

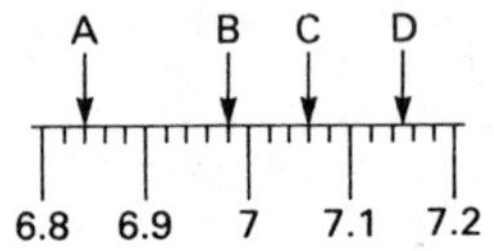

2.

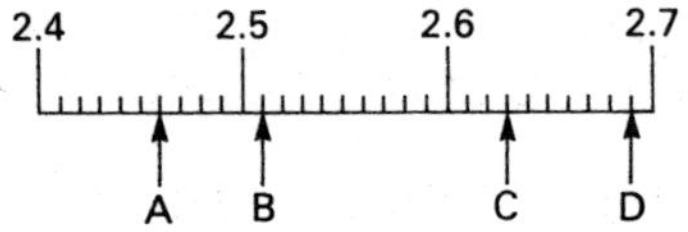

5.

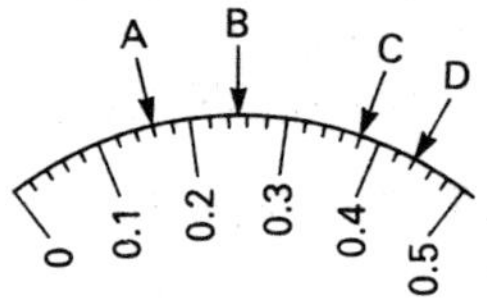

3.

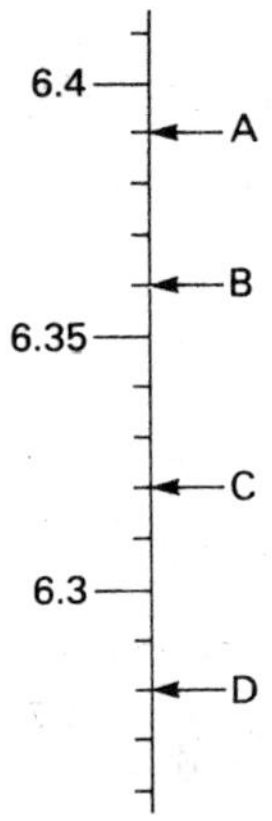

6.

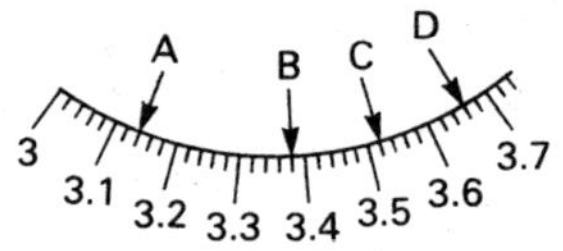

Exercise 4 M

Carefully copy each of the scales given opposite.
Only some of the points have been numbered.
Number all the points on your drawings.

1.

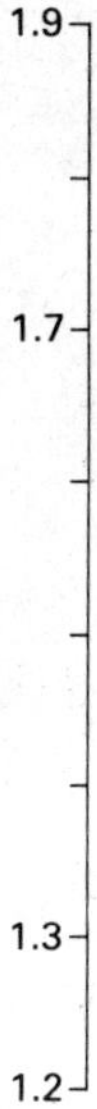

2.

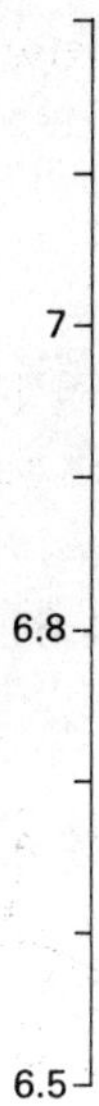

3.

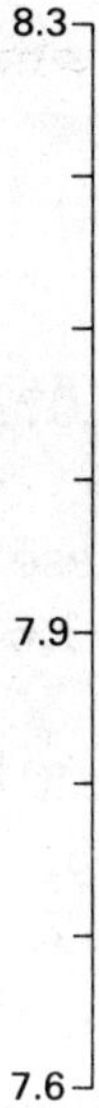

4.

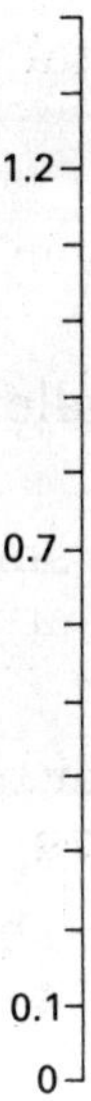

5.

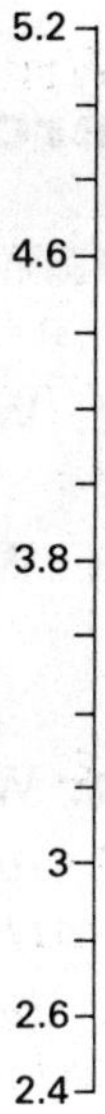

6.

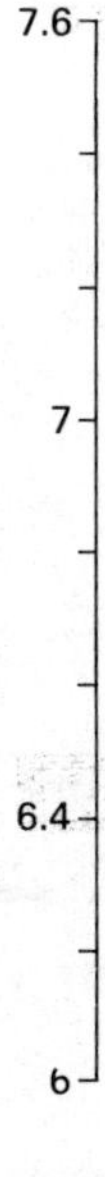

7.

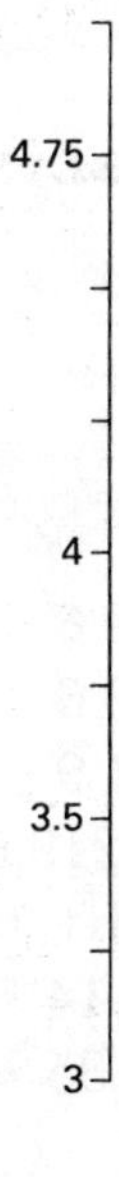

8.

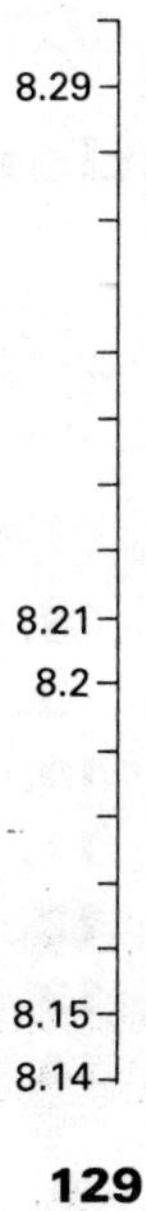

9.

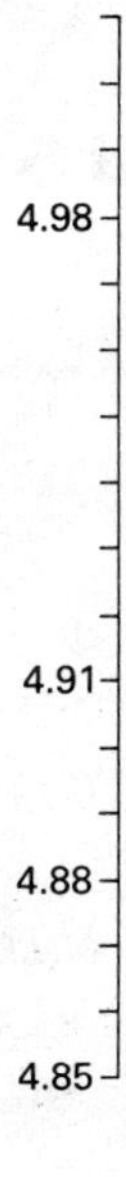

10.

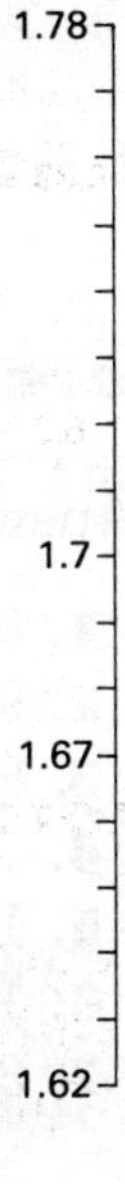

Miscellaneous Work on Decimals

Exercise 5

1. Which is larger, 4.03 or 4.3?

2. Which is smaller, 2.70 or 2.07?

3. Which is the smallest of these numbers?

14.6, 4.16, 16.4, 1.64, 1.46, 4.61

4. Write in order of size, largest first:

(*a*) 4.62, 0.78, 41.8, 4.26, 24.6
(*b*) 7.9, 2.4, 8.1, 6.2, 4.7, 9.3, 8.6
(*c*) 0.24, 2.04, 0.42, 0.4, 0.2, 2.4, 4.2

5. Write in order of size, smallest first:

(*a*) 5.3, 3.5, 0.35, 0.53, 0.36, 0.63, 3.6, 6.3
(*b*) 1.1, 0.11, 1.2, 0.12, 0.21, 2.1
(*c*) 30.7, 3.07, 7.03, 7.3, 3.7, 70.3, 30.07
(*d*) 0.708, 0.87, 0.087, 0.078, 0.807, 0.78

Addition and Subtraction of Decimals

Exercise 6

Answer these:

1. 6.7 + 2.5
2. 9.6 − 4.3
3. 8.2 − 3.5
4. 5.4 + 8.9
5. 12.6 + 25.7
6. 19.8 − 8.9
7. 30.7 − 18.4
8. 51.4 + 29.6
9. 51.4 − 29.6
10. 38.3 + 51.9
11. 136.2 + 219.6
12. 4.5 + 2.6 + 1.3
13. 8.25 − 6.14
14. 9.07 + 6.35

15. 7.09 − 2.68
16. 8.01 − 5.77
17. 1.64 + 4.25 + 3.18
18. 76.4 + 52.6 + 65.3
19. 60.3 − 29.7
20. 71.1 + 18.6 + 20.3

Exercise 7

Answer these:

1. 8 + 3.62
2. 34.9 + 82
3. 9 − 2.6
4. 12.9 + 7.3
5. 56.8 + 2.15
6. 7.62 − 3.4
7. 45 − 18.3
8. 62.7 − 35.9
9. 46 + 3.8 + 17.46
10. 5.7 + 2.84 + 19.77
11. 62.5 − 9.8
12. 70.3 − 12.65

Exercise 8

1. Meg cycled 14.8 km on one day and 12.6 km on the next. How far was that altogether?

2. ABCD is a parallelogram. AB = 7.2 cm and AD = 3.4 cm. Find its perimeter.

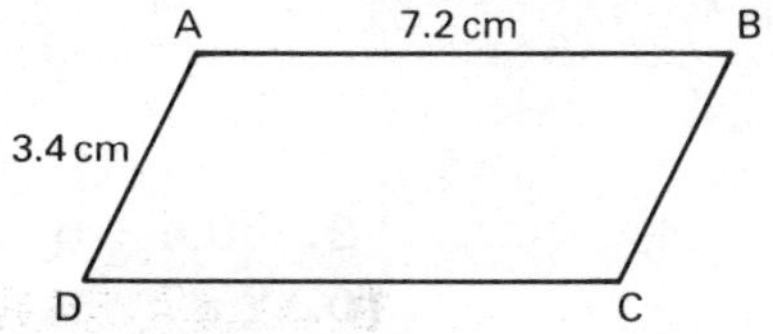

3. A piece of cord is 4.3 m long. If 2.6 m is cut off, what length remains?

4. Mr May lived 6.2 miles from his nearest garage. After travelling 4.9 miles, how much further had he to travel?

5. Find the perimeter of a rectangle that is 5.7 cm long and 3.8 cm wide.

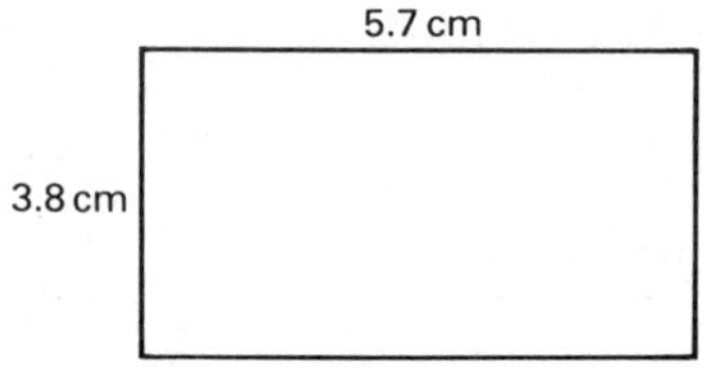

6. A middleweight boxer weighs 73.4 kg while a flyweight weighs 49.8 kg. How much heavier is the middleweight?

7. Mrs Wade's car used 12.6 ℓ of petrol on one day and 11.9 ℓ on the next. How many litres was that altogether?

8. I walked 0.8 km to the bus stop. My bus journey was a distance of 13.7 km. I then walked 1.4 km to work. How far did I travel altogether?

Multiplication and Division of a Decimal by a Single-digit Number

Exercise 9

Work out:

1. 4.7×3
2. $8.4 \div 3$
3. $7.4 \div 2$
4. 0.6×4
5. 8.9×5
6. 24.7×2
7. $14.8 \div 2$
8. $46.8 \div 4$
9. $50.4 \div 6$
10. 2.87×6
11. 9.08×7
12. $21.25 \div 5$
13. $47.25 \div 9$
14. 8×243.7
15. 9×36.51

Exercise 10

1. If a suit is made from 3.2 m of material, what length of material is needed to make 7 suits?

2. If a car travels 14.3 km on 1 ℓ of petrol, how many kilometres should it travel on 9 ℓ?

3. A hiker walked 20.1 km in 3 h. If he walked at a steady rate, how far did he walk each hour?

4. Calculate the area of a rectangle of length 8.6 cm and width 4 cm.

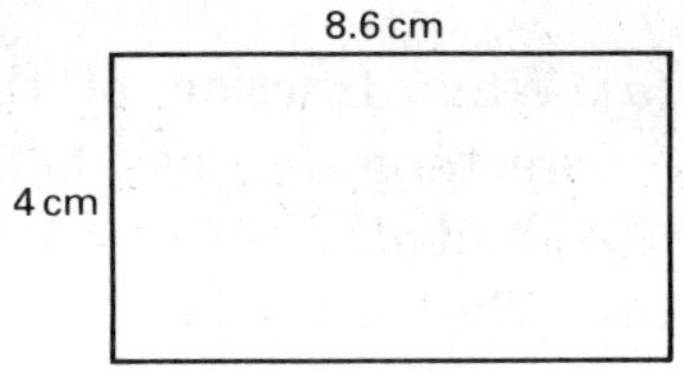

5. The distance around a certain village is 2.8 km. A runner ran around the village 6 times. How far did the runner run altogether?

Decimal Calculations on a Calculator

Exercise 11

Work these out on a calculator:

1. 48.73 + 97.64
2. 74.6 − 13.87
3. 46.8 × 9
4. 71.2 ÷ 4
5. 765.4 + 39.88
6. 47 + 73.68
7. 100 − 37.9
8. 57.6 − 28.92
9. 7.63 × 6
10. 1611.2 ÷ 8
11. 7.86 + 84.9 + 302.7
12. 98.7 − 37.642
13. 34.7 × 2.3
14. 37.196 ÷ 6.8
15. 9.02 × 0.81
16. 47 + 3.9 + 206.5 + 37.21
17. 846.7 − 98.92
18. 8.8 × 3.3
19. 0.74 × 0.95
20. 23.985 ÷ 0.65

Vulgar Fractions

Exercise 12

1. (*a*) What fraction of the large square has been shaded?

(*b*) What fraction remains unshaded?

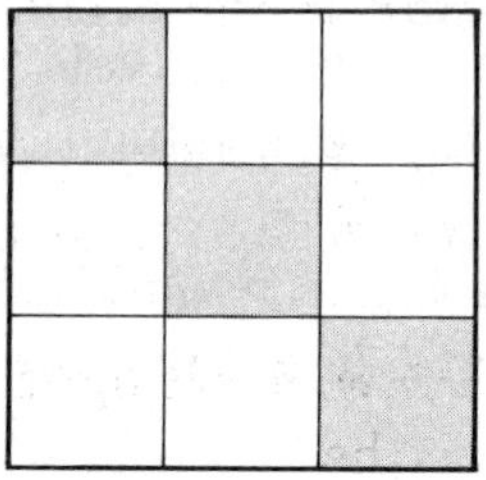

2. (*a*) What fraction of the rectangle has been shaded?

(*b*) What fraction remains unshaded?

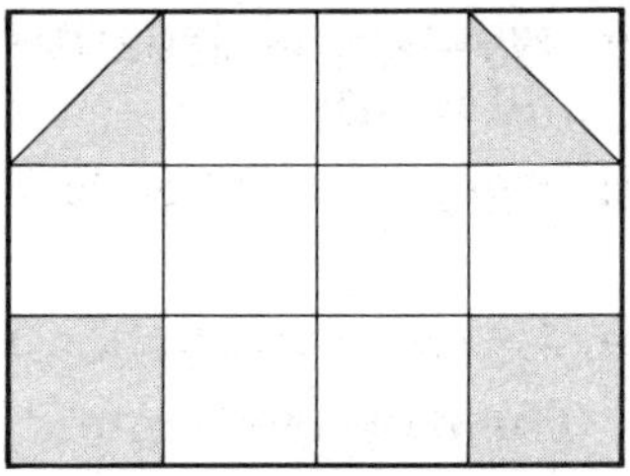

3. (*a*) What fraction of the large square has been shaded?

(*b*) What fraction remains unshaded?

(*c*) How many more small squares would need to be shaded so that altogether $\frac{3}{4}$ of the large square is shaded?

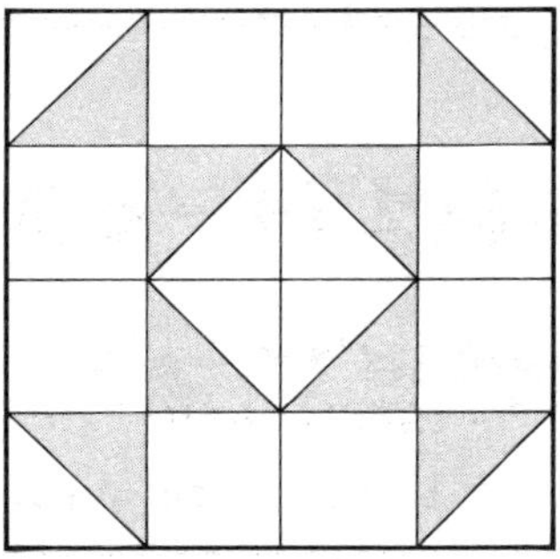

4. The given circle has been divided into 12 equal parts.

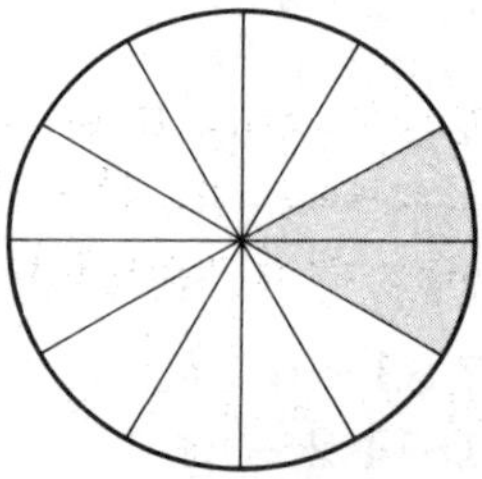

(*a*) What fraction has been shaded?
(*b*) How many more parts need to be shaded so that altogether $\frac{2}{3}$ of the circle is shaded?

5. How many eighths make a whole?

6. (*a*) Write three-fifths in figures.
(*b*) Write $\frac{7}{10}$ in words.

7. Draw a straight line 120 mm in length. Divide it into sixths.

8. How many quarters are there in $2\frac{3}{4}$?

9. Find $\frac{3}{5}$ of 30 girls.

10. Draw a straight line that is $\frac{2}{3}$ the length of the given straight line:

Exercise 13 Equivalent Fractions

A 1. Draw three straight lines, each 8 cm long. The lines should be exactly under each other as shown.

Divide the first line into halves, the second line into quarters and the third line into eighths.
Copy and complete:

(*a*) $\frac{1}{2} = \frac{\boxed{?}}{4} = \frac{\boxed{?}}{8}$ (*b*) $\frac{3}{4} = \frac{\boxed{?}}{3}$

2. Draw five straight lines, each 12 cm long. The lines should be under each other. Divide the first line into halves, the second into sixths, the third into twelfths, the fourth into thirds and the fifth into quarters.
Copy and complete:

(*a*) $\frac{1}{2} = \frac{\boxed{?}}{6} = \frac{\boxed{?}}{12}$

(*b*) $\frac{1}{3} = \frac{\boxed{?}}{6} = \frac{\boxed{?}}{12}$

(*c*) $\frac{2}{3} = \frac{\boxed{?}}{6} = \frac{8}{\boxed{?}}$

(*d*) $\frac{1}{4} = \frac{3}{\boxed{?}}$

(*e*) $\frac{3}{4} = \frac{\boxed{?}}{12}$

B *e.g.1* $\frac{1}{2} \underset{\times 4}{\overset{\times 4}{=}} \frac{4}{8}$

e.g.2 $\frac{16}{24} \underset{\div 8}{\overset{\div 8}{=}} \frac{2}{3}$

Copy and complete:

1. $\frac{1}{2} = \frac{\boxed{?}}{12}$

2. $\frac{1}{3} = \frac{\boxed{?}}{15}$

3. $\frac{12}{16} = \frac{\boxed{?}}{4}$

4. $\frac{12}{18} = \frac{\boxed{?}}{3}$

5. $\frac{2}{5} = \frac{\boxed{?}}{25}$

6. $\frac{15}{24} = \frac{5}{\boxed{?}}$

7. $\frac{4}{7} = \frac{\boxed{?}}{21}$

8. $\frac{70}{100} = \frac{7}{\boxed{?}}$

9. $\frac{15}{18} = \frac{\boxed{?}}{6}$

10. $\frac{16}{20} = \frac{\boxed{?}}{5}$

11. $\frac{3}{8} = \frac{\boxed{?}}{24}$

12. $\frac{1}{6} = \frac{4}{\boxed{?}}$

13. $\frac{3}{12} = \frac{1}{\boxed{?}}$

14. $\frac{27}{30} = \frac{\boxed{?}}{10}$

15. $\frac{1}{12} = \frac{2}{\boxed{?}}$

16. $\frac{21}{35} = \frac{\boxed{?}}{5}$

17. $\frac{28}{32} = \frac{\boxed{?}}{8}$

18. $\frac{13}{52} = \frac{1}{\boxed{?}}$

Addition of Vulgar Fractions

Draw a circle with a radius of 2 cm.

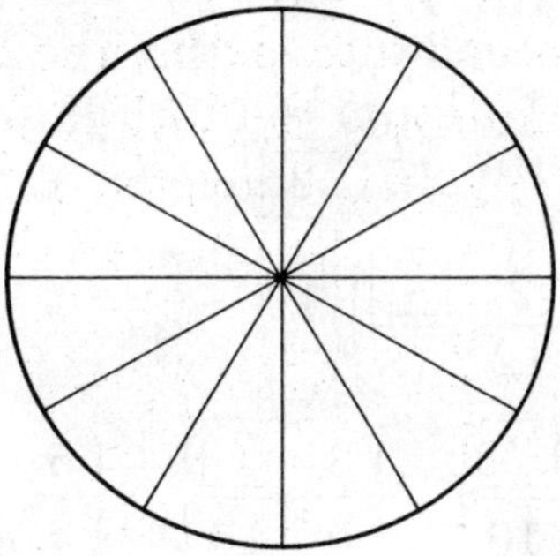

Divide the circle into 12 equal parts.

Shade $\frac{2}{3}$ of the circle.

$\left(\frac{2}{3} = \frac{\boxed{?}}{12}\right)$

Shade a further $\frac{1}{12}$ of the circle.

What fraction has been shaded altogether?

This can be worked out in this way:

$\frac{2}{3} + \frac{1}{12}$	*Notes:*
$= \frac{8}{12} + \frac{1}{12}$	since $\frac{2}{3} = \frac{8}{12}$
$= \frac{9}{12}$	
$= \underline{\underline{\frac{3}{4}}}$	since $\frac{9}{12} = \frac{3}{4}$

Exercise 14 M

A Copy and complete these sets of equivalent fractions:

$\left\{\frac{1}{2}, \frac{2}{4}, \frac{3}{6}, \frac{4}{8}, \frac{\boxed{?}}{10}, \frac{6}{\boxed{?}}, \frac{\boxed{?}}{14}, \frac{\boxed{?}}{16}, \frac{9}{\boxed{?}}, \frac{10}{\boxed{?}}\right\}$

$\left\{\frac{1}{3}, \frac{2}{6}, \frac{3}{9}, \frac{4}{\boxed{?}}, \frac{\boxed{?}}{15}\right\}$ $\left\{\frac{2}{3}, \frac{4}{6}, \frac{6}{9}, \frac{\boxed{?}}{12}, \frac{10}{\boxed{?}}\right\}$

$\left\{\frac{1}{4}, \frac{2}{8}, \frac{3}{12}, \frac{\boxed{?}}{16}, \frac{5}{\boxed{?}}\right\}$ $\left\{\frac{3}{4}, \frac{6}{8}, \frac{9}{12}, \frac{12}{\boxed{?}}, \frac{15}{20}\right\}$

$\left\{\frac{1}{5}, \frac{2}{10}, \frac{3}{15}, \frac{\boxed{?}}{20}\right\}$ $\left\{\frac{2}{5}, \frac{4}{10}, \frac{6}{15}, \frac{\boxed{?}}{20}\right\}$

$\left\{\frac{3}{5}, \frac{6}{10}, \frac{\boxed{?}}{15}, \frac{12}{\boxed{?}}\right\}$ $\left\{\frac{4}{5}, \frac{8}{10}, \frac{12}{\boxed{?}}, \frac{\boxed{?}}{20}\right\}$

$\left\{\frac{1}{6}, \frac{2}{12}, \frac{\boxed{?}}{18}\right\}$ $\left\{\frac{5}{6}, \frac{10}{12}, \frac{\boxed{?}}{18}\right\}$

$\left\{\frac{1}{8}, \frac{\boxed{?}}{16}\right\}$ $\left\{\frac{3}{8}, \frac{\boxed{?}}{16}\right\}$ $\left\{\frac{5}{8}, \frac{10}{\boxed{?}}\right\}$ $\left\{\frac{7}{8}, \frac{\boxed{?}}{16}\right\}$

$\left\{\frac{1}{10}, \frac{\boxed{?}}{20}\right\}$ $\left\{\frac{3}{10}, \frac{6}{\boxed{?}}\right\}$ $\left\{\frac{7}{10}, \frac{\boxed{?}}{20}\right\}$ $\left\{\frac{9}{10}, \frac{\boxed{?}}{20}\right\}$

B Use the above sets of fractions to help you with the following:

e.g. 1 $\frac{3}{8} + \frac{7}{16}$

$= \frac{6}{16} + \frac{7}{16}$

$= \underline{\underline{\frac{13}{16}}}$

e.g.2 $\frac{3}{5} + \frac{1}{15}$

$= \frac{9}{15} + \frac{1}{15}$

$= \frac{10}{15}$

$= \underline{\underline{\frac{2}{3}}}$

1. $\frac{1}{2} + \frac{1}{4}$

2. $\frac{1}{2} + \frac{3}{8}$

3. $\frac{1}{4} + \frac{5}{8}$

4. $\frac{1}{4} + \frac{7}{16}$

5. $\frac{3}{4} + \frac{1}{8}$

6. $\frac{1}{2} + \frac{3}{16}$

7. $\frac{1}{16} + \frac{3}{8}$

8. $\frac{1}{3} + \frac{4}{9}$

9. $\frac{1}{3} + \frac{7}{12}$

10. $\frac{1}{4} + \frac{3}{8}$

11. $\frac{3}{16} + \frac{3}{4}$

12. $\frac{3}{5} + \frac{1}{10}$

13. $\frac{2}{5} + \frac{3}{10}$

14. $\frac{9}{20} + \frac{2}{5}$

15. $\frac{1}{2} + \frac{5}{12}$

Subtraction of Vulgar Fractions

Exercise 15

Carry out these subtractions:

1. $1 - \frac{4}{5}$

2. $1 - \frac{1}{4}$

3. $1 - \frac{1}{8}$

4. $1 - \frac{7}{8}$

5. $1 - \frac{1}{10}$

6. $1 - \frac{3}{10}$

7. $1 - \frac{7}{12}$

8. $1 - \frac{2}{5}$

9. $1 - \frac{11}{16}$

10. $1 - \frac{3}{16}$

11. $1 - \frac{4}{7}$

12. $1 - \frac{9}{20}$

Exercise 16

Carry out these subtractions:

1. $\frac{1}{2} - \frac{1}{4}$

2. $\frac{3}{4} - \frac{1}{2}$

3. $\frac{5}{8} - \frac{1}{4}$

4. $\frac{3}{4} - \frac{3}{8}$

5. $\frac{7}{8} - \frac{1}{2}$

6. $\frac{7}{10} - \frac{2}{5}$

7. $\frac{2}{3} - \frac{4}{9}$

8. $\frac{7}{9} - \frac{1}{3}$

9. $\frac{11}{12} - \frac{5}{6}$

10. $\frac{11}{12} - \frac{1}{2}$

11. $\frac{2}{3} - \frac{7}{12}$

12. $\frac{1}{2} - \frac{3}{16}$

13. $\frac{11}{16} - \frac{3}{8}$

14. $\frac{3}{4} - \frac{5}{16}$

15. $\frac{9}{10} - \frac{2}{5}$

Multiplication of a Vulgar Fraction by a Whole Number

Exercise 17

A Work these out:

1. $2 \times \frac{2}{5}$

2. $3 \times \frac{5}{16}$

3. $5 \times \frac{3}{16}$

4. $5 \times \dfrac{1}{2}$

5. $4 \times \dfrac{2}{3}$

6. $3 \times \dfrac{3}{4}$

7. $4 \times \dfrac{3}{5}$

8. $7 \times \dfrac{3}{8}$

9. $6 \times \dfrac{4}{5}$

10. $3 \times \dfrac{7}{10}$

11. $9 \times \dfrac{3}{10}$

12. $7 \times \dfrac{9}{16}$

B Work these out. Simplify your answers:

1. $4 \times \dfrac{1}{2}$

2. $8 \times \dfrac{3}{4}$

3. $6 \times \dfrac{2}{3}$

4. $6 \times \dfrac{4}{3}$

5. $10 \times \dfrac{2}{5}$

6. $9 \times \dfrac{2}{3}$

7. $5 \times \dfrac{4}{5}$

8. $12 \times \dfrac{5}{6}$

9. $2 \times \dfrac{3}{4}$

10. $6 \times \dfrac{3}{4}$

11. $4 \times \dfrac{5}{6}$

12. $2 \times \dfrac{5}{8}$

Finding a Fraction of a Quantity

Exercise 18

1. Find one-half of 48 h.

2. Find one-half of £8.64.

3. Find $\frac{1}{2}$ of £7.82.

4. (*a*) Find one-quarter of 80 mm.
(*b*) Find three-quarters of 80 mm.

5. (*a*) Find $\frac{1}{3}$ of 75 m.
(*b*) Find $\frac{2}{3}$ of 75 m.

6. (a) Find $\frac{1}{5}$ of 40 km.

   (b) Find $\frac{4}{5}$ of 40 km.

7. (a) Find one-fifth of 35 apples.
   (b) Find three-fifths of 35 apples.

8. (a) Find $\frac{1}{8}$ of 32 weeks.

   (b) Find $\frac{5}{8}$ of 32 weeks.

9. (a) Find $\frac{1}{8}$ of 72 stamps.

   (b) Find $\frac{7}{8}$ of 72 stamps.

10. (a) Find $\frac{1}{10}$ of 60 miles.

    (b) Find $\frac{7}{10}$ of 60 miles.

Exercise 19 Using a Calculator

e.g. Find $\frac{1}{8}$ of £71.52.

Key in [7] [1] [·] [5] [2] [÷] [8] [=]

$\frac{1}{8}$ of £71.52 = $\underline{\underline{\text{£}8.94}}$

Answer the following with a calculator. You need not show which keys you press.

1. Find $\frac{1}{10}$ of £42.80.

2. Find $\frac{1}{5}$ of 328.5 m.

3. Find one-quarter of 112 kg.

4. Find $\frac{1}{8}$ of £192.80.

5. Find $\frac{1}{6}$ of 414 min.

6. Find $\frac{1}{2}$ of 29 376 votes.

7. Find one-third of 1422 yards.

8. Find one-twelfth of £8054.40.

9. Find $\frac{1}{16}$ of 83.2 cm.

10. Find $\frac{1}{20}$ of 19 380 people.

Decimal and Vulgar Fractions

Exercise 20

Use a calculator to change the following vulgar fractions into decimals:

1. $\frac{2}{5}$	**5.** $\frac{37}{50}$	**9.** $\frac{3}{20}$	**13.** $\frac{15}{16}$
2. $\frac{4}{5}$	**6.** $\frac{1}{25}$	**10.** $\frac{13}{20}$	**14.** $\frac{9}{32}$
3. $\frac{5}{8}$	**7.** $\frac{6}{25}$	**11.** $\frac{3}{16}$	**15.** $\frac{3}{40}$
4. $\frac{9}{50}$	**8.** $\frac{19}{25}$	**12.** $\frac{7}{16}$	**16.** $\frac{21}{40}$

Exercise 21

For each question, write the bigger fraction. Use a calculator to help you.

e.g. Which is bigger, $\frac{3}{4}$ or $\frac{2}{3}$?

$\frac{3}{4} = 0.75$ and $\frac{2}{3} = 0.666\,666\,6$ So $\underline{\underline{\frac{3}{4} \text{ is bigger}}}$

1. $\frac{3}{5}$ or $\frac{9}{16}$	**4.** $\frac{7}{20}$ or $\frac{3}{8}$	**7.** $\frac{5}{12}$ or $\frac{7}{16}$
2. $\frac{5}{8}$ or $\frac{7}{10}$	**5.** $\frac{11}{20}$ or $\frac{9}{16}$	**8.** $\frac{7}{8}$ or $\frac{21}{25}$
3. $\frac{14}{25}$ or $\frac{11}{20}$	**6.** $\frac{17}{50}$ or $\frac{5}{16}$	**9.** $\frac{7}{10}$ or $\frac{13}{18}$

11 Circumference of a Circle

Exercise 1

The diagrams in this exercise have not been drawn to size:

1. Write the radius and diameter of each circle:

(*a*)

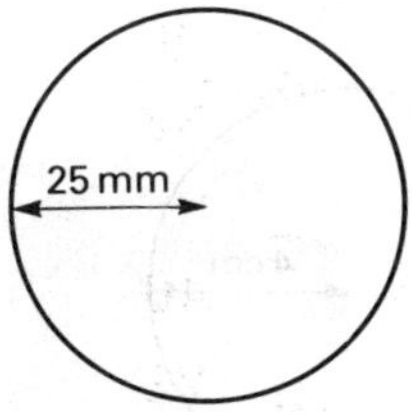

(*b*)

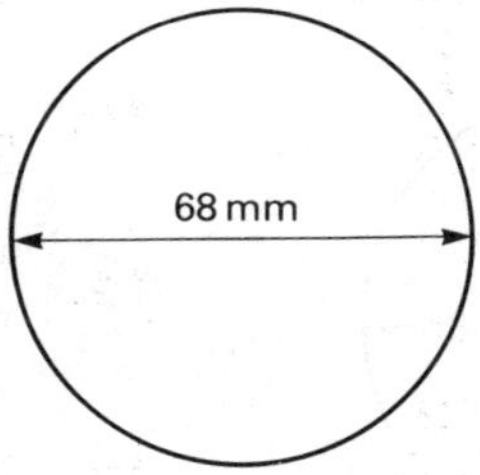

(*c*)

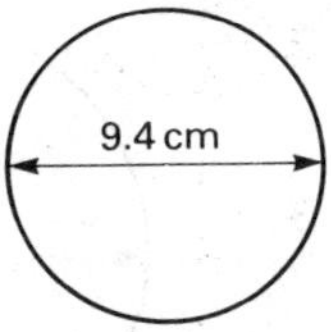

2. Find the diameter of a circle when its radius is:
(*a*) 65 mm (*b*) 3.9 cm (*c*) 7.3 m (*d*) 1.7 ft

3. Find the radius of a circle when its diameter is:
(*a*) 46 mm (*b*) 7 m (*c*) 5.6 in (*d*) 23.8 cm

4. A circular flower bed is shown.
Its radius is 1.8 m.
Find its diameter.

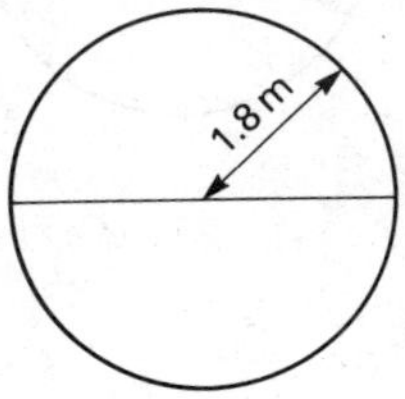

5. A semicircular flower bed is shown. Its radius is 3.2 m.
Find the distance AB along its straight side.

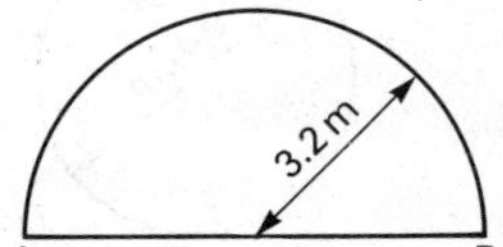

Reminder:

The circumference of a circle is approximately 3 times the diameter.
We can write, $C \approx 3d$

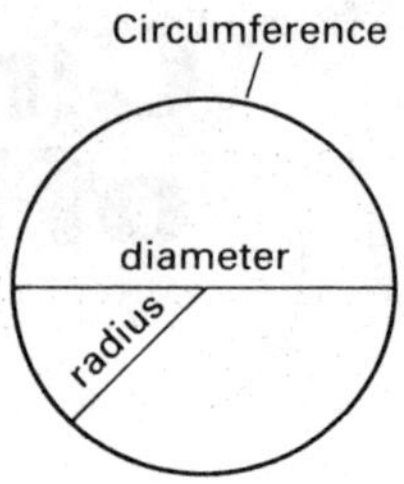

Exercise 2

Find the circumference of each circle. Use $C \approx 3d$.

1.

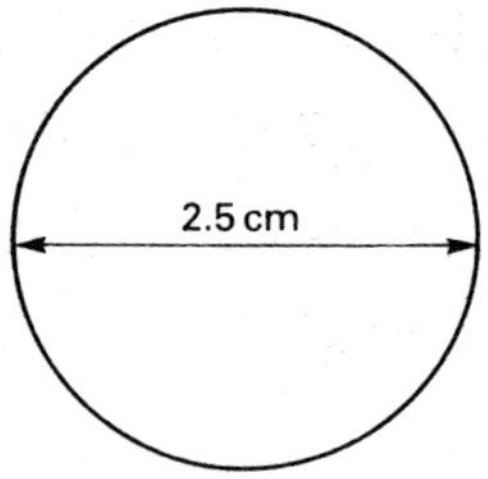

2.

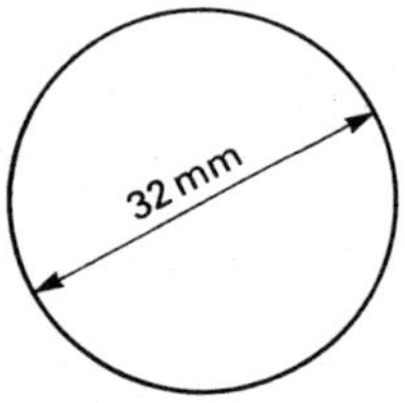

3.

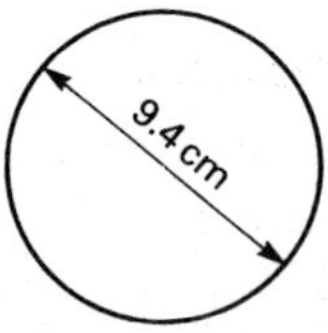

4.

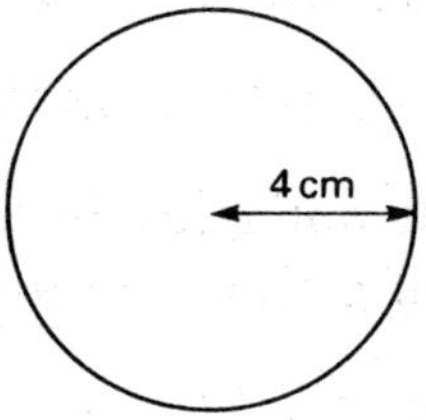

5.

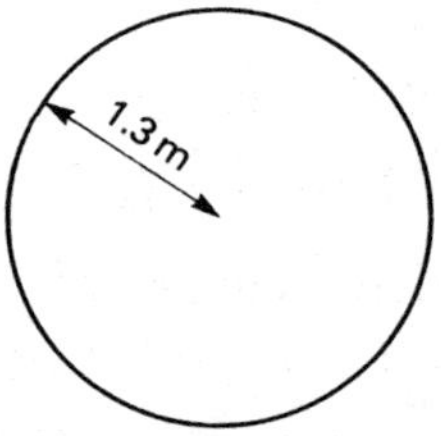

6.

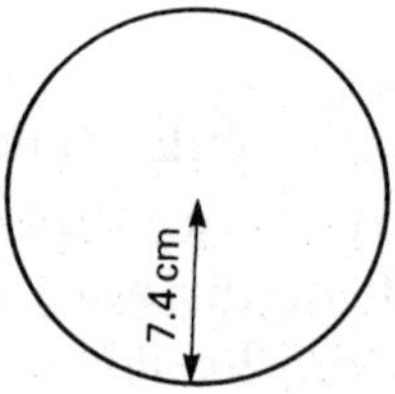

7.

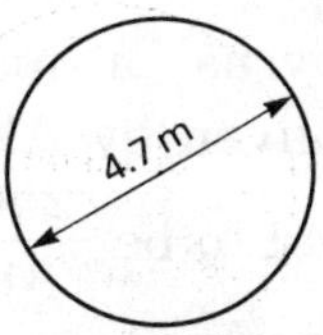

8.

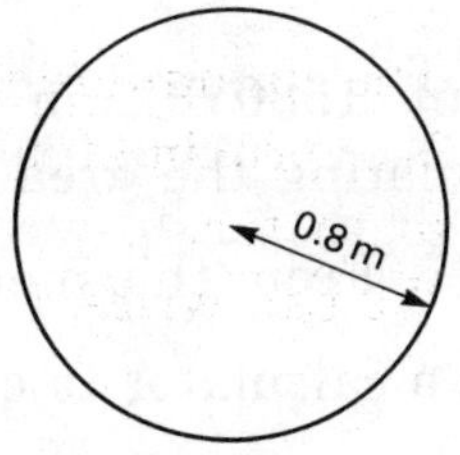

The circumference of a circle happens to be slightly more than 3 times the diameter.

Instead of writing the circumference of a circle $C \approx 3d$ we normally write, the circumference of a circle $C = \pi d$.

The Greek letter π (pi) stands for a number that is slightly more than 3. It does not have an exact value. Here is its value to 39 decimal places:

$\pi = 3.141\,592\,653\,589\,793\,238\,462\,643\,383\,279\,502\,884\,197\ldots$

You will not need to use many decimal places. The decimal values of π that you are most likely to use are

$$\pi = 3.14 \quad \text{or} \quad \pi = 3.142 \quad \text{or} \quad \pi = 3.1.$$

Some calculators have a π key on them. Calculators are most likely to use $\pi = 3.141\,592\,7$ (7 decimal places).

The world record for memorising π is held by Rajan Mahadevan of India who memorised 31 811 decimal places.

The British record of memorising 20 013 decimal places of π is held by Creighton Carvello of Redcar.

To remember π to 4 decimal places, just remember

Yes	I	like	a	cuddle
3 letters	1 letter	4 letters	1 letter	6 letters
3 .	1	4	1	6

How	I	wish	I	could	recollect	pi	easily	today
3 .	1	4	1	5	9	2	6	5

gives 8 decimal places.

Exercise 3

1. About 1580 BC, in the Rhind Papyrus, a method for calculating the area of a circle was given by Ahmes, the scribe. From this method, pi was found to be $\frac{256}{81}$.
Use a calculator to change this to a decimal.

Compare its accuracy with the decimal value given on p.145.

2. Over 2000 years ago, Archimedes (287–212 BC), a Greek mathematician, showed that the value of pi lies between $3\frac{10}{71}$ and $3\frac{1}{7}$. Use a calculator to change these two values of pi into decimals. Compare their accuracy using the decimal value given on p.145.
(Note that $3\frac{1}{7}$ is still used as a suitable approximation to π.)

3. About AD 150, Ptolemy, another Greek, found pi to have the value of 3°8′30″. This is $3\frac{17}{120}$ or $\frac{377}{120}$.

Change $\frac{377}{120}$ to a decimal and compare it with the decimal value given on p.145.

4. In China, about AD 480, Tsu Ch'ung-Chih found quite an accurate value of pi. He gave the value $\frac{355}{113}$.

Change $\frac{355}{113}$ to a decimal.
Compare it with the decimal value given on p.145.

Some Notes on π

1. About AD 1430, the Persian mathematician, Jemshid Al-Kashi, gave pi to 16 decimal places.

2. Ludolf van Ceulen (1540–1610), a German, found pi to 35 decimal places. He had worked on the problem for most of his life.

3. William Shanks (1853) found π to 707 decimal places. However, the last 180 digits were incorrect. It took him about 20 years of calculating!

4. In 1949, the ENIAC computer calculated π to 2035 decimal places in 80 h of machine time.

5. In 1958, G. E. Felton used a Ferranti Pegasus computer to find π to 10 021 places in 33 machine hours. (This would have taken about 100 years to work out on a calculator!)

6. Dr Daniel Shanks and Dr John W. Wrench Jr in New York in 1961 used an IBM 7090 computer to calculate π to 100 265 places in 8 h 43 min.

7. In 1981, Dr Kazunori Miyoshi of the University of Tsukuba in Japan worked out π to two million decimal places in 137 h on a FACOM M200 computer.

8. In 1983, two other Japanese mathematicians, Yoshiaki Tamura and Yasumasa Kanada succeeded in working out π to eight million decimal places in 7 h using a HITAC M280H computer.

Exercise 4

In Exercise 3, several ratios were given for π.

Here are some more. Use a calculator to change these to decimals. In each case, state the number of decimal places that are correct, by comparing with the value given on p.145.

1. $\frac{333}{106}$ **3.** $\frac{531}{169}$ **5.** $\frac{732}{233}$ **7.** $\frac{1043}{332}$

2. $\frac{421}{134}$ **4.** $\frac{688}{219}$ **6.** $\frac{1021}{325}$ **8.** $\frac{1354}{431}$

Exercise 5

In Exercises 3 and 4 there are a number of different ratios that give good approximations for π.

Try to find some ratios of your own that give approximations for π.

Exercise 6

Find the circumference of each of the following circles.
Use the formula:

$$\text{Circumference of a circle, } C = \pi d$$

and take π to be 3.14. (As a check, estimate the answers using $\pi = 3$.)

A **1.**

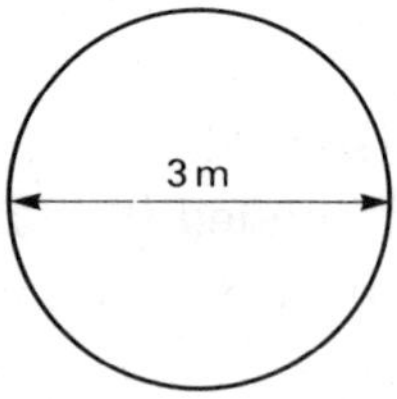

2.

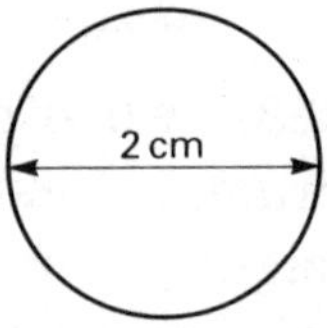

3.

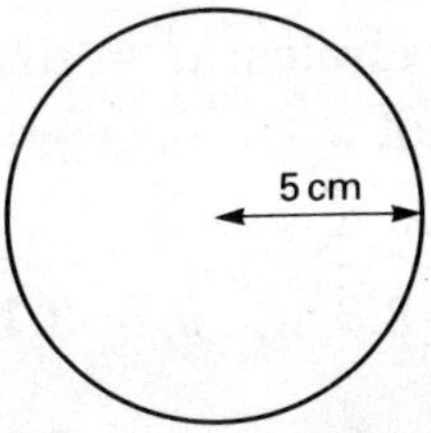

6.

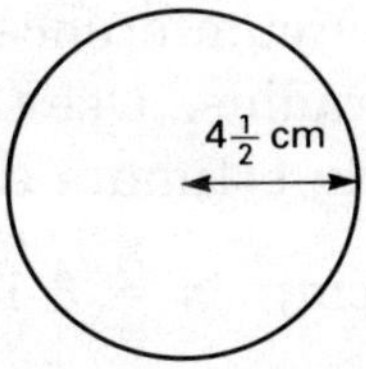

4.

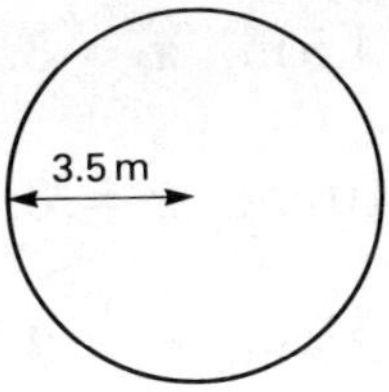

7.

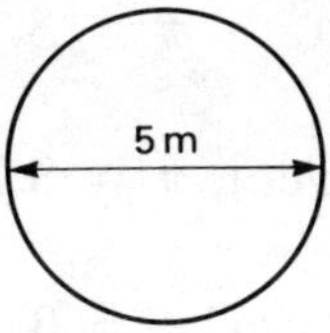

5.

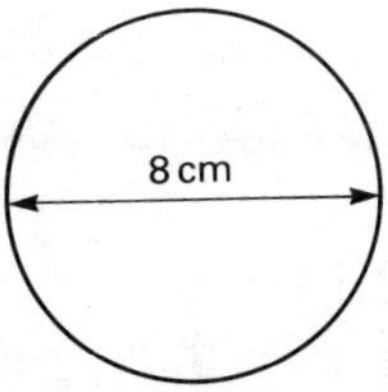

8.

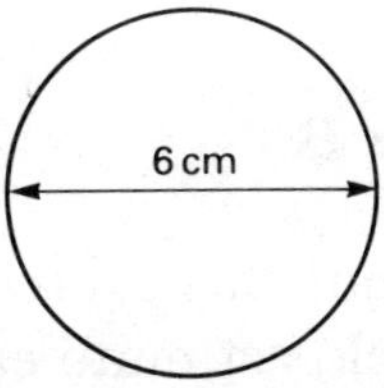

B

1.

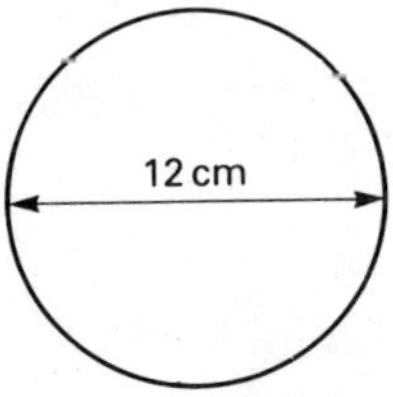

3.

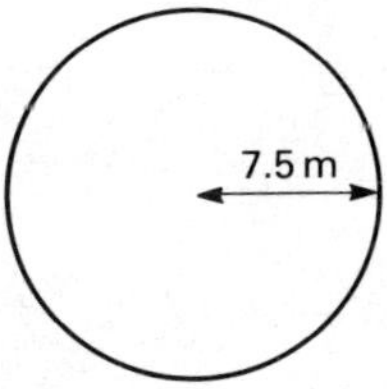

2.

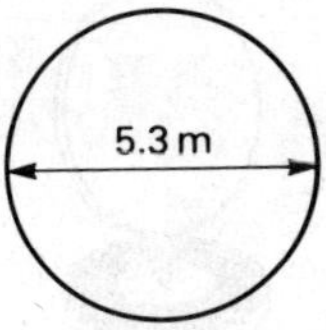

4.

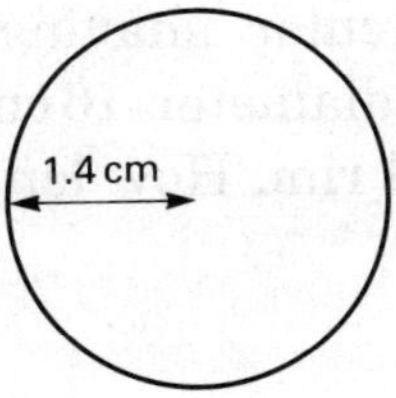

Exercise 7

Find the circumference of the following circles. d = diameter and r = radius. Use the given values of π.
As a check, estimate each answer using $\pi = 3$.

1. $d = 5\text{ cm}$, $\pi = 3.142$
2. $d = 9\text{ ft}$, $\pi = 3.142$
3. $r = 3\text{ m}$, $\pi = 3.142$
4. $d = 8.3\text{ cm}$, $\pi = 3.1$
5. $r = 21\text{ mm}$, $\pi = 3.1$
6. $r = 3.3\text{ m}$, $\pi = 3.14$
7. $d = 68\text{ in}$, $\pi = 3.1$
8. $d = 4.5\text{ m}$, $\pi = 3.142$
9. $r = 10\text{ cm}$, $\pi = 3.142$
10. $d = 2.9\text{ m}$, $\pi = 3.1$

Exercise 8

Throughout this exercise, take π to be 3.14.
As a check, estimate each answer using $\pi = 3$.

1. A cylindrical waste-paper bin has a diameter of 20 cm. Calculate the distance around the bin.

2. A circular shaving-mirror with diameter 16 cm has a metal rim. How long is the rim?

3. The sketches show some kitchen containers with two metal bands around each of them:

The diameter of each container is:
tea 9 cm, coffee 11 cm, sugar 14 cm and flour 17 cm.

(*a*) Calculate the length of each metal band.
(*b*) Find the total length of all eight metal bands giving the answer in metres.

4. A dart-board is edged with a metal strip. If the dart-board has a diameter of 430 mm,

(*a*) Find the length of the metal strip to the nearest millimetre.
(*b*) Change the length of the metal strip into metres.

5. The radius of a circular flower bed is $2\frac{3}{4}$ m. Find its perimeter.

6. The sketch shows a semi-circular sand pit. Its radius is 3.1 m.

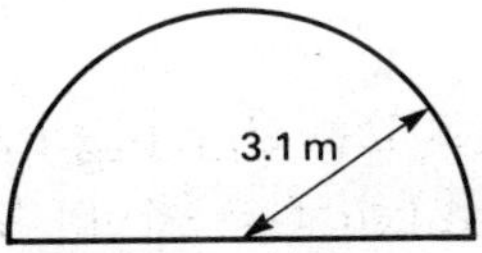

(*a*) Find the length of the straight part of its border.
(*b*) Find the length of the curved part of its border.
(*c*) Find the perimeter of the sand pit.

7. A motorbike is racing around a circular track of radius 150 m. How far does the motorbike go in:

(*a*) 1 lap? (*b*) 5 laps? (*c*) 30 laps?

8. The diameter of the roller on Mrs Green's lawn mower is 21 cm.

(*a*) Calculate the circumference of the roller.

(*b*) What is the length of Mrs Green's lawn if the roller makes 20 full turns in mowing one strip? (Give the answer in metres.)

(*c*) How far will Mrs Green walk in mowing the lawn if the roller makes 500 full turns?

12 Area and Volume

Area of a Rectangle

Area of a rectangle $= l \times b$

$$A = lb$$

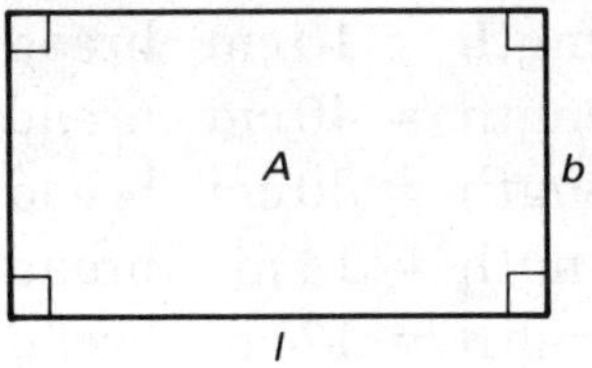

Exercise 1

A Find the areas of the following rectangles:

1.

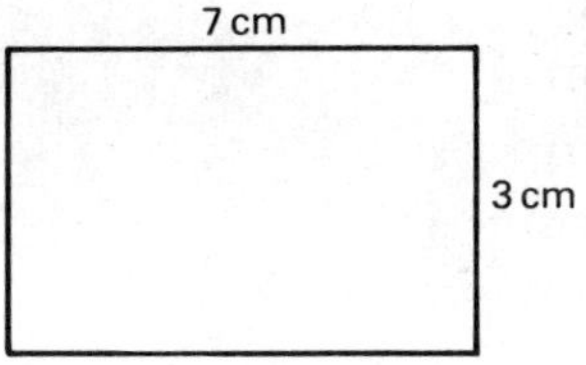

4.

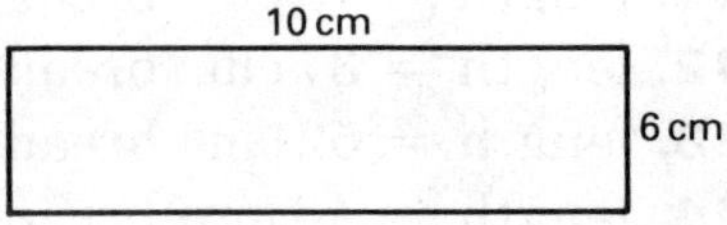

2.

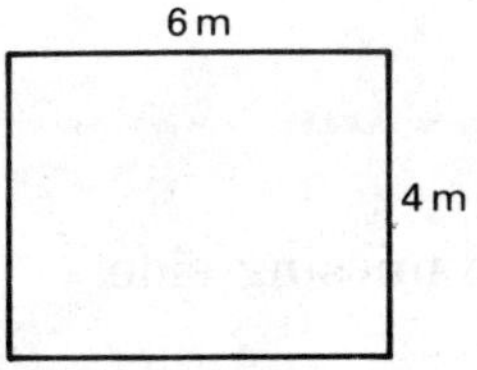

5.

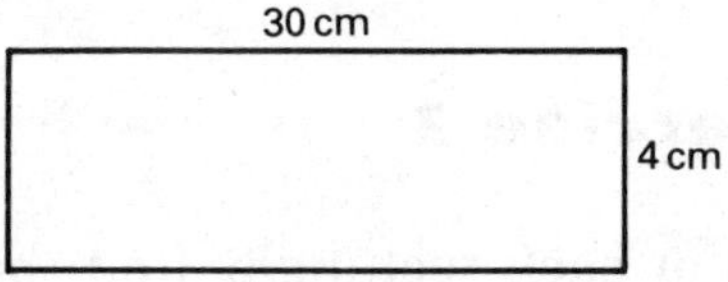

3.

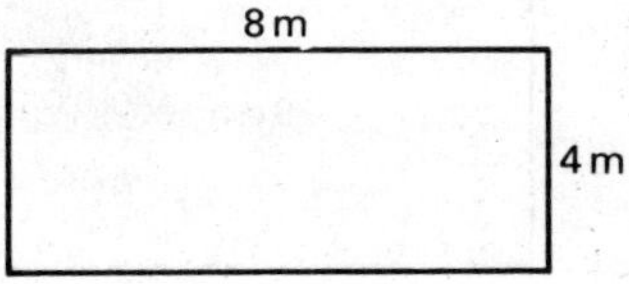

6.

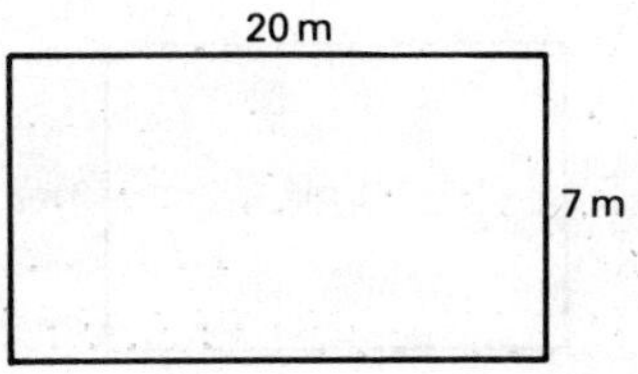

7.

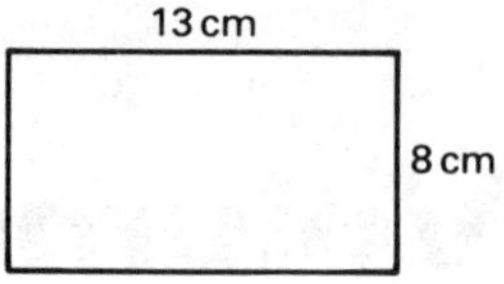

8.

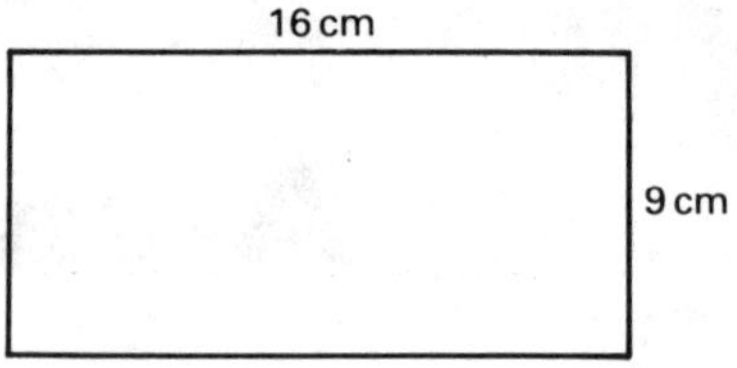

B Find the area of rectangles having the following lengths and breadths:

1. length = 10 cm breadth = 3 cm
2. length = 40 cm breadth = 2 cm
3. length = 30 cm breadth = 8 cm
4. length = 14 m breadth = 3 m
5. length = 17 cm breadth = 5 cm
6. length = 12 m breadth = 9 m
7. length = 34 cm breadth = 7 cm
8. length = 4.1 cm breadth = 8 cm
9. length = 2.3 m breadth = 6 m
10. length = 29 cm breadth = 10 cm
11. length = 18 cm breadth = 14 cm
12. length = 37 cm breadth = 25 cm
13. length = 52 cm breadth = 36 cm
14. length = 46 mm breadth = 42 mm
15. length = 79 mm breadth = 33 mm

Exercise 2

For each rectangle, find the length of the missing side:

A 1.

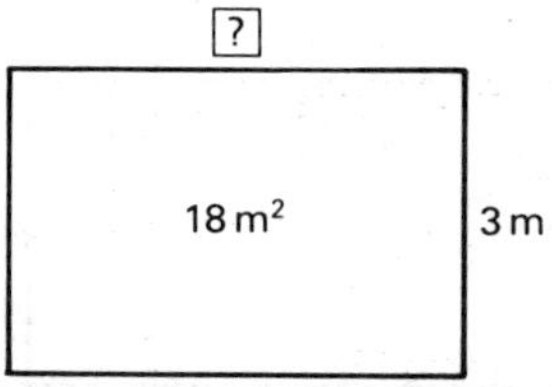

2.

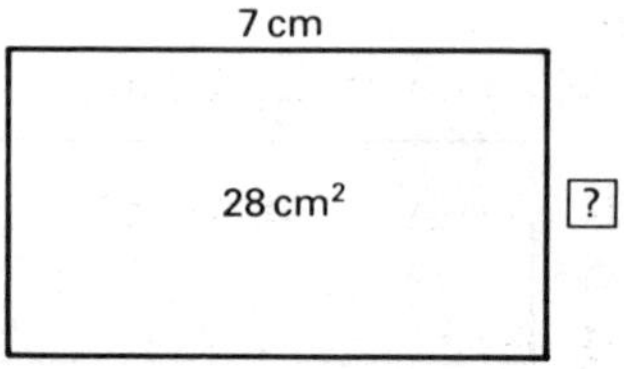

3.

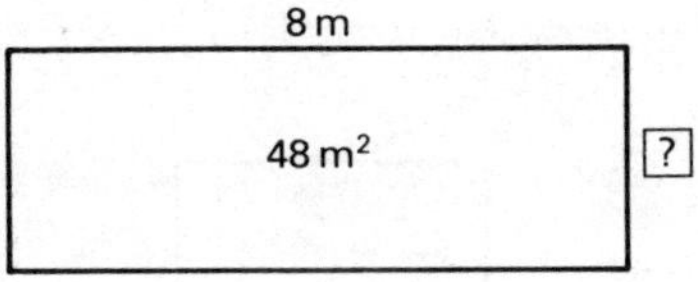

6.

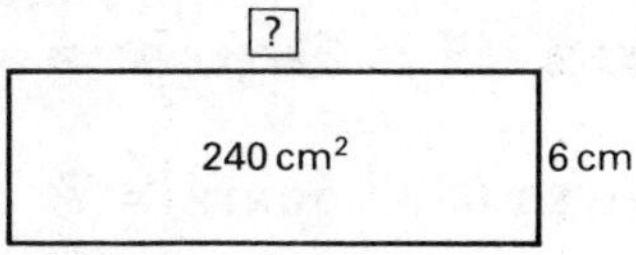

4.

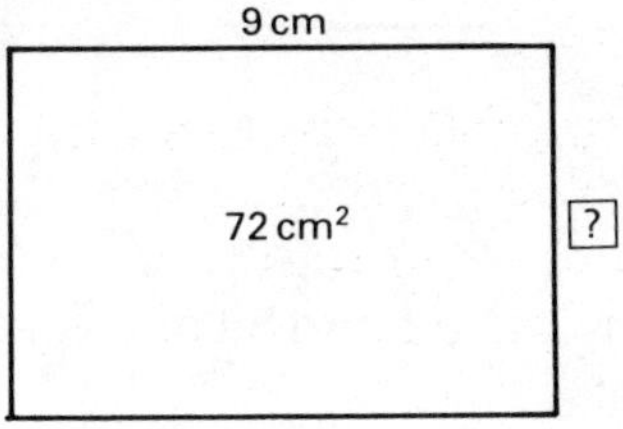

7.

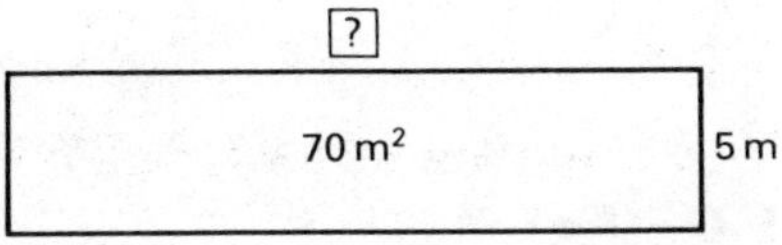

5.

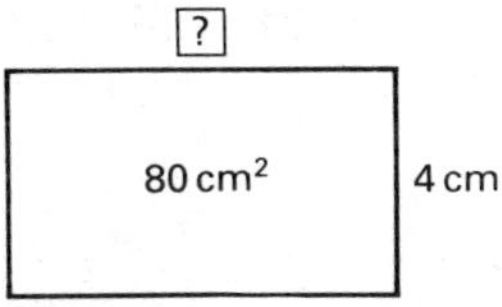

8.

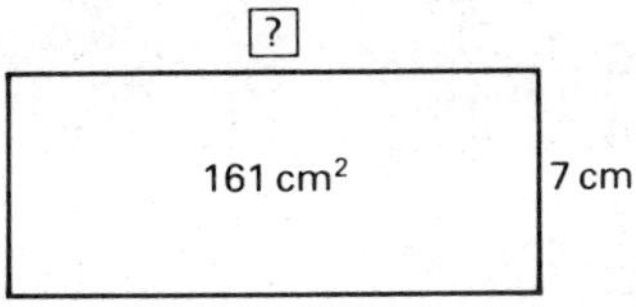

B

1. Area = $56\,m^2$ length = 8 m breadth = ?

2. Area = $63\,cm^2$ length = 9 cm breadth = ?

3. Area = $84\,m^2$ breadth = 7 m length = ?

4. Area = $280\,cm^2$ breadth = 4 cm length – ?

5. Area = $120\,cm^2$ length = 30 cm breadth = ?

6. Area = $510\,cm^2$ breadth = 6 cm length = ?

7. Area = $34\,m^2$ breadth = 5 m length = ?

8. Area = $27.2\,cm^2$ length = 8 cm breadth = ?

9. Area = $364\,cm^2$ breadth = 13 cm length = ?

10. Area = $777\,m^2$ breadth = 21 m length = ?

Area of a Square

Area of a square $= l^2$

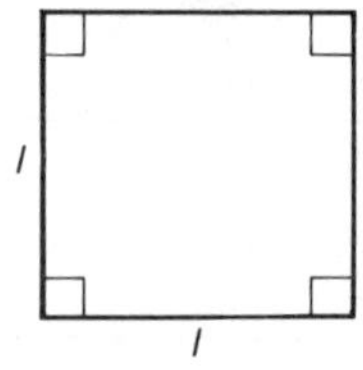

Exercise 3

Find the areas of the squares for which the lengths of the sides are:

1. 6 cm	**4.** 50 cm	**7.** 4.7 m	**10.** 9.1 m
2. 8 m	**5.** 30 m	**8.** 38 cm	**11.** 28 cm
3. 11 cm	**6.** 70 mm	**9.** 62 mm	**12.** 53 cm

Exercise 4

The areas of several squares are given. For each one, use a calculator to help you find the length of each side:

1. $9\,cm^2$	**4.** $400\,cm^2$	**7.** $8100\,cm^2$	**10.** $676\,cm^2$
2. $49\,m^2$	**5.** $1600\,mm^2$	**8.** $441\,cm^2$	**11.** $14.44\,cm^2$
3. $100\,cm^2$	**6.** $3600\,cm^2$	**9.** $2.89\,cm^2$	**12.** $5184\,mm^2$

Area of a Parallelogram

Exercise 5

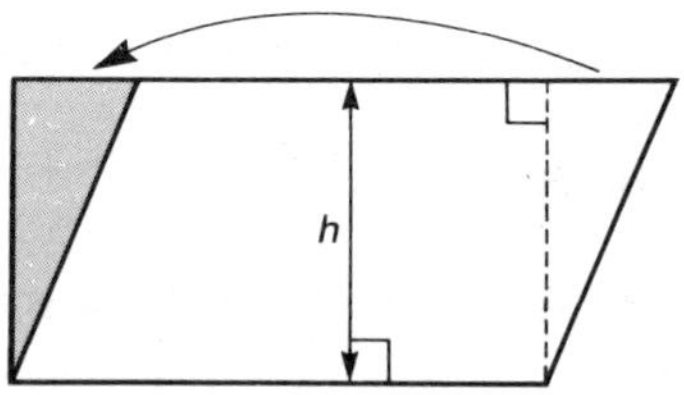

Draw a parallelogram on a piece of paper. Cut it out. Using one straight cut, as shown by the dotted line in the sketch, cut the parallelogram into two pieces. Move the triangular part as shown to form a rectangle. This shows that

Area of a parallelogram $= b \times h$

Note that h is the perpendicular height, not the slant height.

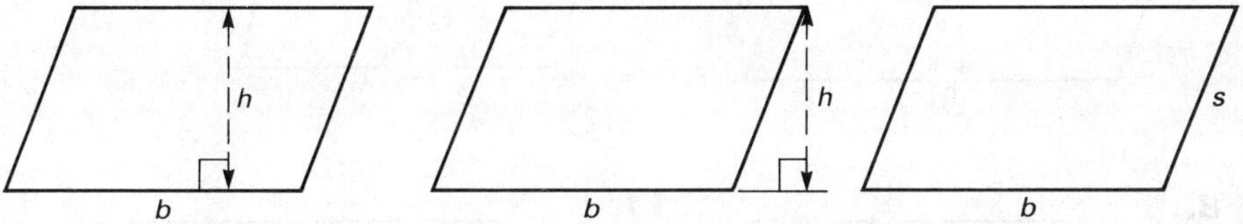

In the first two diagrams above, h is the perpendicular height. In the third diagram, s is the slant height.

In the formula, Area of a parallelogram $= b \times h$, h is the perpendicular height.

Exercise 6

Calculate the area of each parallelogram:

1.

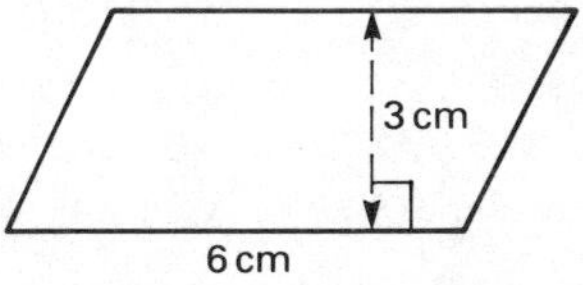

2.

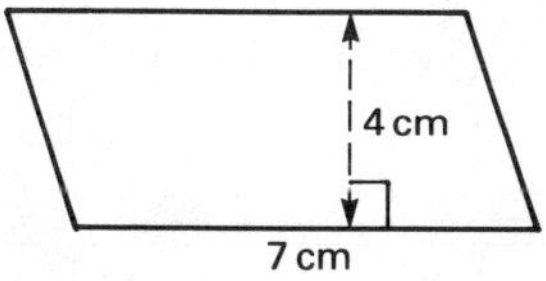

3.

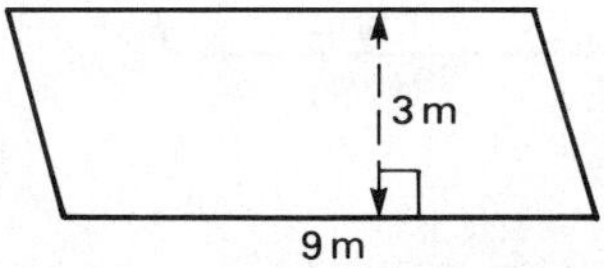

4.

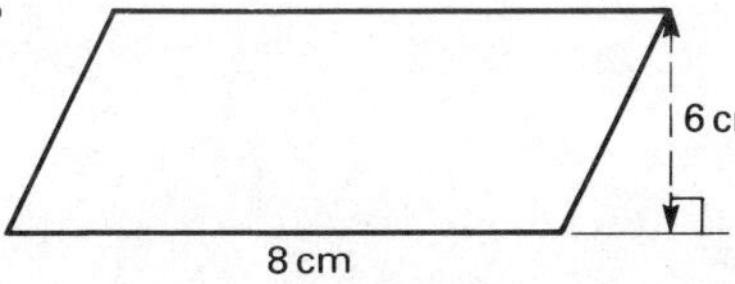

5.

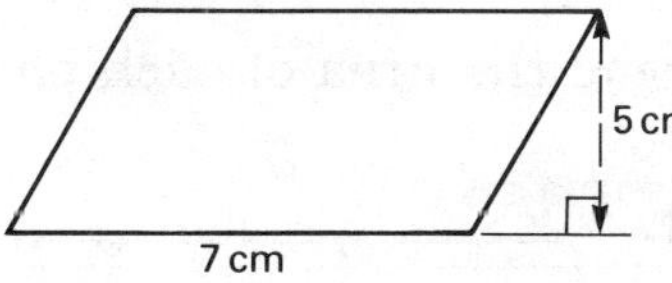

6.

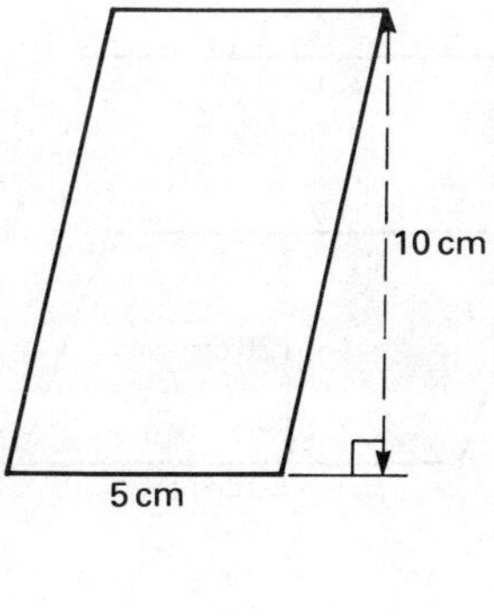

7.
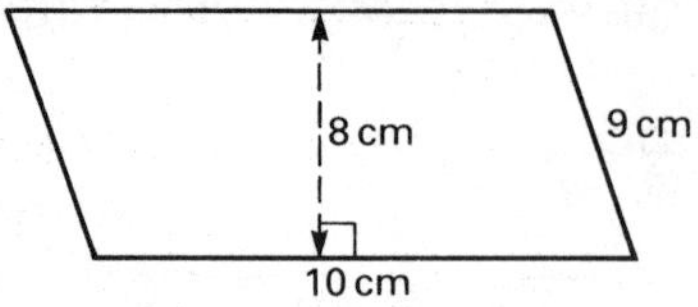

10.
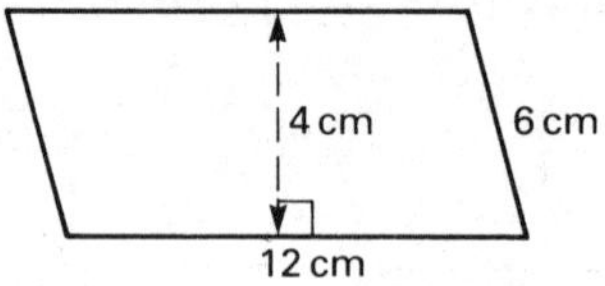

8.
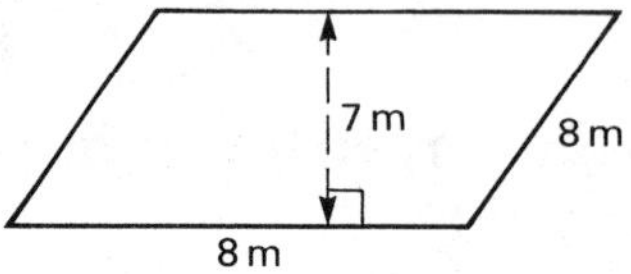

11.
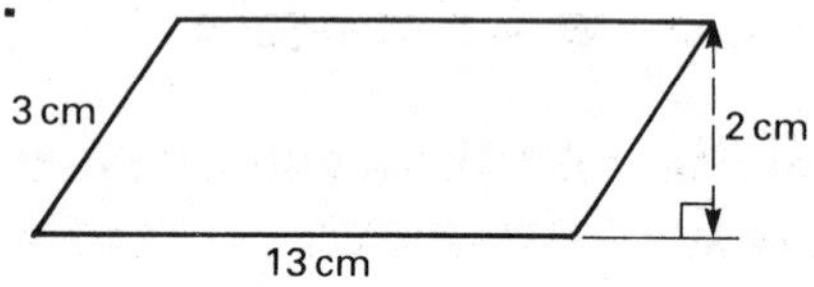

9.
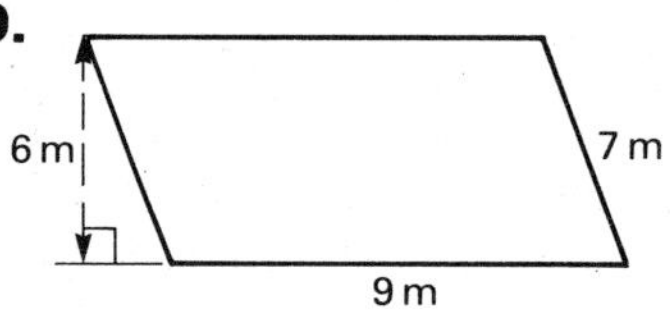

12.
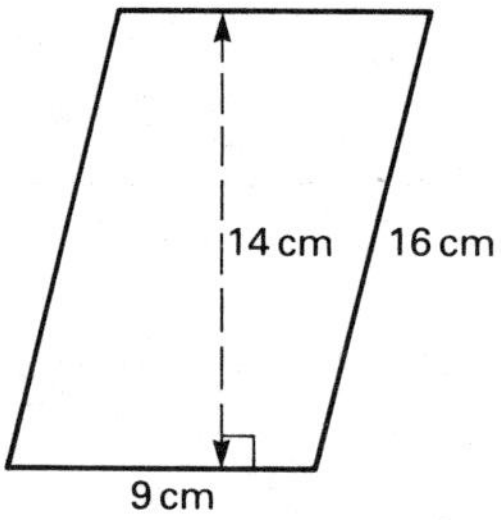

Exercise 7

Find the area of each parallelogram:

1.
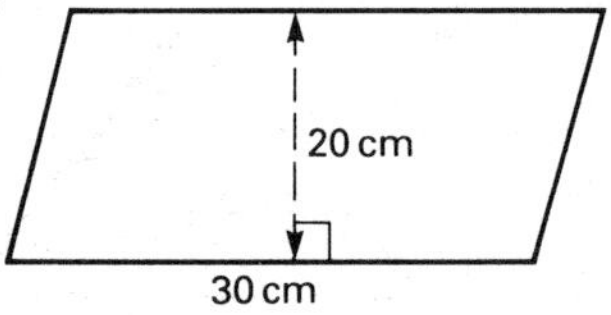

3.
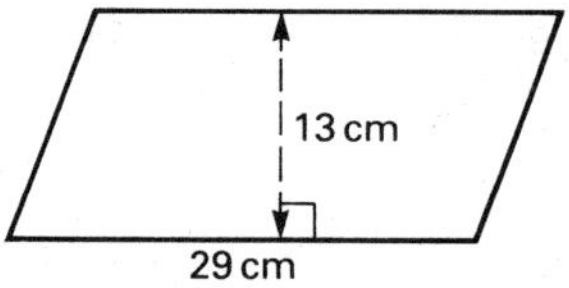

2.
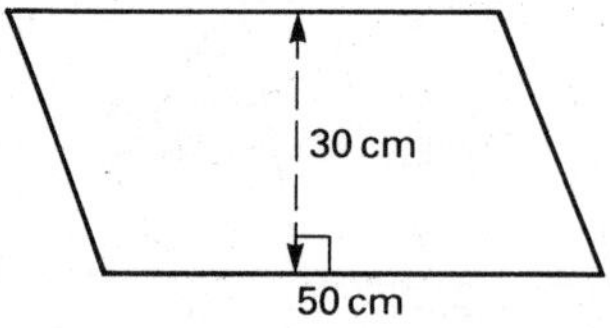

4.
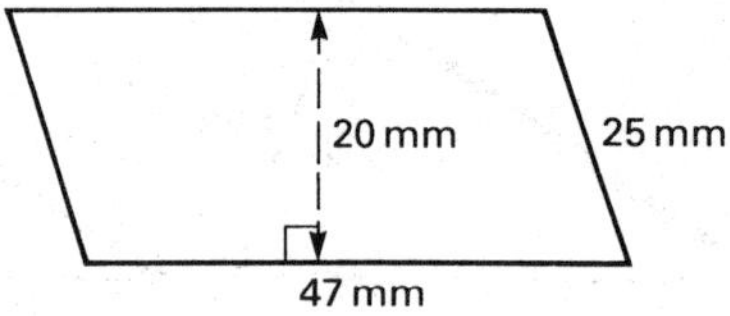

5.

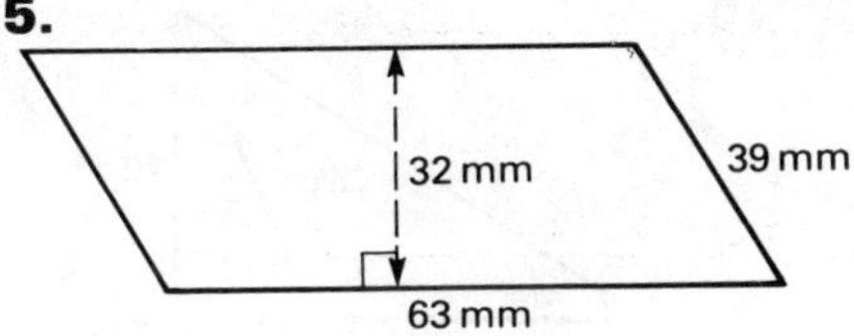

6.

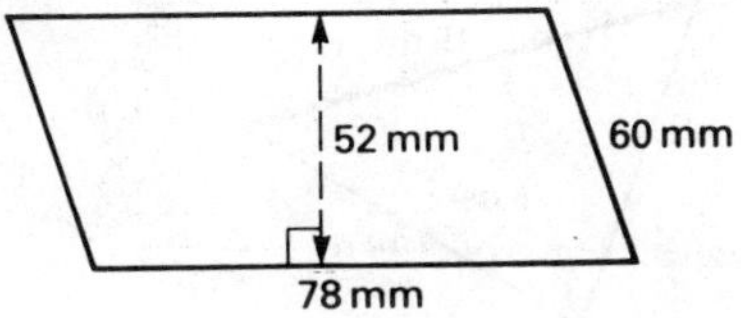

Area of a Triangle

In the following triangles, the base is labelled *b*, and the perpendicular height is labelled *h*.

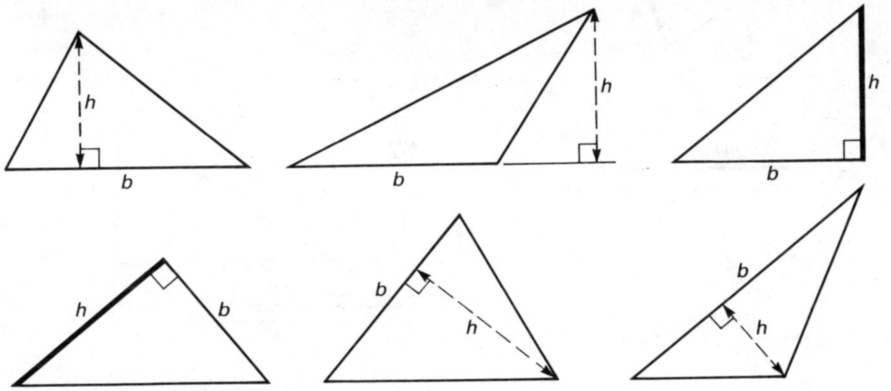

Exercise 8

In each of the following triangles a base is given. Write the size of the perpendicular height.

1.

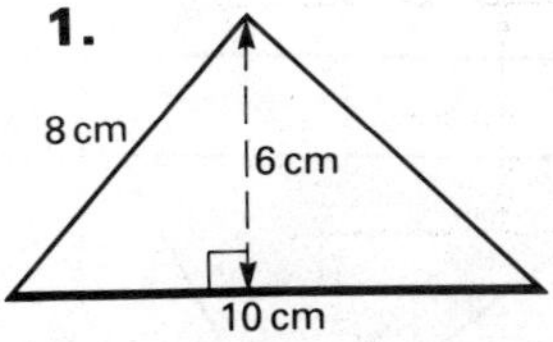

3.

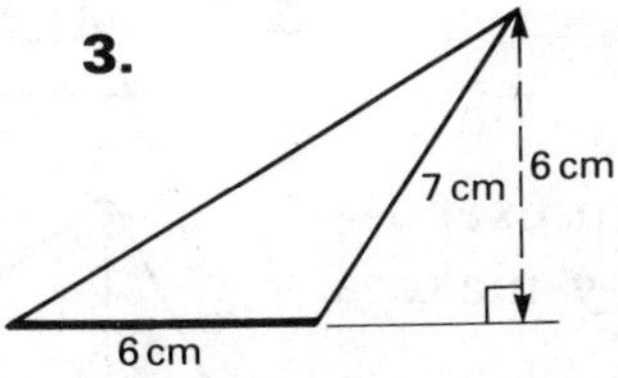

5.

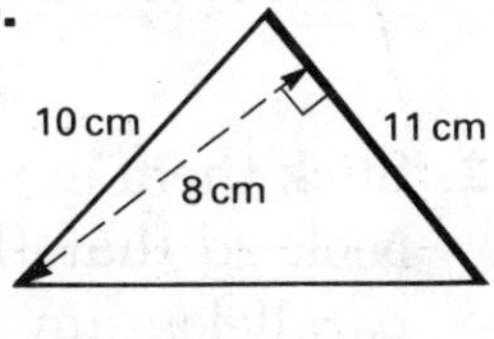

2.

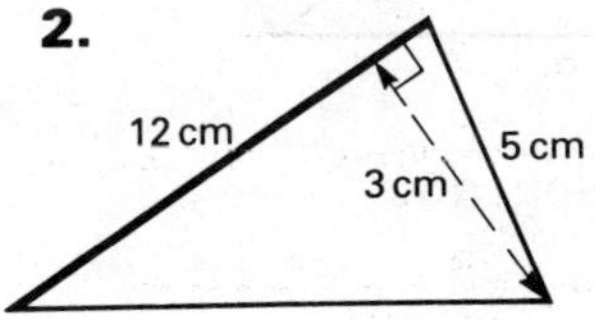

4.

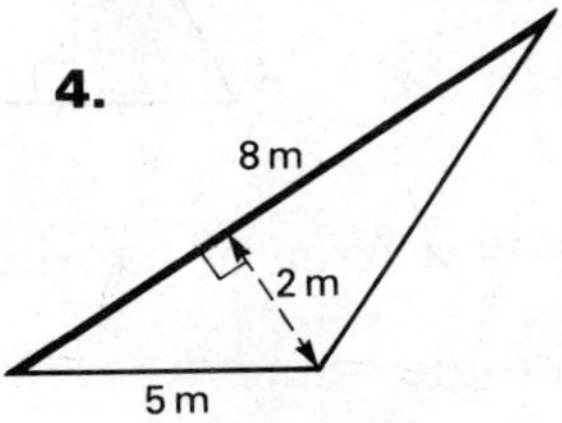

6.

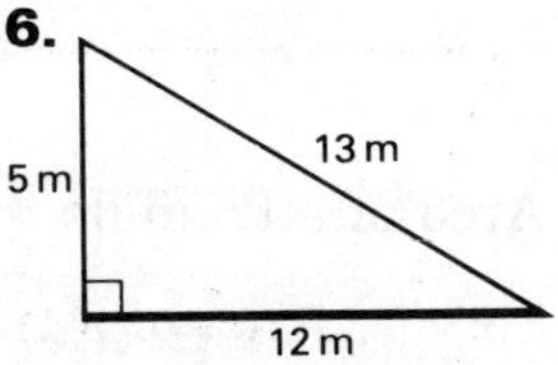

7.

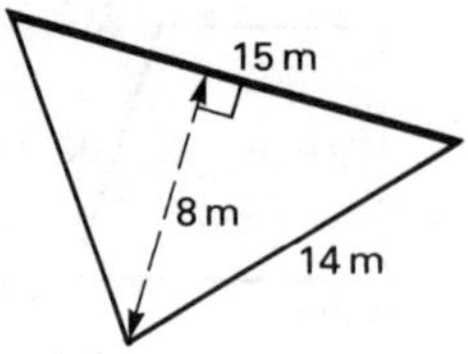

10.

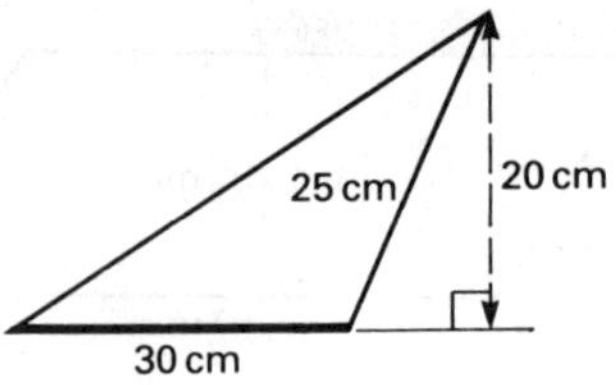

8.

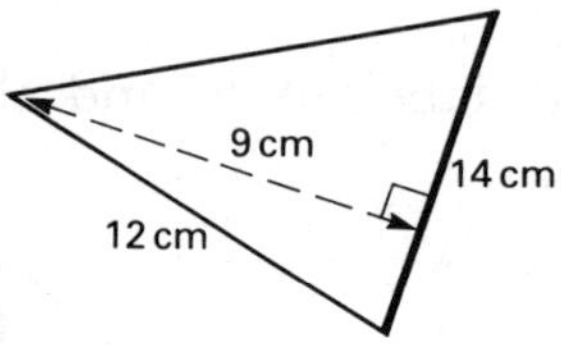

11.

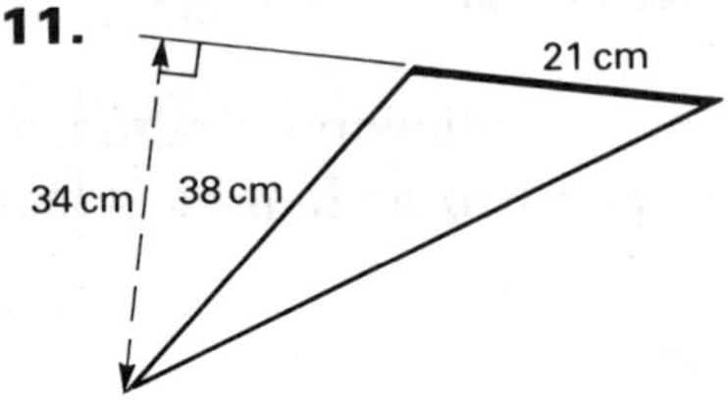

9.

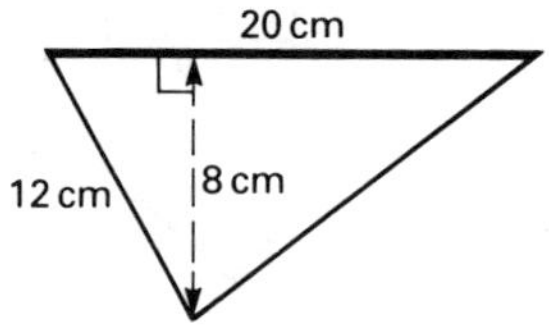

12.

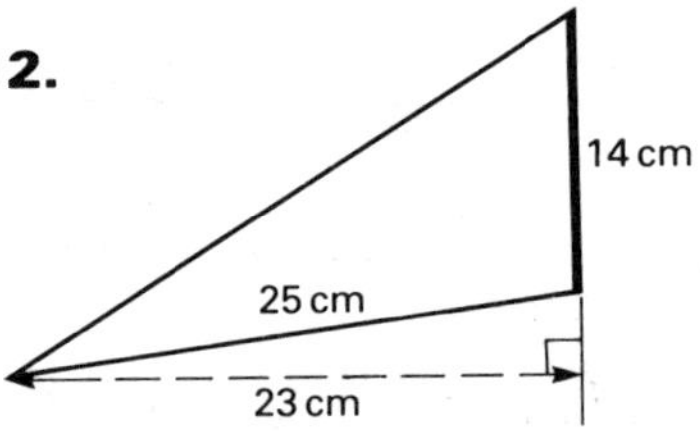

Exercise 9

1. Cut out two identical paper triangles.

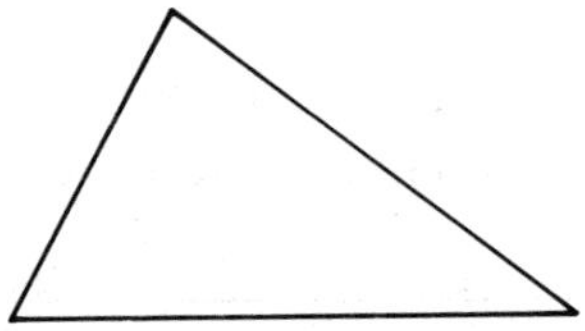

2. Stick them in your exercise book so that they make a parallelogram.

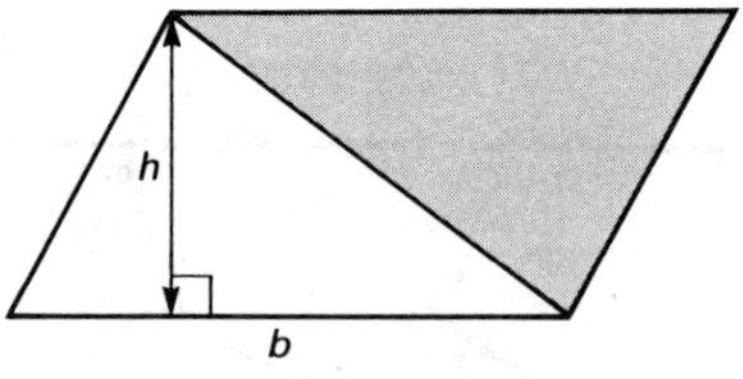

Area of a triangle $= \frac{1}{2} \times$ Area of parallelogram

So: Area of a triangle $= \frac{1}{2}bh$

Exercise 10

Calculate the area of each triangle:

1.

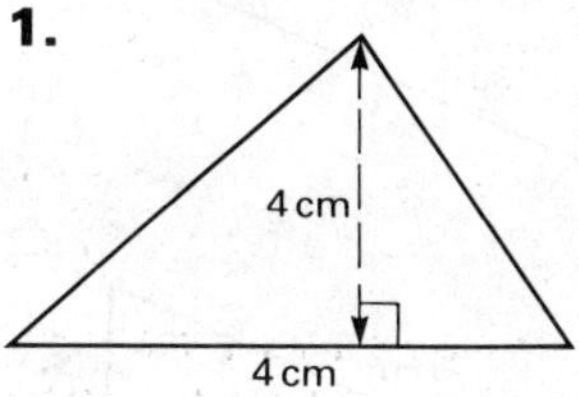

2.

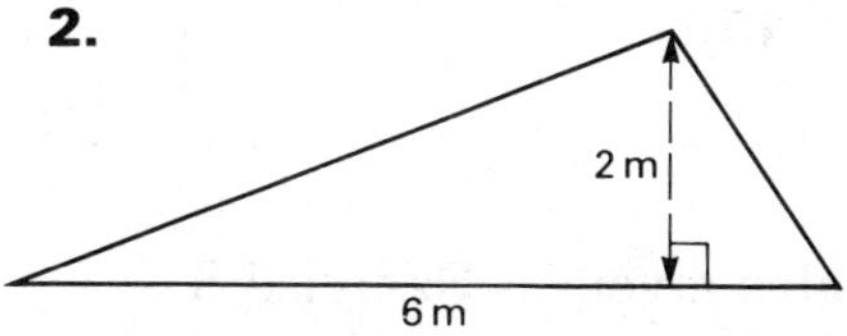

3.

4.

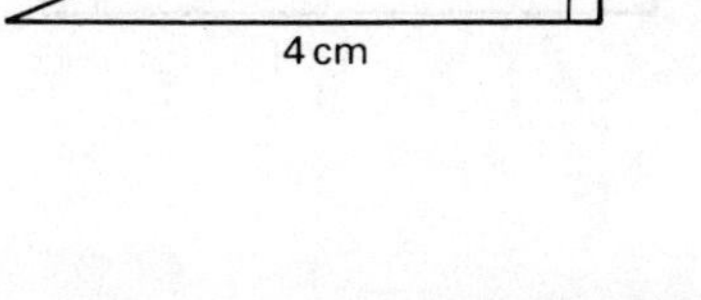

5.

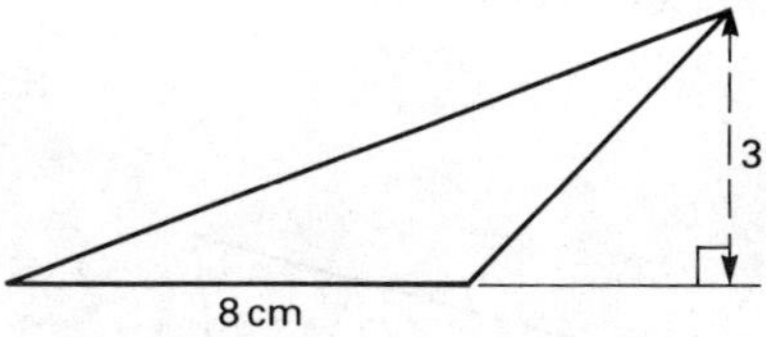

6.

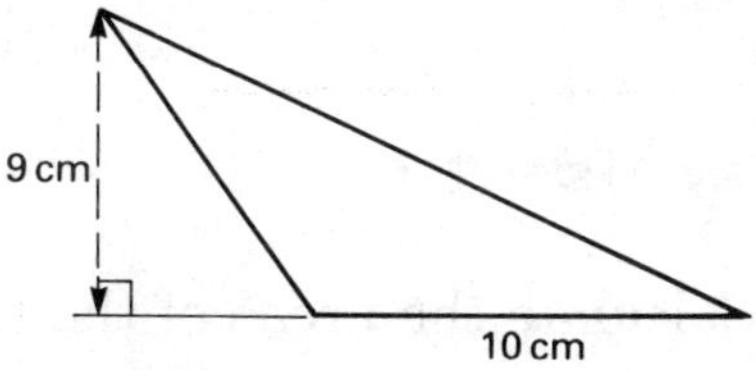

7.

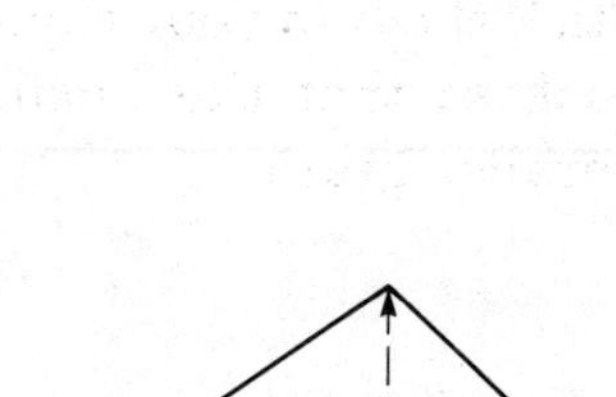

8.

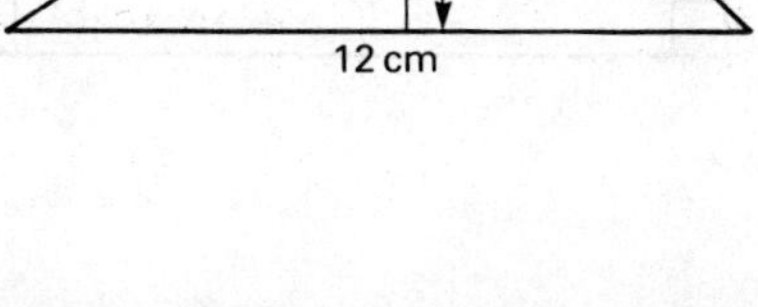

9.

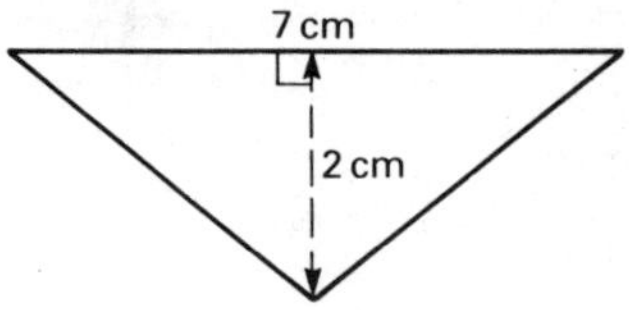

11.

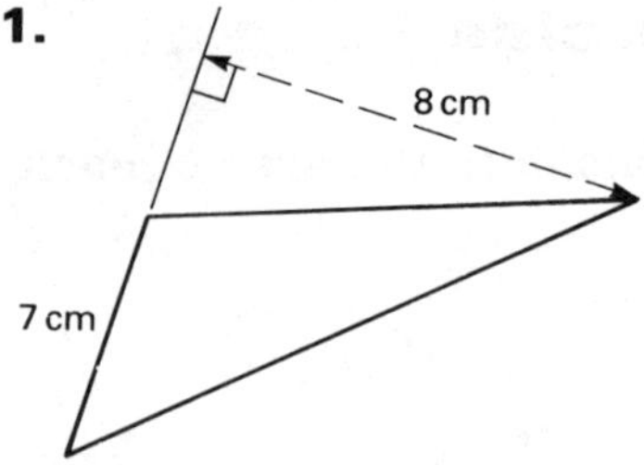

10.

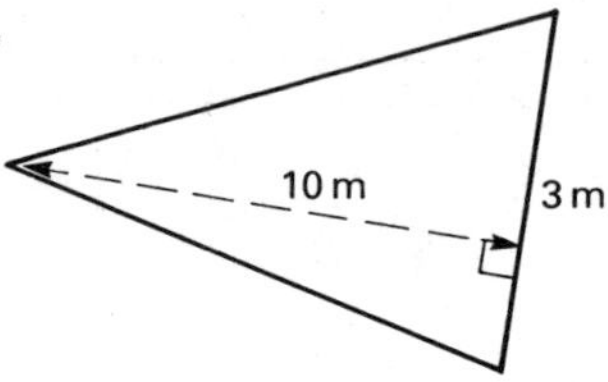

12.

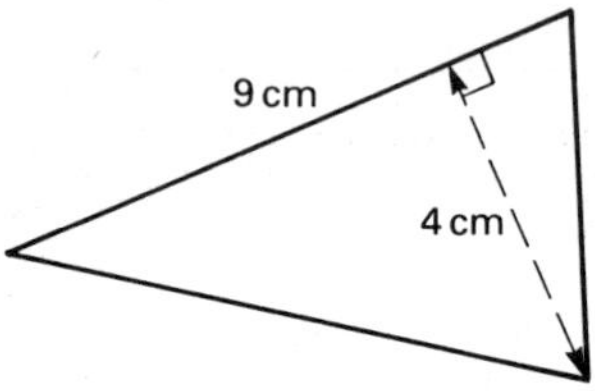

Exercise 11

Calculate the areas of the triangles given in Exercise 8.

Measurements Necessary to Calculate Areas

Exercise 12

By measuring, then calculating, find the area of each of the following shapes:

1.

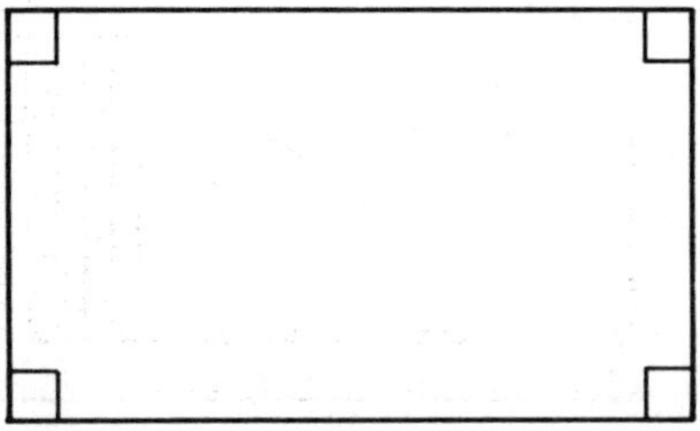

2.

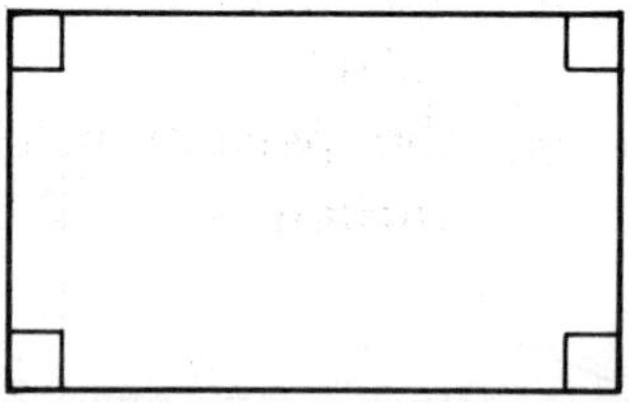

3.

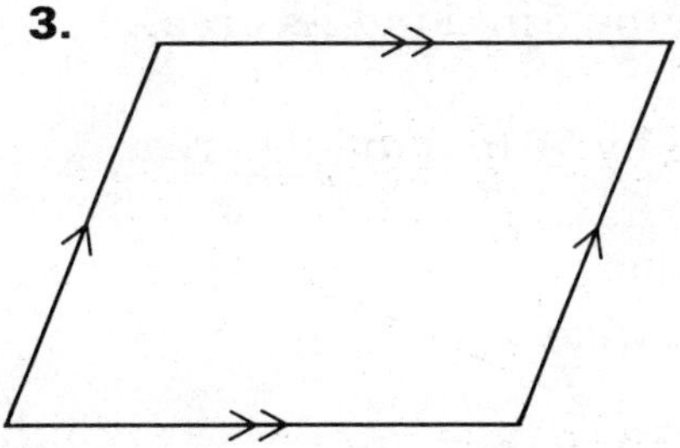

4.

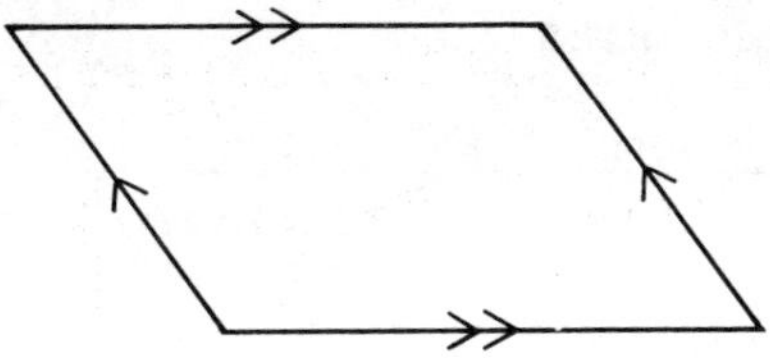

5.

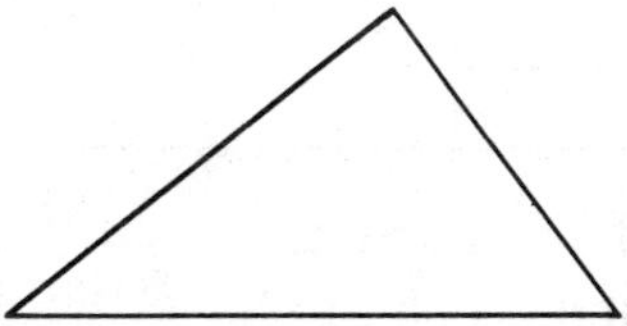

6.

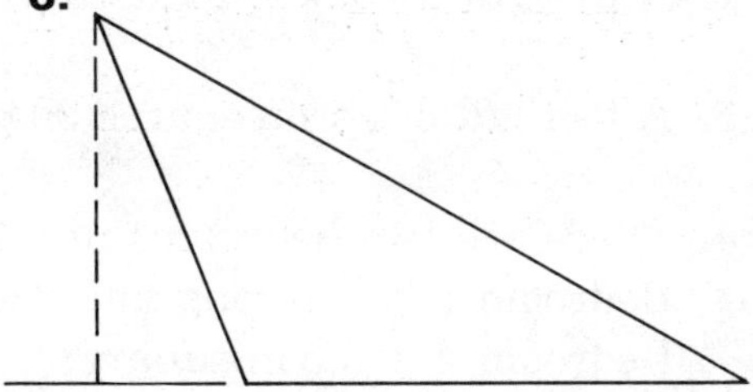

7.

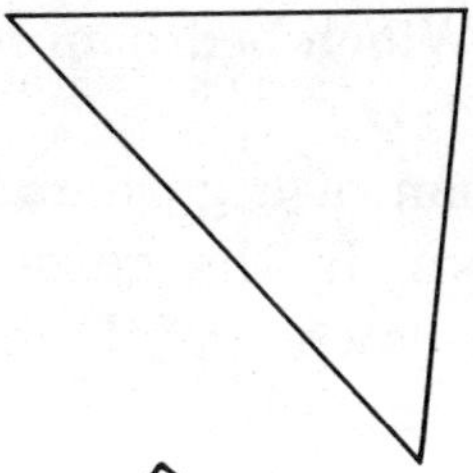

8.

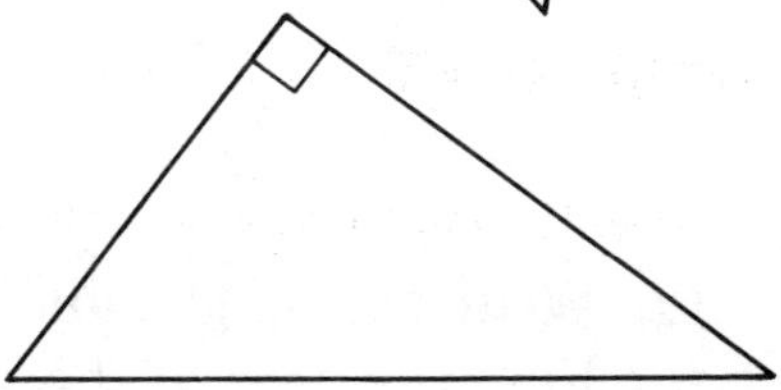

Miscellaneous Problems

Exercise 13

1. The diagram is of a garden. Part of it is lawn. The rest is paved. Find:

(*a*) the area of the lawn,
(*b*) the paved area,
(*c*) the total area of the garden,
(*d*) the perimeter of the garden.

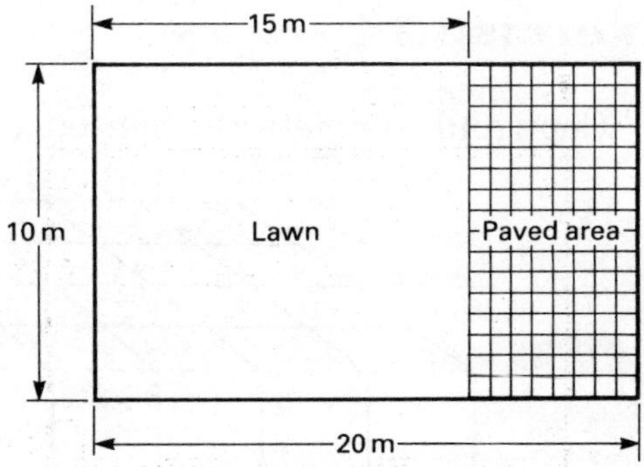

2. A page of a book measures 8 in by 5 in. Find its area.

3. A football pitch measures 90 m by 70 m. Find its area.

4. There are two bedrooms in a house.
Bedroom 1 is 4 m long and 3 m wide.
Bedroom 2 is 3.5 m square.
(*a*) What is the perimeter of each bedroom?
(*b*) What is the area of each bedroom?
(*c*) Which bedroom has the larger area?

5. A plan of a garden is shown. In the garden are a lawn, a path and a triangular flower bed.

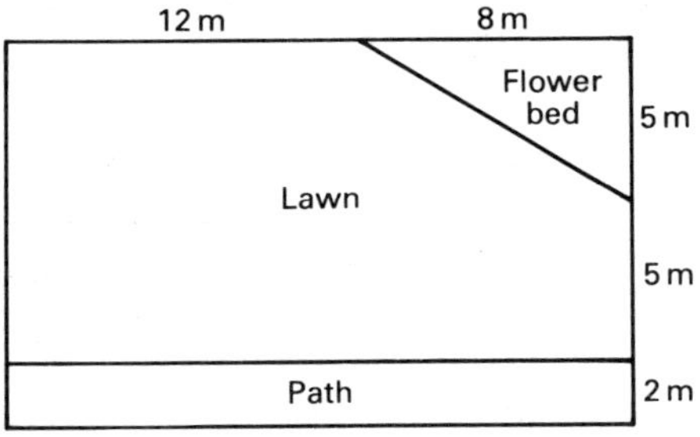

(*a*) Write the length and breadth of the garden.
(*b*) Find the area of the whole garden.
(*c*) Find the perimeter of the whole garden.
(*d*) What is the area of the path?
(*e*) Find the area of the flower bed.
(*f*) Find the area of the lawn.

Volume of a Cuboid

Exercise 14

Copy and complete the sentences:

1.

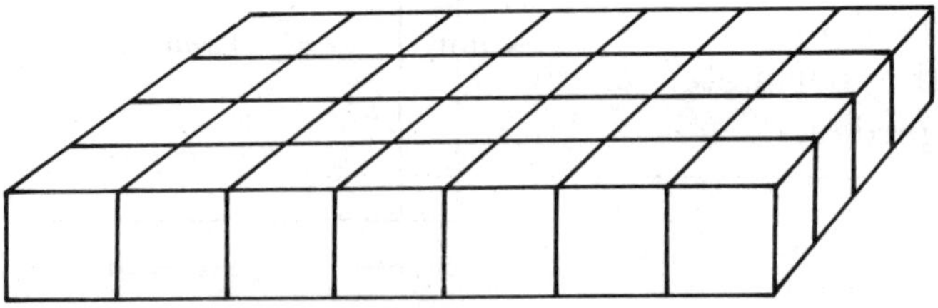

(*a*) There are [?] small cubes in the diagram on p.164.

(*b*) If there were 2 layers there would be [?] small cubes.

(*c*) If there were 5 layers there would be [?] small cubes.

(*d*) If each small cube has a volume of 1 cm^3, the volume of the cuboid in the diagram on p.164 is [?] cm^3.

2.

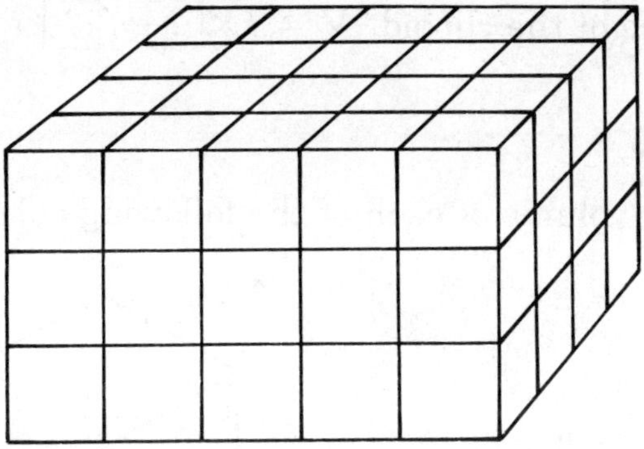

(*a*) There are [?] small cubes in each layer of the cuboid above.

(*b*) There are [?] layers.

(*c*) There are [?] small cubes in the cuboid.

(*d*) If each small cube has a volume of 1 cm^3, the volume of the cuboid is [?] cm^3.

Exercise 15

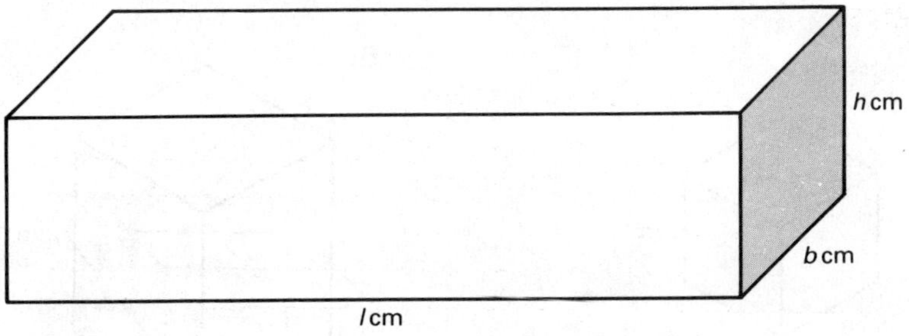

If a cuboid was *l* cm long, *b* cm wide and *h* cm high, then:

1. The number of 1 cm cubes in the length $= \boxed{?}$.

2. The number of 1 cm cubes in the bottom layer of the cuboid $= l \times \boxed{?}$.

3. The total number of 1 cm cubes $= l \times \boxed{?} \times \boxed{?}$.

4. The volume of the cuboid, $V = l \times \boxed{?} \times \boxed{?}$.

Exercise 16

Calculate the volume of each of the following cuboids:

1.

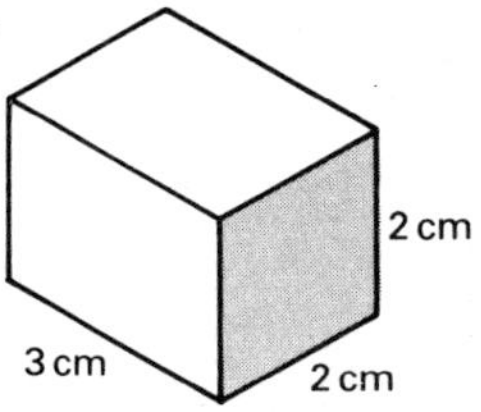

2.

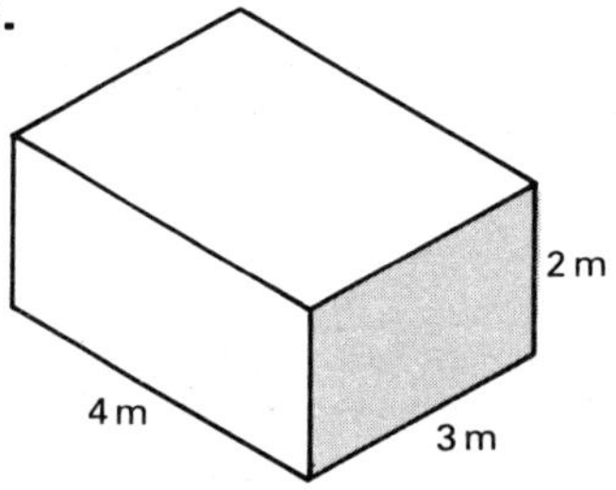

3.

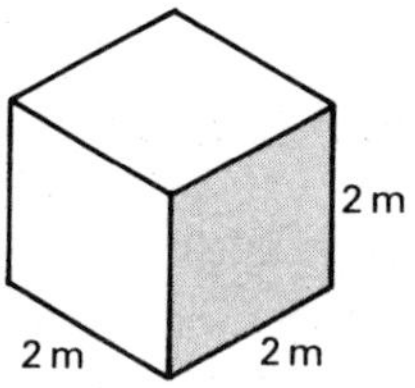

4.

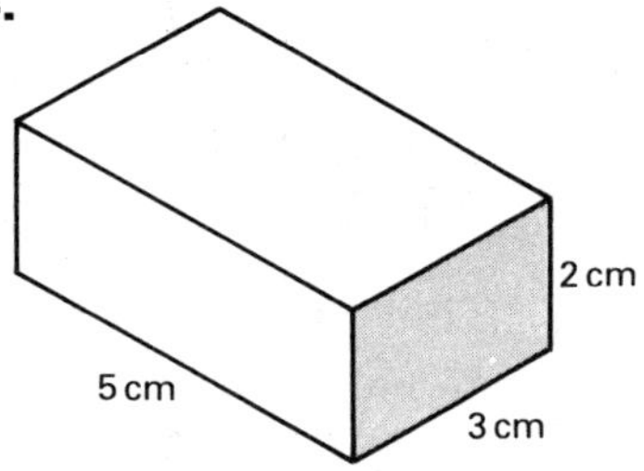

5.

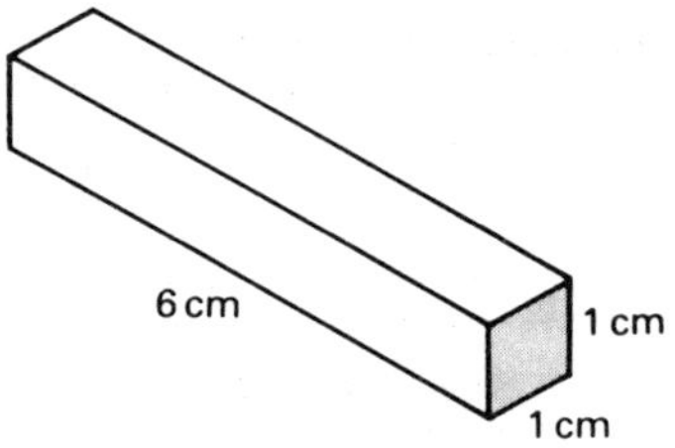

6.

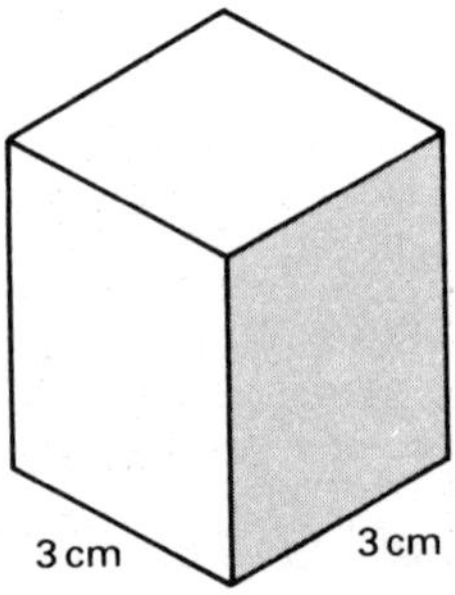

7.

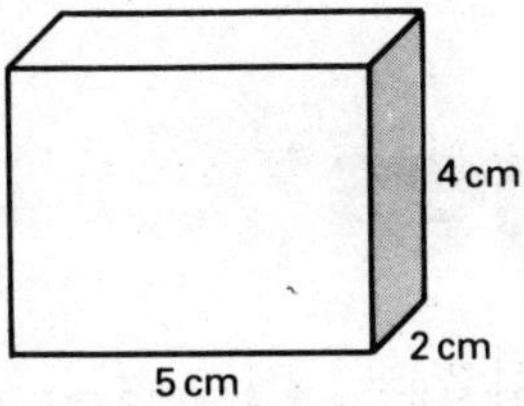
4 cm
2 cm
5 cm

10.

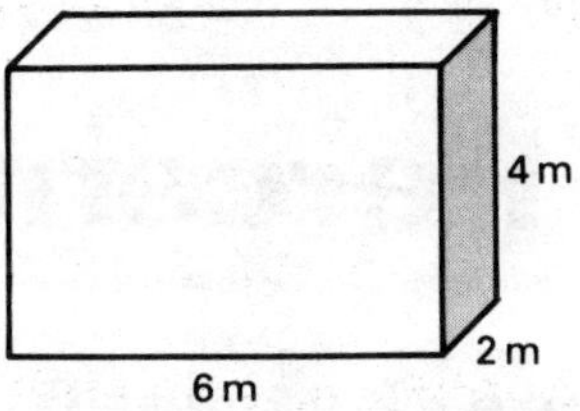
4 m
2 m
6 m

8.

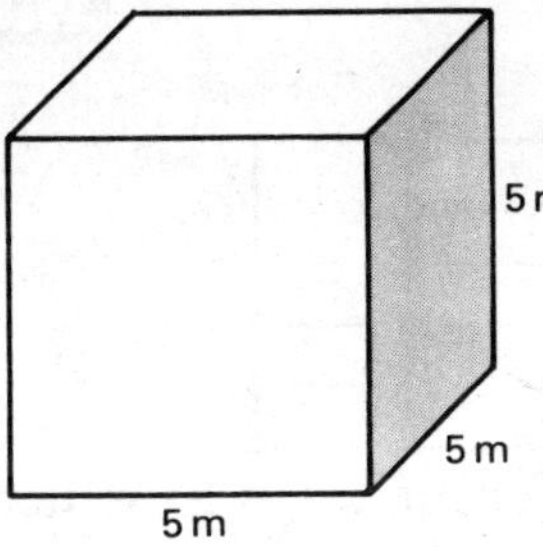
5 m
5 m
5 m

11.

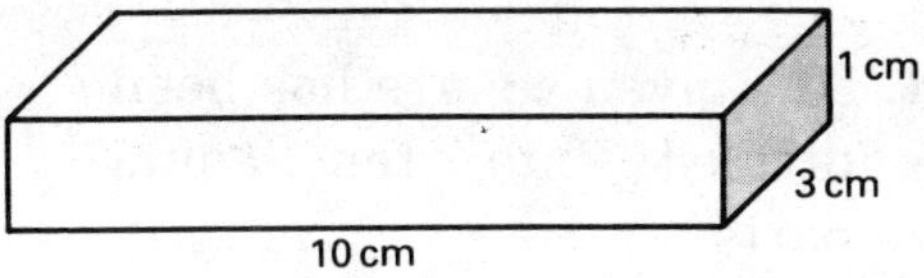
1 cm
3 cm
10 cm

9.

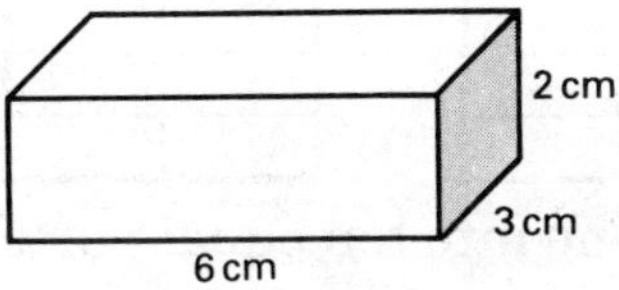
2 cm
3 cm
6 cm

12.

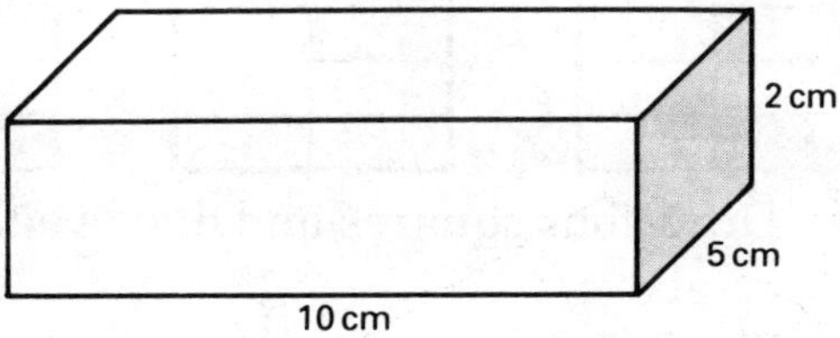
2 cm
5 cm
10 cm

13 Percentages

Equivalences between Vulgar, Decimal and Percentage Fractions

Exercise 1 **M**

1. The given square has been divided into ten equal parts.

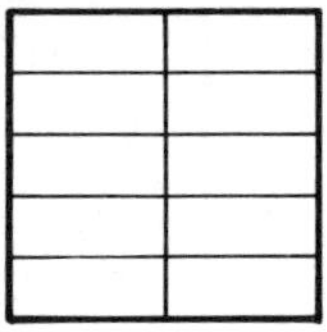

The following diagrams show four different ways in which 50% of the square can be shaded:

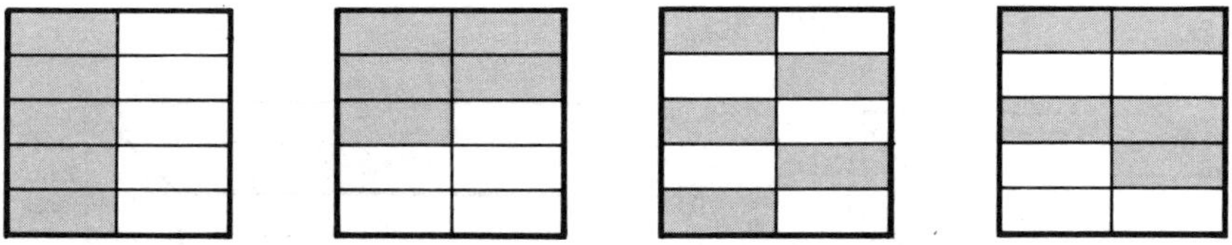

Draw four squares and divide each one into ten equal parts as above.
Shade 50% of each of the four squares in different ways and in different ways from those given above.

2. The squares have been divided into eight equal parts and 25% of each has been shaded. All three squares have been shaded in exactly the same way.

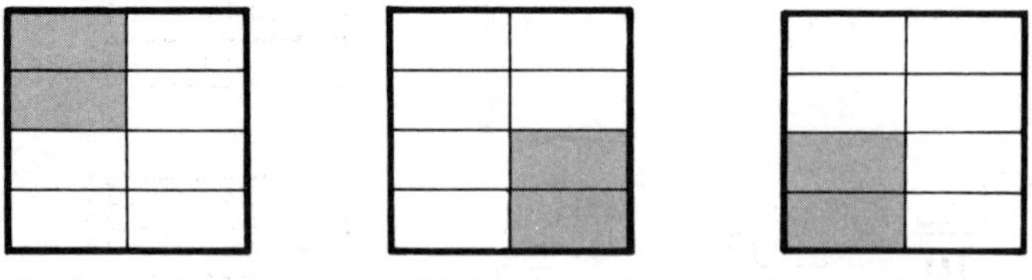

Draw four squares and divide each one into eight equal parts.
Shade 25% of each of the four squares in different ways.

Exercise 2 M

Copy and complete:

$$100\% = 1$$

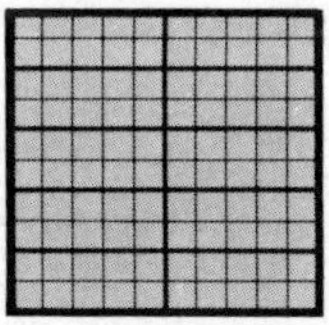

$$\frac{1}{10} = 10\% = 0.1$$

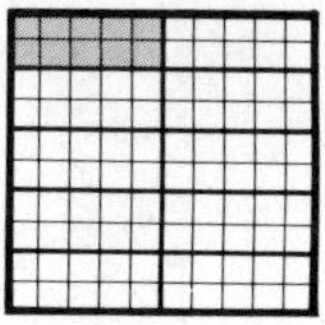

$$\frac{\boxed{?}}{10} = \boxed{?}\,\% = 0.\boxed{?}$$

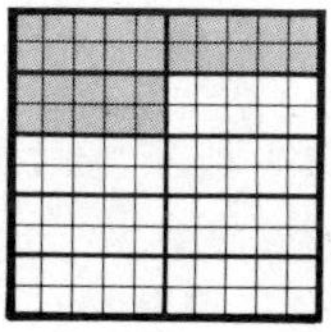

$$\frac{1}{2} = \frac{\boxed{?}}{10} = \boxed{?}\,\% = 0.\boxed{?}$$

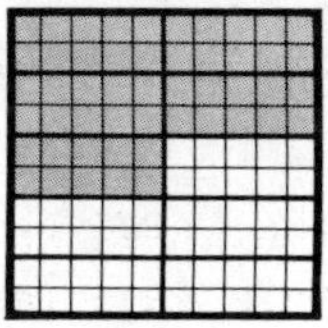

$$\frac{\boxed{?}}{10} = \boxed{?}\,\% = 0.\boxed{?}$$

$$\frac{\boxed{?}}{10} = \boxed{?}\,\% = 0.\boxed{?}$$

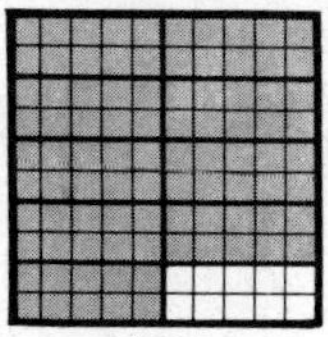

Exercise 3 **M**

Copy and complete:

$$\frac{1}{10} = 10\% = 0.1$$

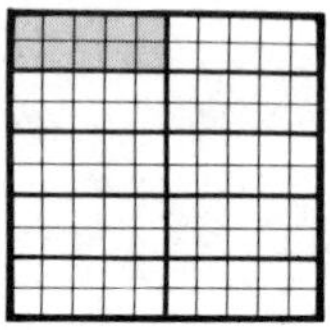

$$\frac{1}{5} = \frac{2}{10} = \boxed{?}\,\% = 0.2$$

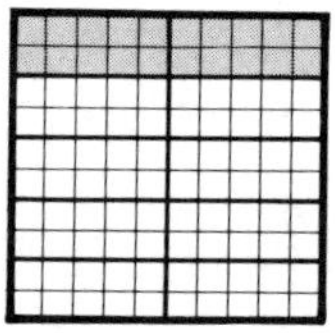

$$\frac{\boxed{?}}{5} = \frac{\boxed{?}}{10} = 40\% = 0.\boxed{?}$$

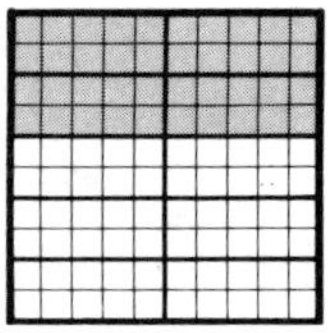

$$\frac{\boxed{?}}{5} = \frac{\boxed{?}}{10} = \boxed{?}\,\% = 0.\boxed{?}$$

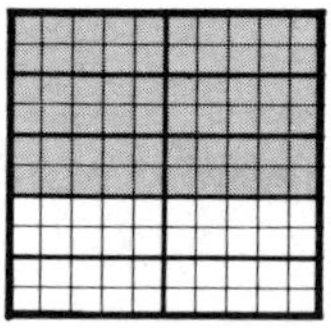

$$\frac{\boxed{?}}{5} = \frac{\boxed{?}}{10} = \boxed{?}\,\% = 0.\boxed{?}$$

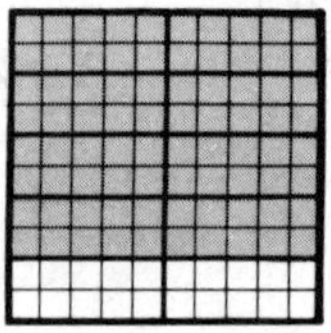

Exercise 4 M

Copy and complete:

$$100\% = 1$$

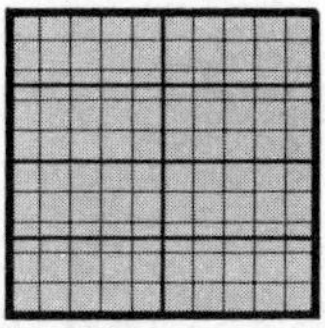

$$\frac{1}{2} = 50\% = 0.\boxed{?}$$

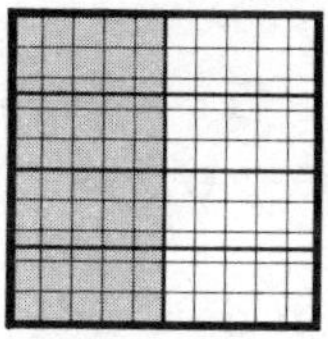

$$\frac{1}{\boxed{?}} = \boxed{?}\% = 0.25$$

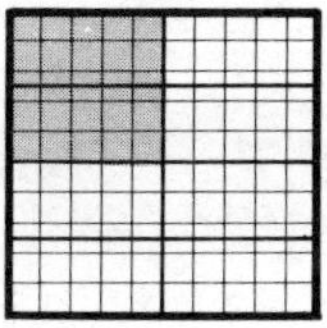

$$\frac{1}{\boxed{?}} = \boxed{?}\% = 0.125$$

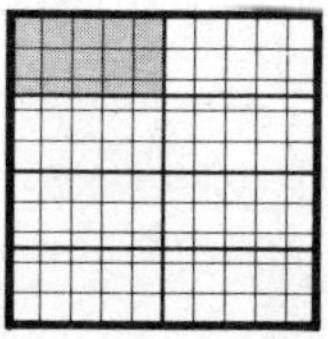

Since $\frac{1}{4} = \boxed{?}\% = 0.25$

$$\frac{3}{4} = \boxed{?}\% = 0.\boxed{?}$$

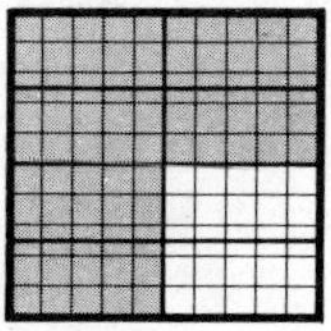

Exercise 5 Conversion from a Percentage to a Decimal

Write each percentage as a decimal:

e.g. 1 84% = 0.84 *e.g. 2* 7% = 0.07

1. 43%	**6.** 62%	**11.** 34%	**16.** 58%
2. 94%	**7.** 36%	**12.** 17%	**17.** 76%
3. 24%	**8.** 72%	**13.** 61%	**18.** 2%
4. 19%	**9.** 40%	**14.** 29%	**19.** 80%
5. 81%	**10.** 55%	**15.** 93%	**20.** 8%

Exercise 6 Conversion from a Percentage to a Vulgar Fraction

Write each percentage in Exercise 5 as a vulgar fraction in its lowest terms.

Exercise 7 Percentages on a Calculator

Show how to key in the following percentages as decimals on a calculator:

e.g. 1 84% = [·] [8] [4]

(Although we normally write 84% as 0.84 rather than .84, there is probably no need to key in the zero. Calculators normally insert the zero for you.)

e.g. 2 70% = [·] [7]

e.g. 3 7% = [·] [0] [7]

1. 25%	**5.** 39%	**9.** 50%	**13.** 1%
2. 64%	**6.** 78%	**10.** 90%	**14.** 5%
3. 83%	**7.** 41%	**11.** 30%	**15.** 4%
4. 12%	**8.** 97%	**12.** 10%	**16.** 6%

Percentage of a Sum of Money

To find 50% of £7.80

Method 1	*Method 2*
(without a calculator)	(using a calculator)
50% of £7.80	Key in:
$= \frac{1}{2}$ of £7.80	[·] [5] [×] [7] [·] [8] [=] (·5 = 50%, 7·8 = £7.80)
= £3.90	50% of £7.80 = £3.90

Exercise 8

A Find 10% of each sum of money:

1. £30
2. £50
3. £500
4. £200
5. £600
6. £180
7. £320
8. £32
9. £94
10. £6
11. 80 p
12. £1
13. £0.90
14. £1.20
15. £39.90
16. £60.40

B Find:

1. 50% of £10
2. 50% of £3
3. 25% of £16
4. 25% of £52
5. 75% of £12
6. 50% of £38
7. 25% of £18
8. 75% of £30
9. 20% of £15
10. 60% of £20
11. 70% of £6
12. 40% of £100
13. 90% of £400
14. 30% of £250
15. 80% of 75 p
16. 20% of £0.95

Exercise 9

Find:

1. 50% of £3.60
2. 25% of £7
3. 75% of £2.40
4. 10% of £42.70
5. 40% of £378
6. 80% of £19

7. 30% of £9.20
8. 15% of £80
9. 65% of £12.60
10. 36% of £6.75
11. 24% of £2
12. 78% of £3.50
13. 57% of £25
14. 42% of £71.50
15. 91% of £100
16. 28% of £40
17. 84% of £95
18. 16% of £12.75
19. 9% of £15
20. 1% of £79

VAT

VAT stands for *Value Added Tax.*

It is a tax on most goods or services and is set by the Government.

Sometimes VAT is added to the price of single items.

Sometimes prices are given without VAT being added. Then the total cost of the goods is found and the VAT on the total amount is added at the end.

The VAT rate is given as a percentage.

Exercise 10

In this exercise, the VAT rate is 15%.
The price before VAT is given. Find the VAT.

1. £1
2. £5
3. £16
4. £30
5. £54
6. £91
7. £100
8. £300
9. £420
10. £235
11. £111
12. 80 p
13. £1.60
14. £4.80
15. £19.40
16. £25.20

Often the amount of VAT payable can work out to be a decimal of 1 p. Then, the amount needs to be written to the nearest penny. For example, 11.85 p is not a sensible answer (12 p is better). Also £1.4475 is not sensible. (This should be written as £1.45.)

Exercise 11

Write each amount of money correct to the nearest penny:

1. 4.7 p
2. 9.2 p
3. 3.45 p
4. 9.75 p
5. 36.95 p
6. £1.4355
7. £9.2885
8. £5.739
9. £12.477
10. £15.918
11. £46.7505
12. £81.098
13. £55.006
14. £39.1515
15. £63.8585
16. £20.9025
17. £98.7095
18. £15.997
19. £32.9965
20. £4.995

Exercise 12

In this exercise, VAT is at 15%.
The price before VAT is given.
Find the VAT (it should be rounded to the nearest penny):

1. A rubber costing 35 p.
2. A pencil sharpener costing 60 p.
3. A dance ticket costing £4.
4. A cinema ticket costing £2.17.
5. A dress costing £25.
6. A meal costing £6.04.
7. A TV set costing £243.
8. A cassette costing £3.90.

Exercise 13

In this exercise the pre-VAT price is given (that is, the price before VAT is added). Find, to the nearest penny:
(*a*) the amount of VAT if the rate is 15%,
(*b*) the price inclusive of VAT.

1. A pack of pencils at £1.73.
2. A writing pad at 95 p.
3. A fishing-rod at £26.
4. A telephone bill at £55.25.
5. A theatre ticket at £6.96.
6. A carpet at £140.
7. A holiday at £251.30.
8. A pair of jeans at £17.38.
9. A computer at £138.26.
10. A TV repair at £40.

Discount

Exercise 14

In this exercise, give your answers correct to the nearest penny.

1. Personal stereo.
10% off.
(*a*) Find the discount.
(*b*) Find the sale price.

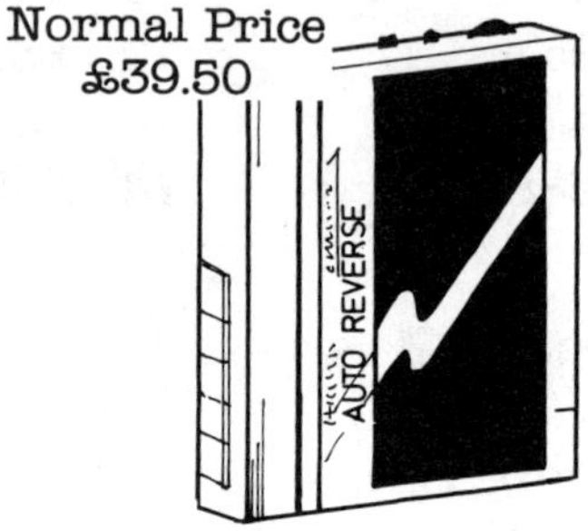

2. Snooker table.
Normal price £135.
Discount 10%.
(*a*) Find the discount.
(*b*) Find the sale price.

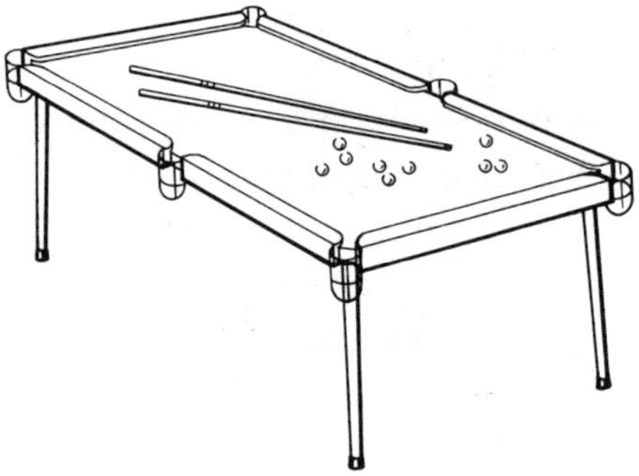

3. Sleeping bag.
Normal price £18.50.
Special offer 10% discount.
(*a*) Find the discount.
(*b*) Find the offer price.

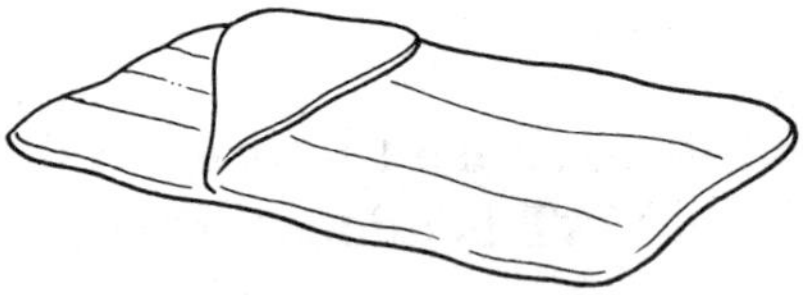

4. A pair of binoculars.
Normal price £49.99.
Sale offer 10% discount.
(*a*) Find the discount.
(*b*) Find the sale price.

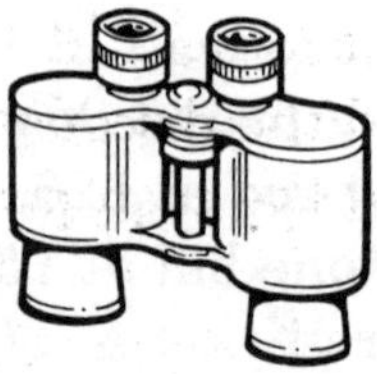

5. Tennis-racket.
Normal price £20.
Special offer 5% discount.
(*a*) Find the discount.
(*b*) Find the sale price.

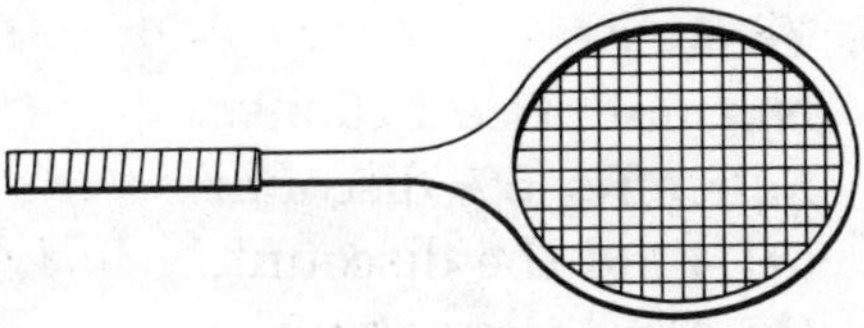

6. Computer.
Was £149.99.
Sale price 5% discount.
(*a*) Find the discount.
(*b*) Find the sale price.

7. Disco skate boots.
Normally £25.99.
Special offer 5% discount.
(*a*) Find the discount.
(*b*) Find the special offer price.

8. Set of 3 pans.
Normal price £24.99.
Discount for cash 5%.
(*a*) Find the discount.
(*b*) Find the cash price.

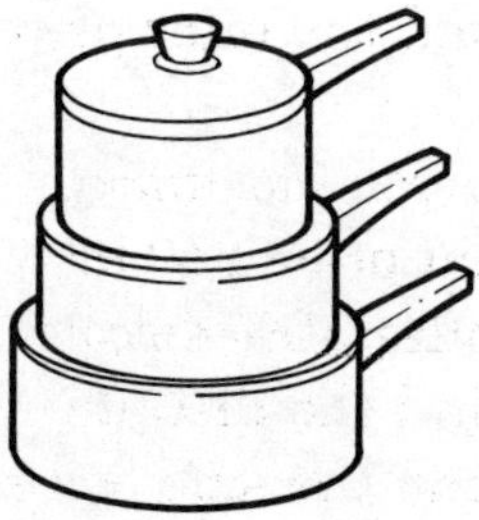

9. Toaster.
Normal price £15.99.
Sale offer 5% discount.
(*a*) Find the discount.
(*b*) Find the sale price.

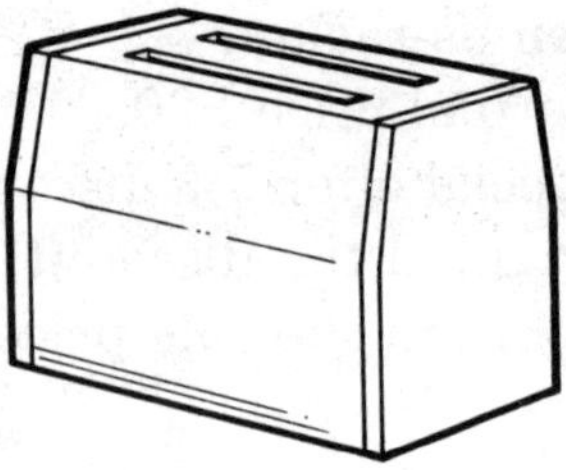

10. Record deck.
- Belt driven
- Auto return
- Was £108

Special offer
$12\frac{1}{2}\%$ discount.
(*a*) Find the discount.
(*b*) Find the special offer price.

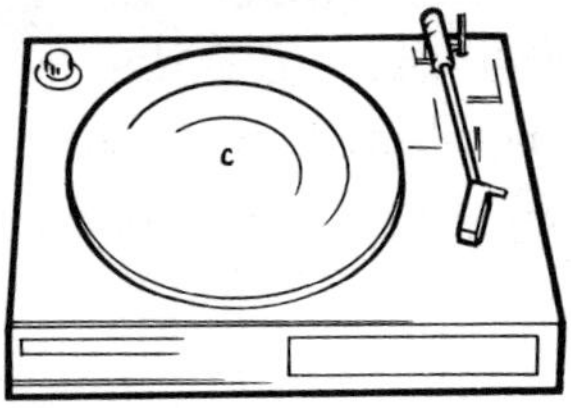

Revision Exercises VII to XIII

Revision Exercise VII

1. A triangular-based prism is shown.

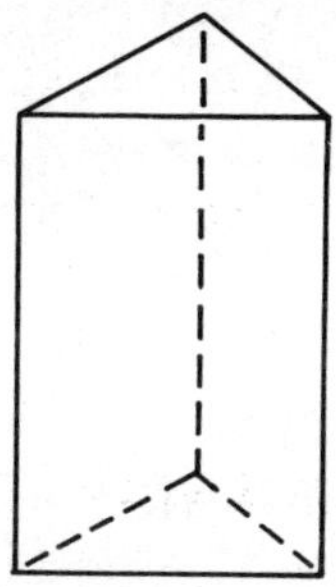

(*a*) How many edges does it have?
(*b*) How many faces does it have?
(*c*) How many vertices does it have?

2. Part of a cuboid is shown. Copy and complete it.

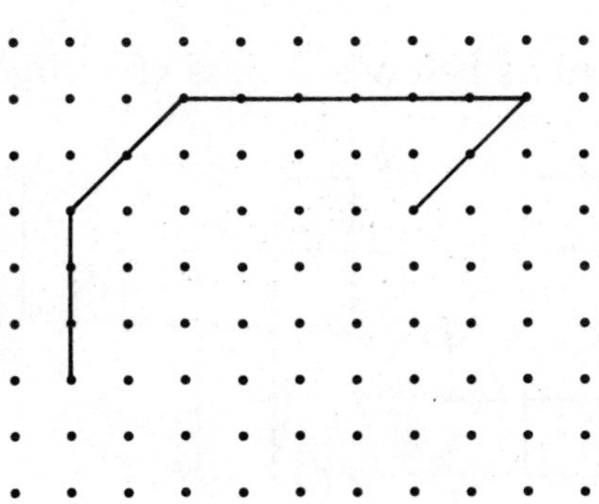

3. Part of a cuboid is shown. Copy and complete it.

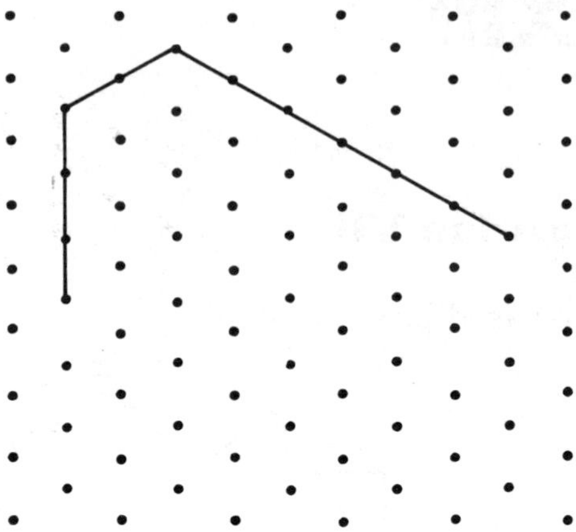

4. What is the smallest number of cubes that could be used to make the given shape?

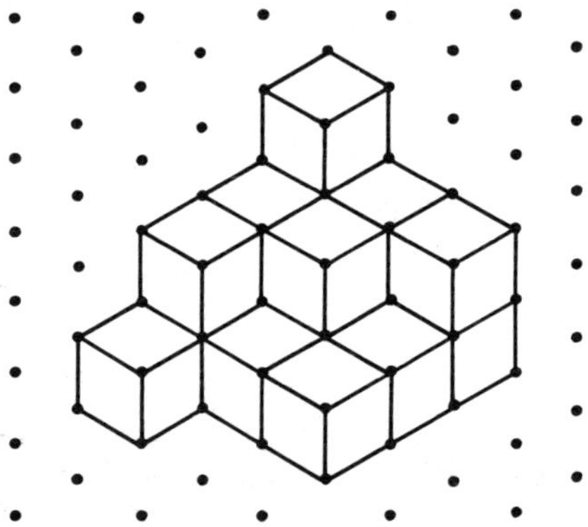

5. Which of the drawings below are nets of a cube?

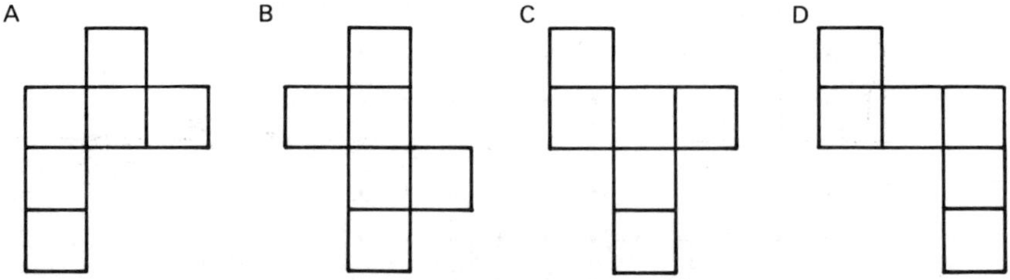

6. Here is a net of a cube. When the cube has been made, which face would be opposite face E?

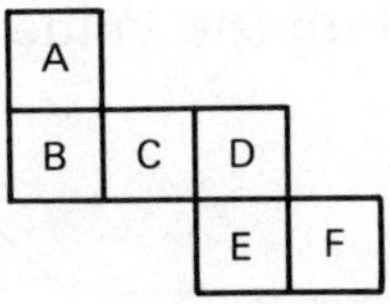

7. Is the given drawing the net of a cuboid?

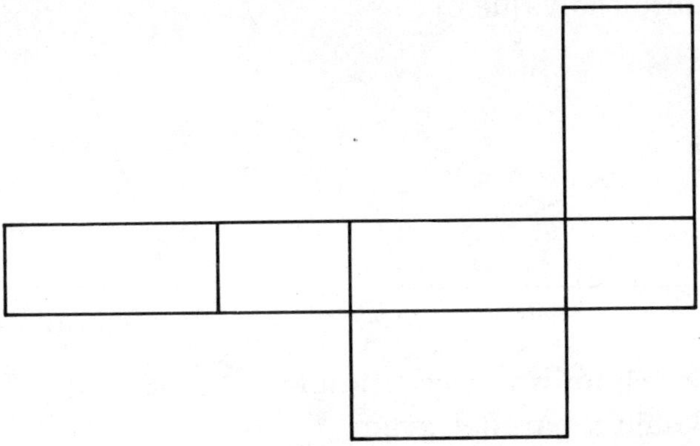

8. What sort of angle is shown?

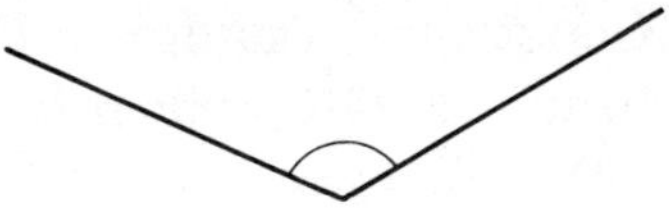

9. In $\triangle PQR$, $\angle PQR = 60°$.
(*a*) Measure $\angle PRQ$. (*b*) Calculate $\angle QPR$.

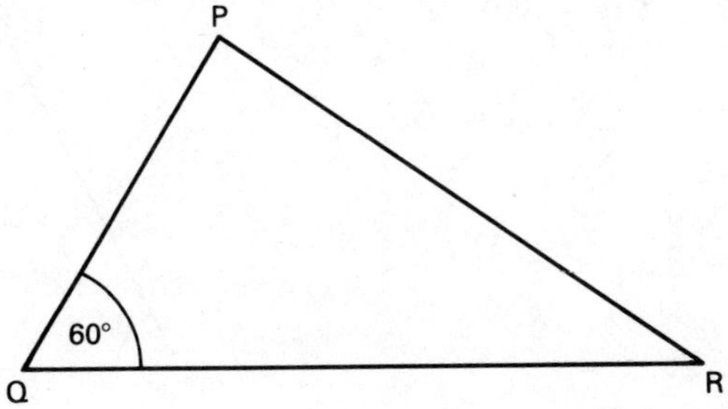

10. Calculate the value of x:

(*a*)

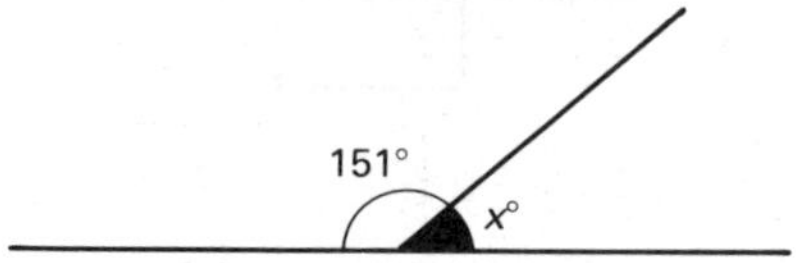

(*b*)

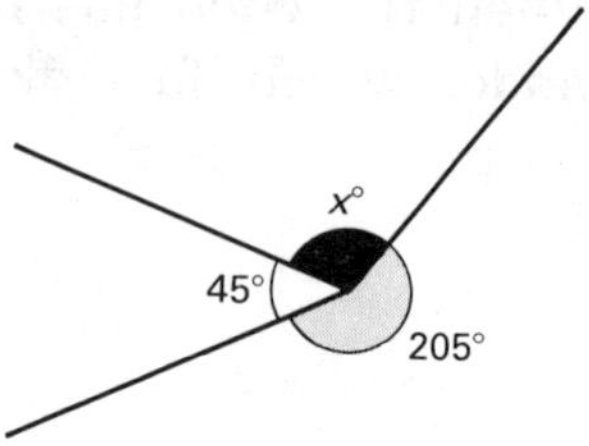

11. Calculate the value of x:

(*a*)

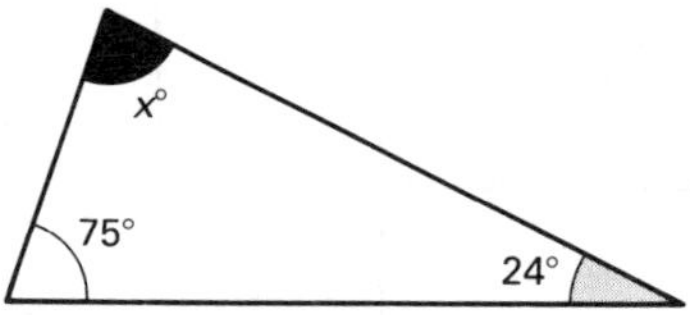

(*b*)

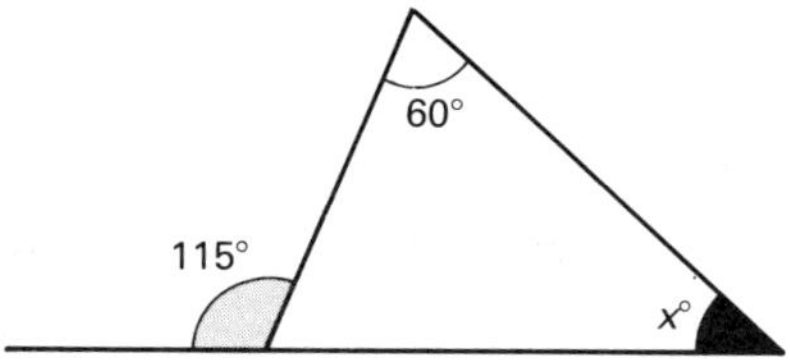

12. (*a*) Sketch an isosceles triangle.
(*b*) Sketch a parallelogram.

Revision Exercise VIII

1. Construct triangle PQR in which QR = 95 mm, QP = 45 mm and RP = 110 mm.
(*a*) Measure $P\hat{Q}R$.
(*b*) Measure $P\hat{R}Q$.

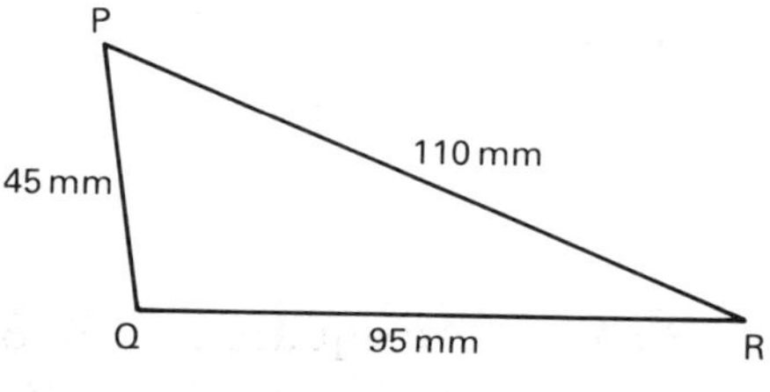

2. Construct an isosceles triangle XYZ where YZ = 5 cm, ∠XYZ = 65° and ∠XZY = 65°. Measure XY.

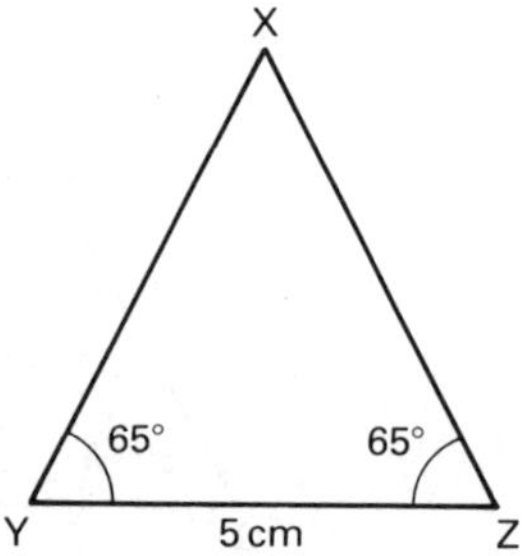

3. Draw a straight line, BC. Make it 9 cm long. Using a protractor, draw an angle ABC equal to 55°. Make AB 4 cm long. Using a set square and ruler, draw a line AQ parallel to BC. Using a pair of compasses, bisect BC. Let the bisector meet AQ at P. Measure AP.

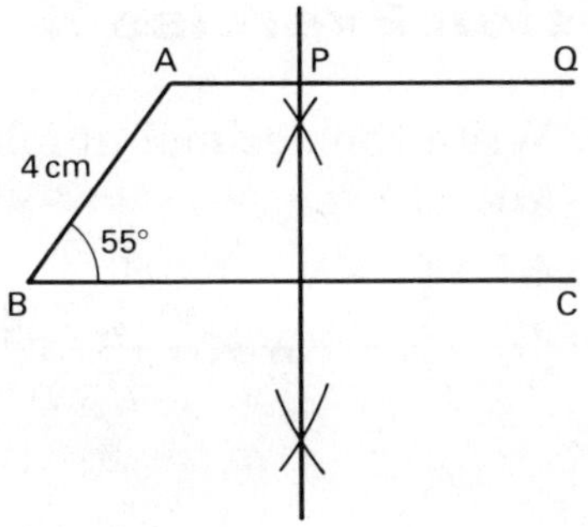

4. (*a*) Draw a circle with a radius of 35 mm. On the circle, draw a radius.
(*b*) Draw a circle with a radius of 3 cm. Show a segment on that circle.

Revision Exercise IX

1. Find the value of 8^2.

2. Draw a dot pattern to show the square number 16.

3. Find the value of 5^3.

4. Find $\sqrt{36}$.

5. Find the square root of 81.

6. Find the two closest whole numbers between which the square root of 41 lies.

7. Using a calculator, find (*a*) 3.7^2 (*b*) 26.8^2.

8. Using a calculator, find the following square roots. For each answer shown in the display, write only the first three digits.
(*a*) $\sqrt{39}$ (*b*) $\sqrt{500}$ (*c*) $\sqrt{76.2}$ (*d*) $\sqrt{9306}$

Revision Exercise X

1. Write the decimal number to which each arrow points:

(*a*)

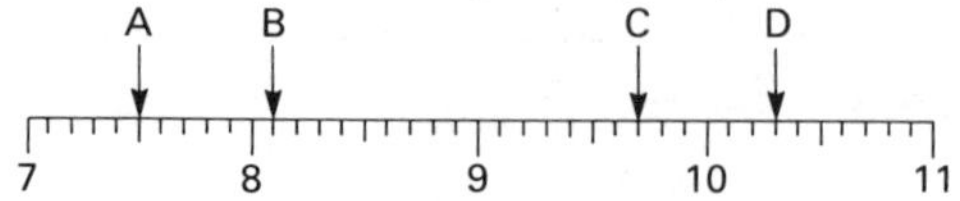

(*b*)

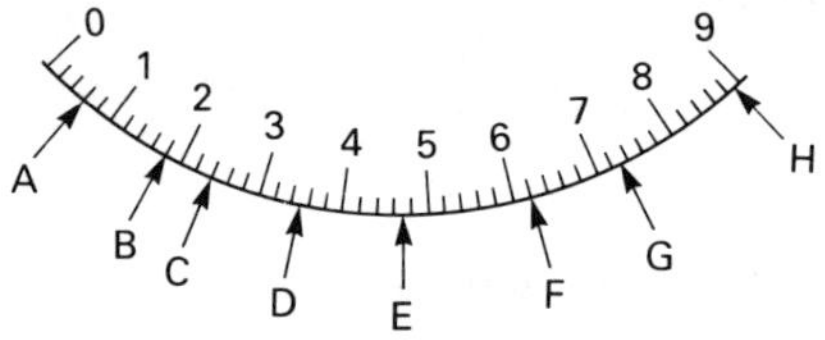

(*c*)

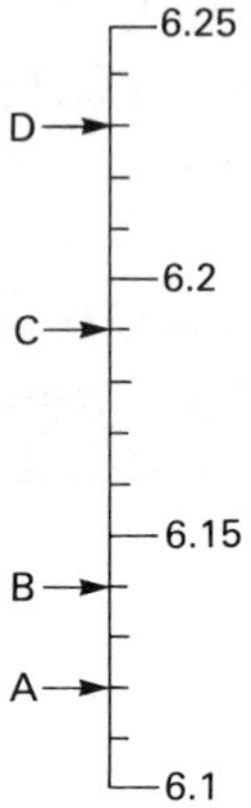

2. Write in order of size, largest first:
6.1, 1.6, 0.16, 0.61, 6.01, 10.6, 1.06

3. Write in order of size, smallest first:
7.83, 78.3, 3.87, 3.78, 37.8, 38.7, 73.8

4. Work out:
(*a*) 76.7 + 19.6 (*b*) 7.16 + 8.53 + 5.72

5. Work out:
(*a*) 48.6 − 23.9 (*b*) 8.07 − 5.79

6. Work out:
(*a*) 7 + 5.68 (*b*) 4.69 + 24 + 15.7

7. Work out:
(*a*) 19.2 − 8.76 (*b*) 73 − 49.5

8. A rectangle measures 29.6 cm by 17.8 cm. Calculate its perimeter.

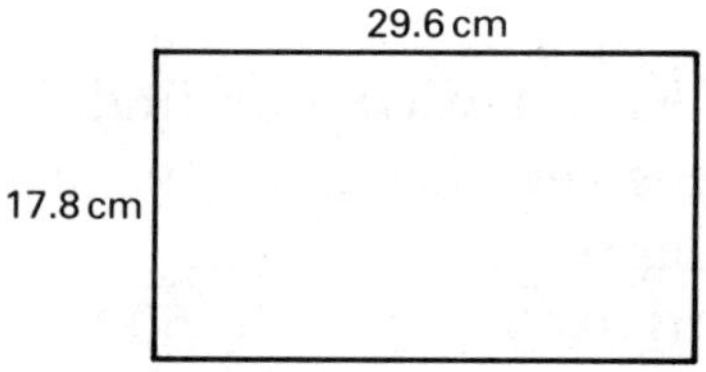

9. Work out:
(*a*) 3.92×4 (*b*) 0.7×6 (*c*) $32.65 \div 5$

10. A car used 8ℓ of petrol in travelling 111.2 km. How many kilometres is that per litre?

11. Work out the following on a calculator:
(*a*) $31.6 + 298 + 7.69$ (*c*) 4.17×7.6
(*b*) $807.3 - 89.62$ (*d*) $17.64 \div 3.6$

12. (*a*) What fraction of the rectangle has been shaded?
(*b*) What fraction remains unshaded?

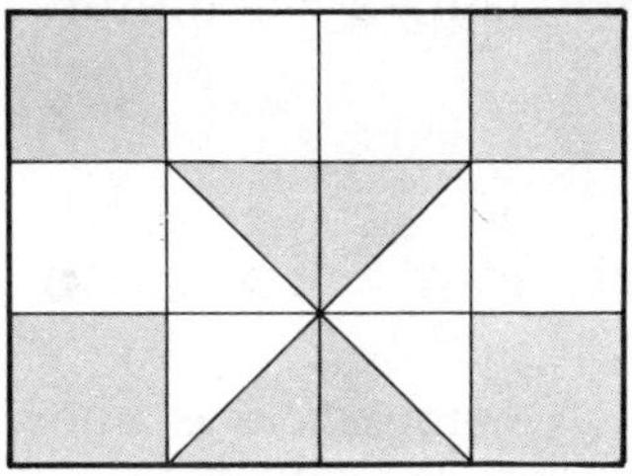

13. How many fifths make a whole?

14. Copy and complete:

(*a*) $\frac{2}{3} = \frac{\boxed{?}}{15}$ (*b*) $\frac{6}{24} = \frac{1}{\boxed{?}}$

15. Carry out these calculations:

(*a*) $\frac{1}{2} + \frac{1}{20}$ (*c*) $1 - \frac{1}{8}$ (*e*) $\frac{5}{6} - \frac{2}{3}$

(*b*) $\frac{3}{8} + \frac{5}{16}$ (*d*) $1 - \frac{9}{16}$ (*f*) $\frac{3}{4} - \frac{5}{12}$

16. Work out $7 \times \frac{5}{8}$.

17. Work out $4 \times \frac{3}{10}$ and simplify your answer.

18. (*a*) Find $\frac{1}{4}$ of 28 m.

(*b*) Find $\frac{3}{4}$ of 28 m.

19. Use a calculator to change the following to decimals:

(*a*) $\frac{17}{20}$ (*b*) $\frac{8}{25}$ (*c*) $\frac{9}{16}$ (*d*) $\frac{1}{40}$

20. Using a calculator, find $\frac{1}{8}$ of £188.64.

Revision Exercise XI

1. A semi-circle is shown. Its radius is 5.7 cm. Find straight line distance CD.

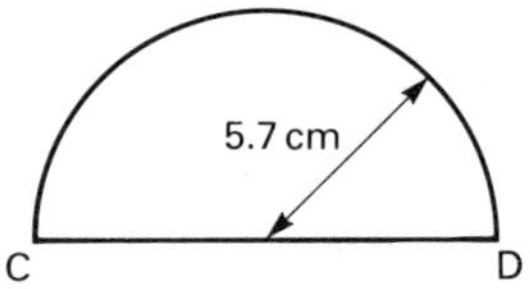

2. Find the circumference of the circle. Use the formula:

$$C = \pi d$$

(C is the circumference of a circle, and d is the diameter. Take π to be 3.142).

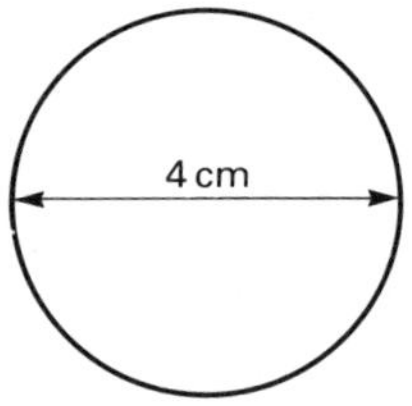

3. (*a*) By measuring, find the diameter of the given circle.

(*b*) By calculating, find the circumference of the circle. Use the formula $C = \pi d$ and take π to be 3.14.

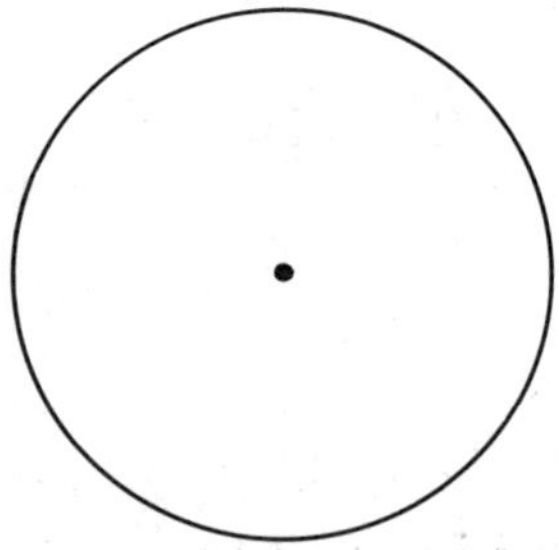

4. Donna's bicycle wheels have a diameter of 40 cm.
   (*a*) Calculate the circumference of the rear wheel, taking π to be 3.14. (Give your answer in metres.)
   (*b*) How far does Donna cycle when the rear wheel makes 10 full turns?
   (*c*) How far does Donna cycle when the rear wheel makes 80 full turns?
   (*d*) How far does Donna cycle when the rear wheel makes 100 full turns?
   (*e*) How far does Donna cycle when the rear wheel makes 2000 full turns? (Give the answer in kilometres.)

Revision Exercise XII

1. A rectangle measures 14 cm by 5 cm. Calculate its area.

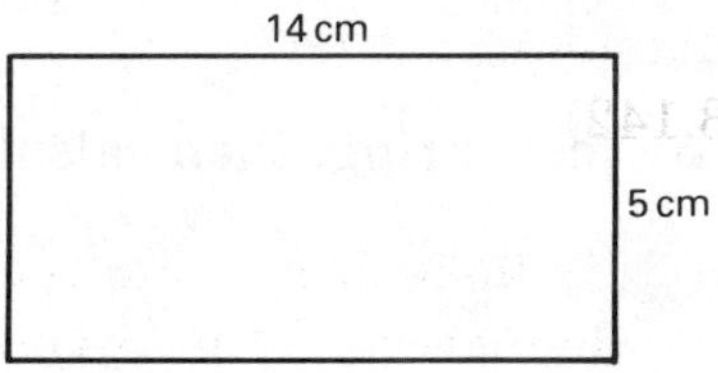

2. A rectangle has an area of 45 m^2. If its length is 9 m, calculate its breadth.

3. A rectangle has an area of 22.8 cm^2. If its breadth is 3 cm, calculate its length.

4. A square has an area of $25\,cm^2$.

(*a*) Find the length of each side of the square.

(*b*) Find the perimeter of the square.

5. Find the area of a square if the length of each side is 70 cm.

6. Find the areas of the parallelograms:

(*a*)

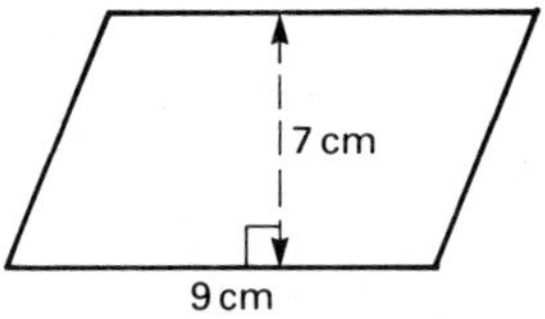

(*b*)

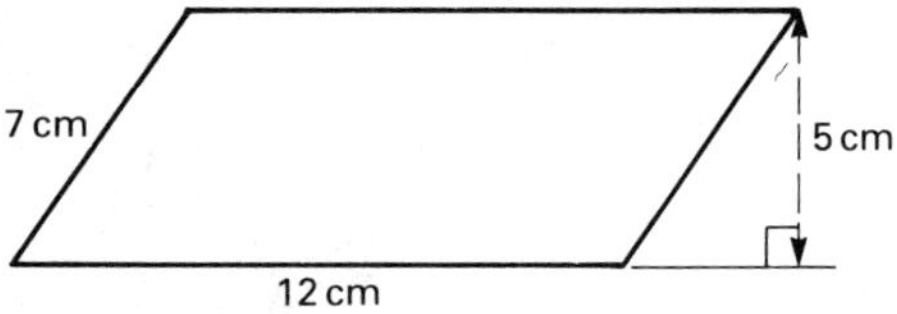

7. Find the areas of the triangles:

(*a*)

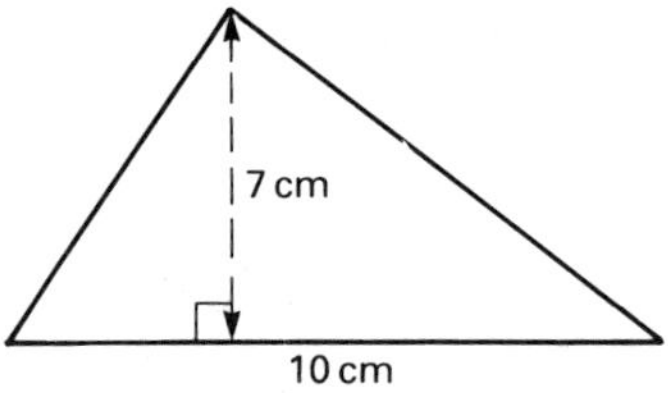

(*b*)

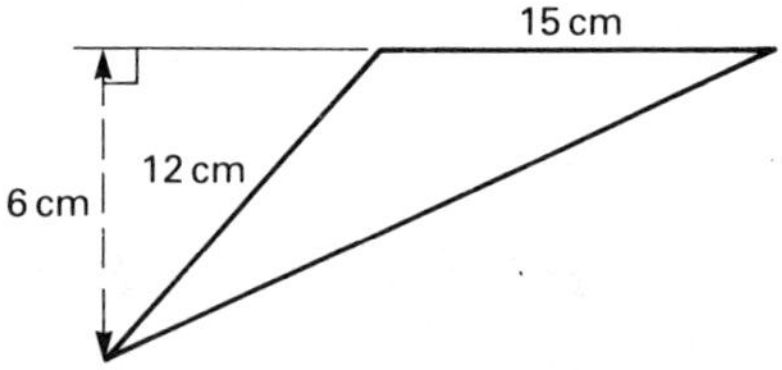

8. By measuring, then calculating, find the area of $\triangle ABC$.

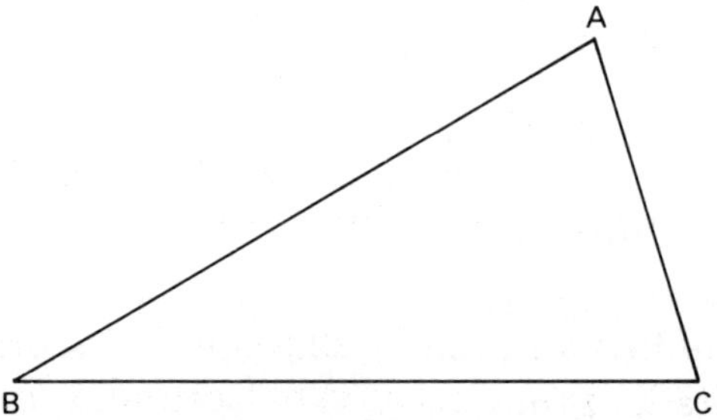

9. Calculate the volumes of these cuboids:

(*a*)

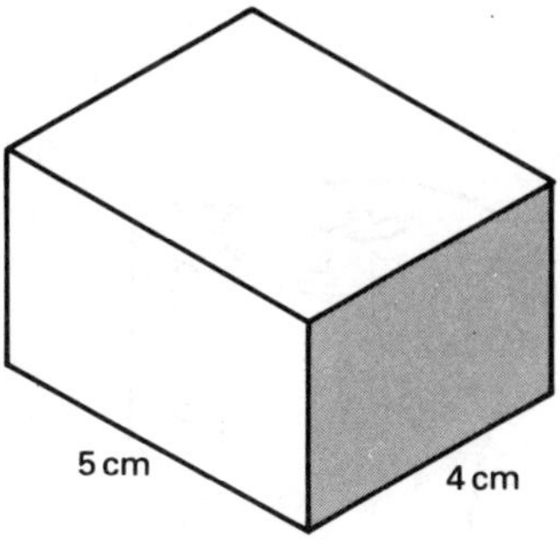

(*b*)

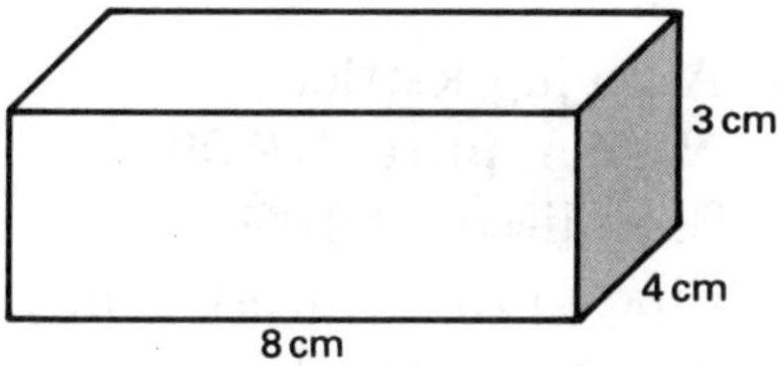

Revision Exercise XIII

1. Here is a set of decimal and vulgar fractions:

$$\{\tfrac{1}{10}, \tfrac{1}{8}, 0.2, 0.25, 0.3, \tfrac{2}{5}, 0.5, 0.6, 0.7, \tfrac{3}{4}, 0.8, \tfrac{9}{10}\}$$

Write the following percentages. Next to each one, write the decimal or vulgar fraction that has the same value. Choose from the set above.

(*a*) 50%	(*d*) 10%	(*g*) 90%	(*j*) 30%
(*b*) 25%	(*e*) 20%	(*h*) 80%	(*k*) 70%
(*c*) 75%	(*f*) 60%	(*i*) 40%	(*l*) $12\frac{1}{2}$%

2. Write as decimals:

(*a*) 49% (*b*) 91% (*c*) 60% (*d*) 6%

3. (*a*) Find 10% of £85.
(*b*) Find 10% of £5.40.
(*c*) Find 25% of £28.
(*d*) Find 40% of £35.

4. (*a*) Find 20% of £6.90.
(*b*) Find 54% of £32.
(*c*) Find 18% of £29.50.

5. A plumber's bill came to £50 + VAT at 15%. Calculate:
(*a*) the amount of VAT,
(*b*) the total bill including VAT.

6. Auto jug kettle.
Normal price £19.99.
Sale discount 10%.
Find to the nearest penny:
(*a*) the discount,
(*b*) the sale price.

14 Symmetry

Bilateral and Rotational Symmetry

Exercise 1

For each shape, write its number of lines of bilateral symmetry (there may not be any). If you need to do so, first make a rough copy of the shape in your exercise book.

1.

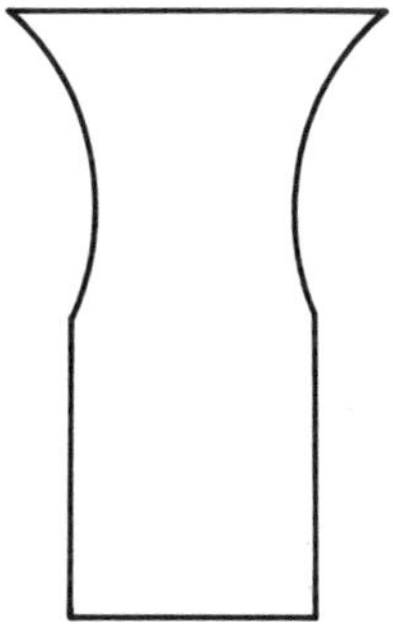

2.

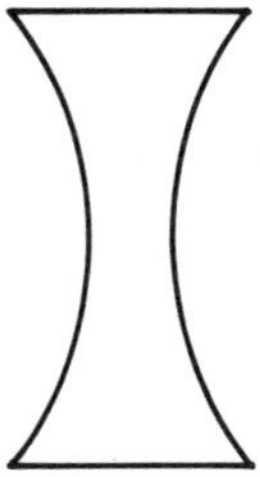

3.

4.

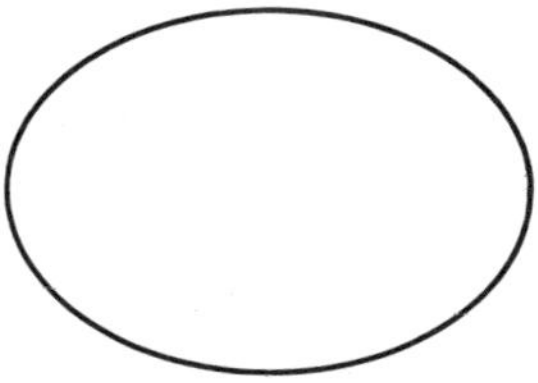

5.

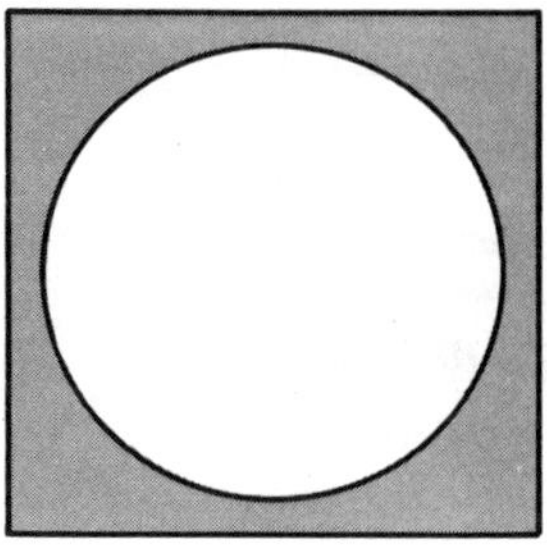

6.

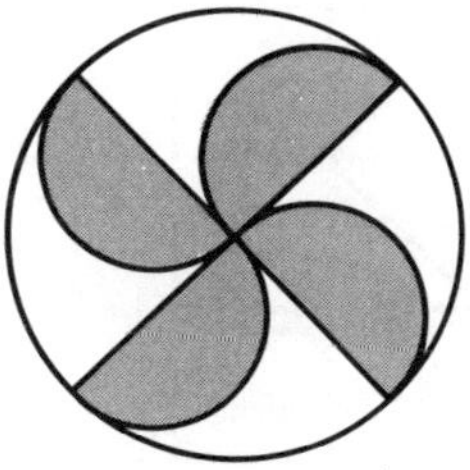

7.

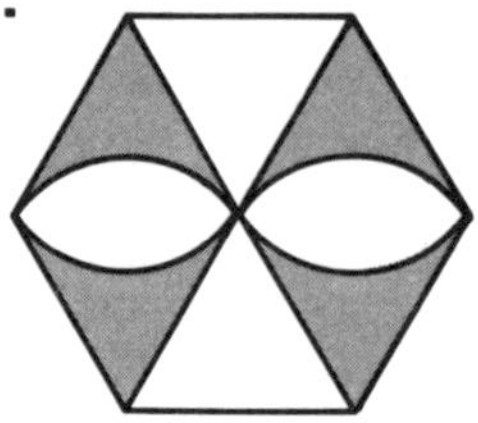

9.

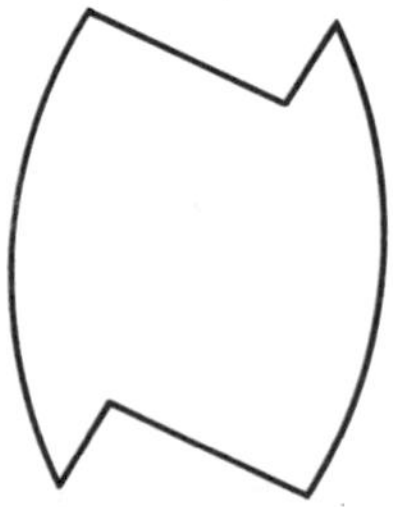

8.

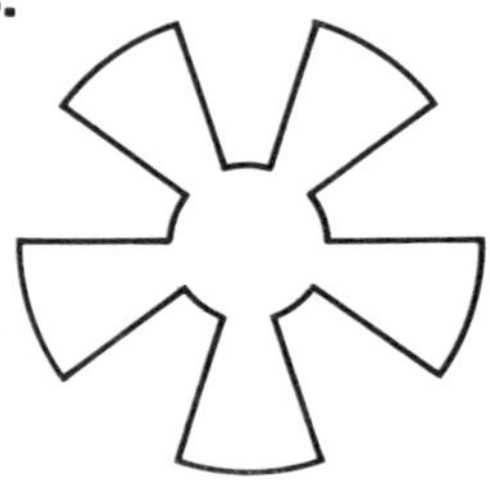

10.

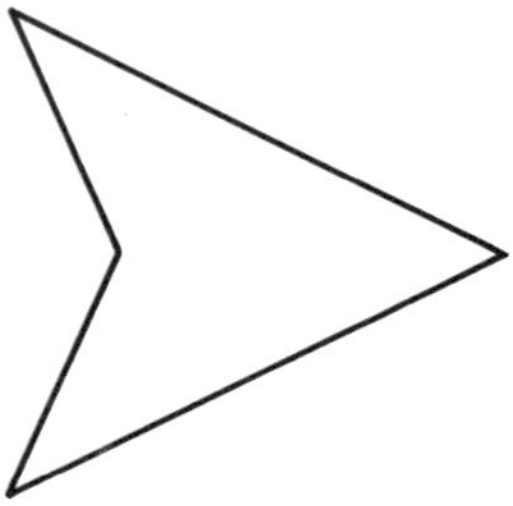

Exercise 2

For each shape in Exercise 1, write whether or not it has rotational symmetry.

Exercise 3

M

A Copy each shape on to squared paper.
Complete each one so that the broken line is a line of symmetry.

1.

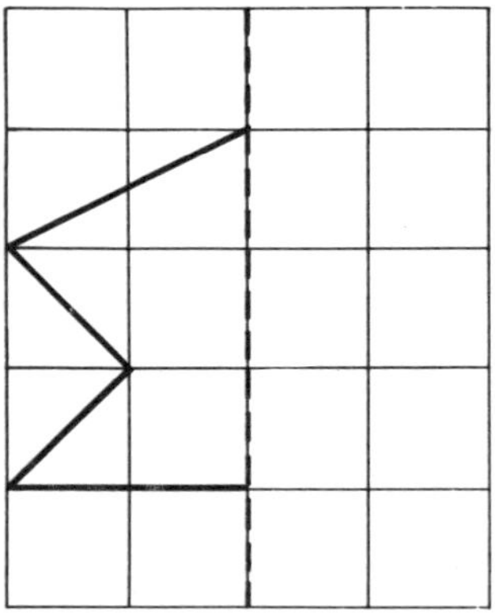

2.

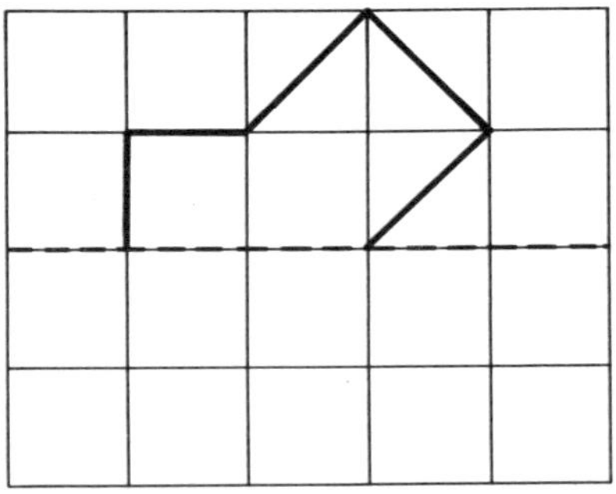

3.

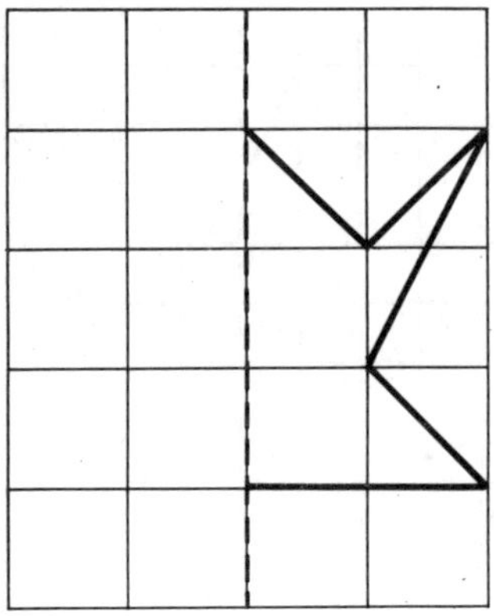

5.

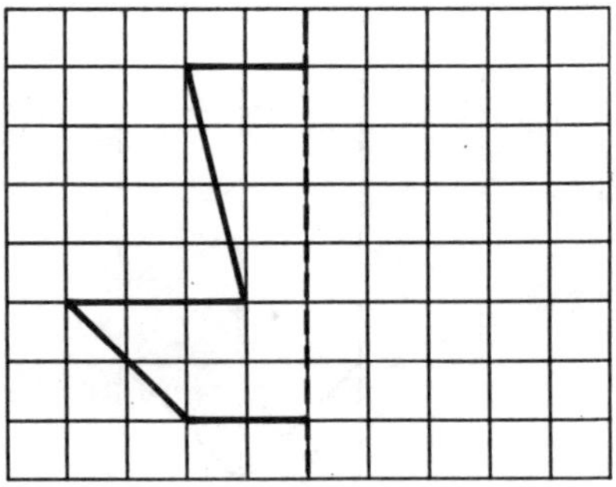

6.

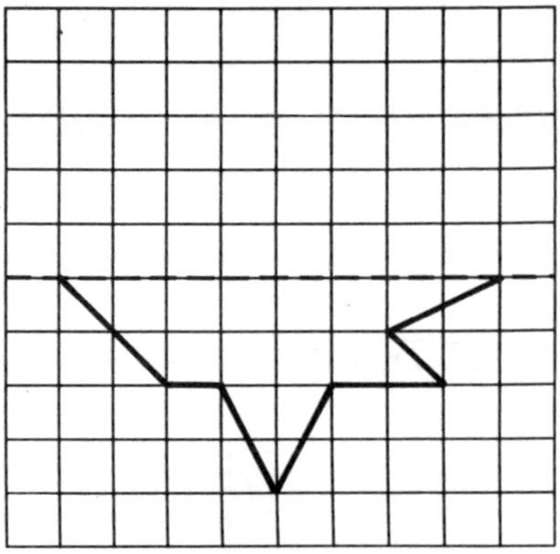

4.

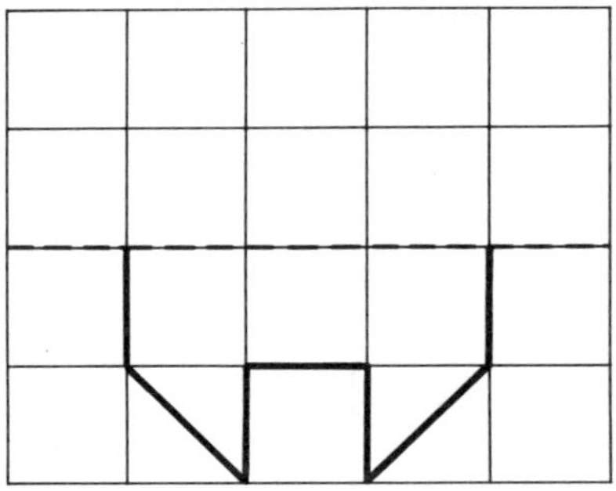

B Copy the following shapes.
Complete each one so that it has a horizontal and a vertical line of symmetry as shown by the broken lines.

1.

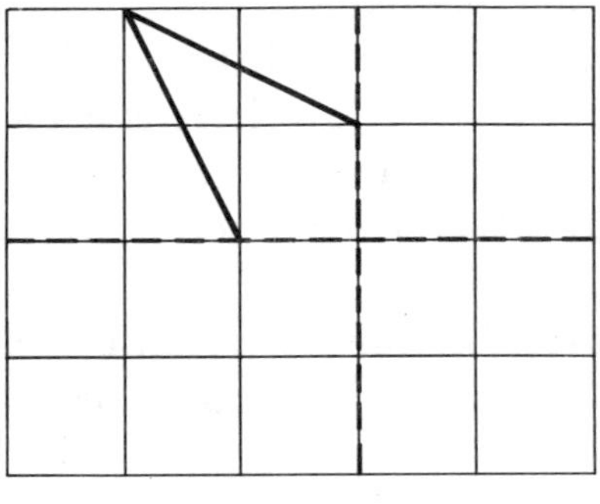

2.

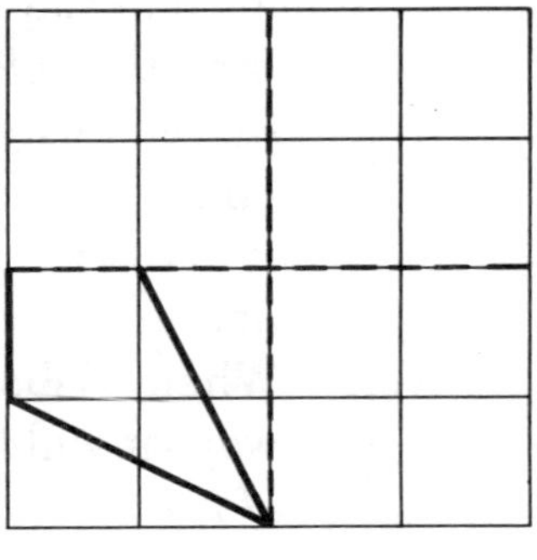

3.

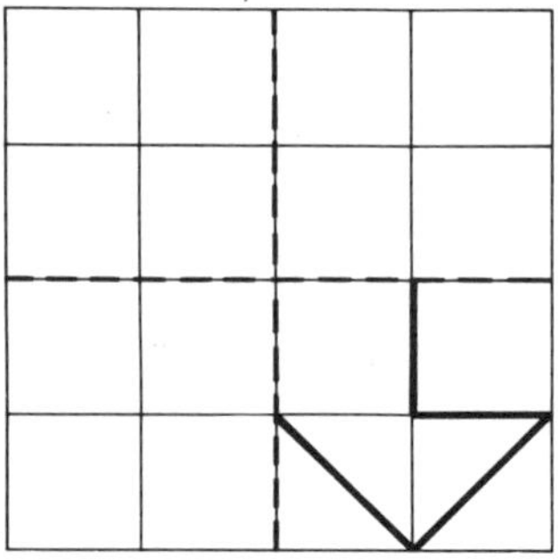

4.

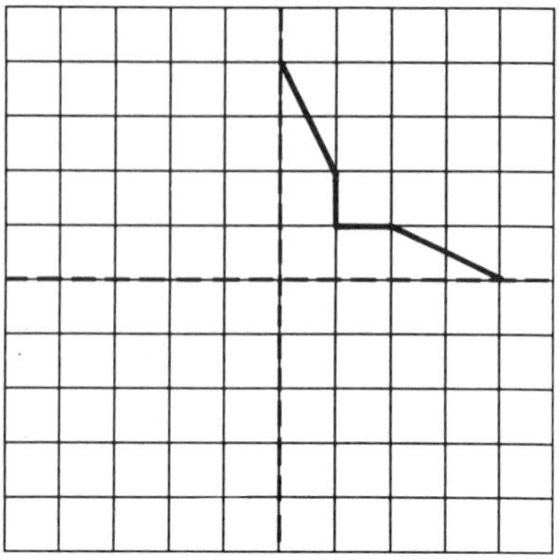

Order of Rotational Symmetry

Exercise 4

1. (*a*) Draw a rectangle as shown.

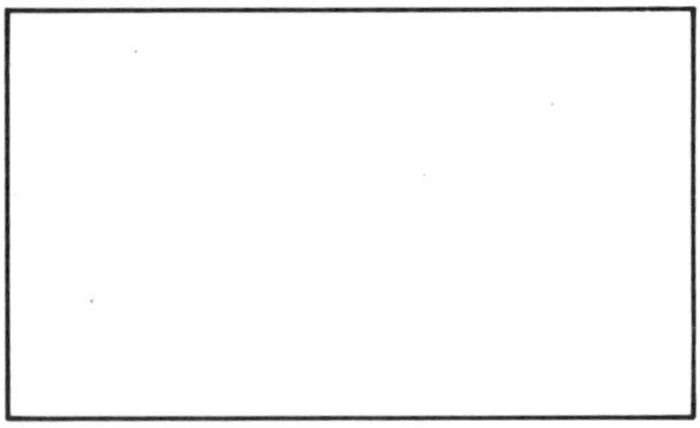

(*b*) Draw an identical rectangle on a piece of card and cut it out.

(*c*) Colour one side of the cut-out rectangle.

(*d*) Keeping the coloured side uppermost, find the number of different ways the cut-out rectangle will fit exactly on top of the rectangle you have drawn.

(*e*) What is the order of rotational symmetry of a rectangle?

(*Note* When a shape is rotated, the number of different ways it will fit into its own outline is called *the order of rotational symmetry*.)

2. Repeat question 1 for a regular hexagon. Make each side of the hexagon 3 cm long.

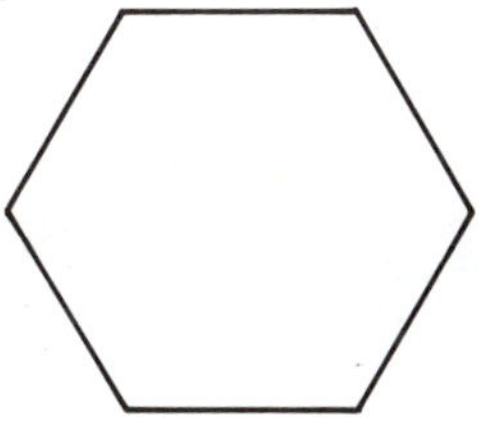

Note that a kite will fit into its own outline in only one way, and so the order of rotational symmetry of a kite is 1.

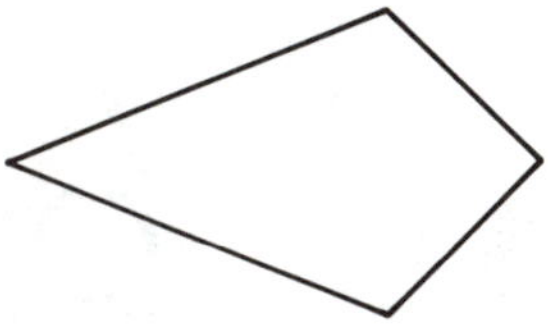

Exercise 5

Write the order of rotational symmetry of each shape given in Exercise 1.

Exercise 6

Throughout this exercise, check whether or not a given item has bilateral symmetry; if it has, write the number of axes. Check also whether or not the item has rotational symmetry; if it has, give its order of rotational symmetry.

A Road signs:

1.

National speed limit applies

2.

No motor vehicles except solo motorcycles scooters or mopeds

3.

No entry for vehicular traffic

4.

No stopping (clearway)

5.

Keep left

6.

Vehicles may pass either side

7.

Mini-roundabout

8.

Uneven road

9.

Crossroads

10.

Roundabout

11.

Dual carriageway ends

12.

Road narrows on both sides

13.

Two-way traffic straight ahead

14.

Hump bridge

15.

Cattle

16.

Low-flying aircraft or sudden aircraft noise

B Ordnance Survey Symbols:

1.  Church or chapel with spire

2.  Church or chapel with tower

3.  Church or chapel without tower or spire

4.  Youth hostel

5.  Bus or coach station

C Flags (ignore the edging for hanging the flags):

1.

Canada

2.

Ireland

3.

India

4.

Jamaica

5.

Turkey

6.

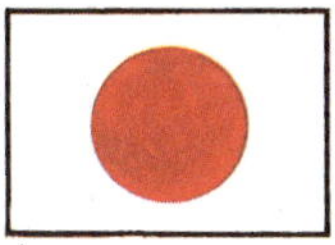

Japan

7.

Denmark

8.

Switzerland

9.

Trinidad & Tobago

10.

Israel

11.

Thailand

12.

Netherlands

13.

Morocco

14.

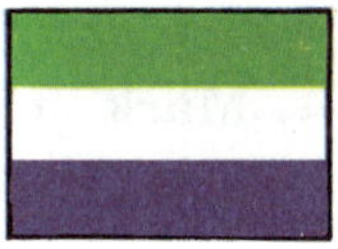

Sierra Leone

15.

Belgium

16.

United Kingdom

Exercise 7 Properties of Triangles Related to their Symmetries

A **1.** Copy the given diagram.

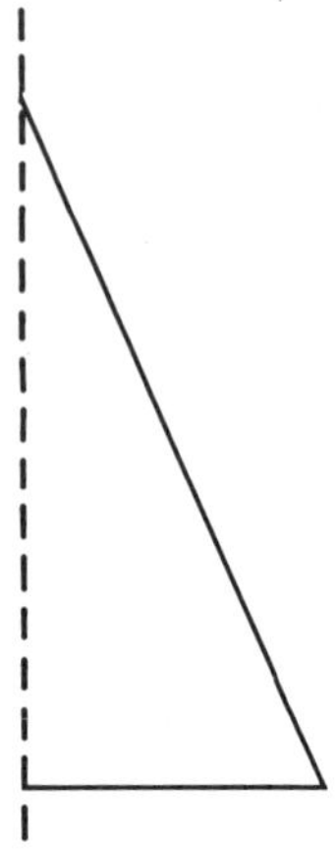

2. Complete it so that the broken line is a line of symmetry.

3. What sort of triangle have you obtained?

4. Mark on your triangle the sides that are equal.

5. Mark on your triangle the angles that are equal.

B **1.** Construct an equilateral triangle with sides that measure 50 mm.

2. (*a*) Draw the lines of bilateral symmetry.
(*b*) How many lines of bilateral symmetry has an equilateral triangle?

3. Mark the equal sides of the triangle.

4. Mark the equal angles of the triangle.

C **1.** Draw a right-angled triangle that has a line of symmetry.

2. Write what is special about a right-angled triangle that has a line of symmetry.

15 Functions

Exercise 1 M

Copy each of these. Complete each one by filling in the missing numbers.

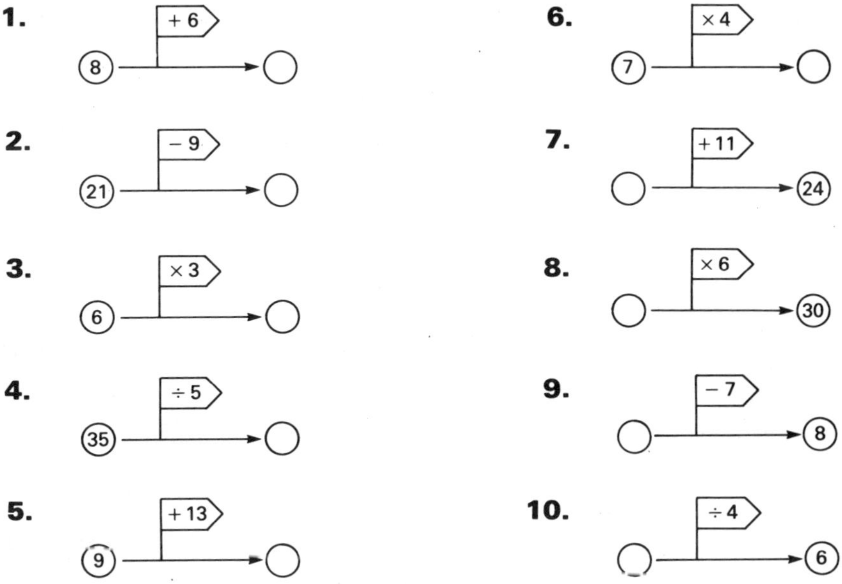

Exercise 2 M

Copy and complete:

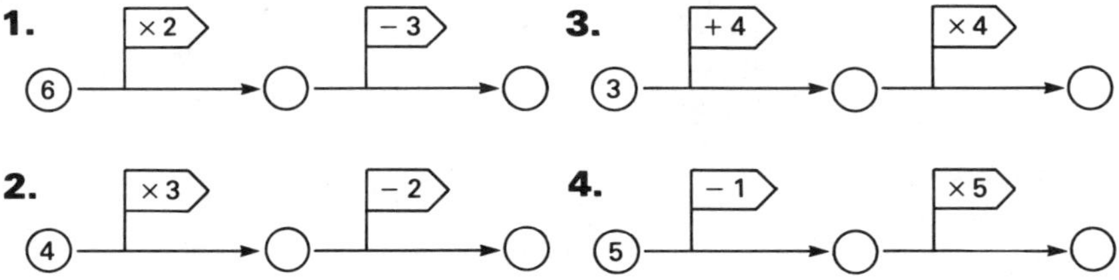

5.

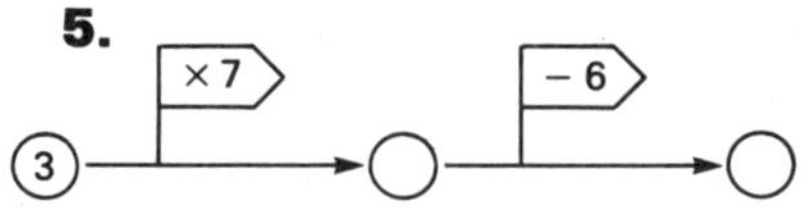

6.

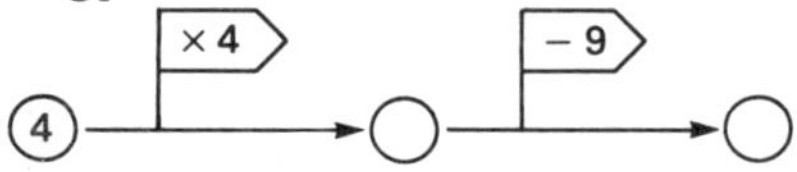

7.

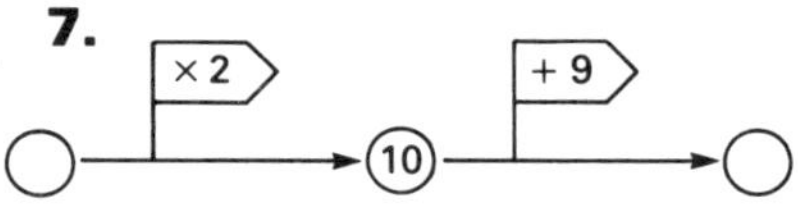

8.

9.

10.

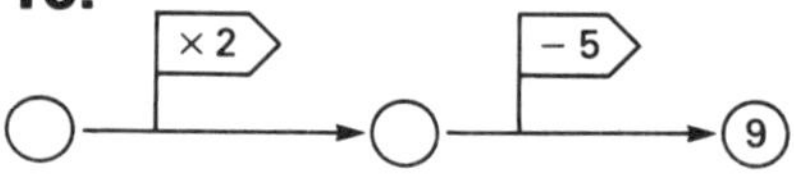

11.

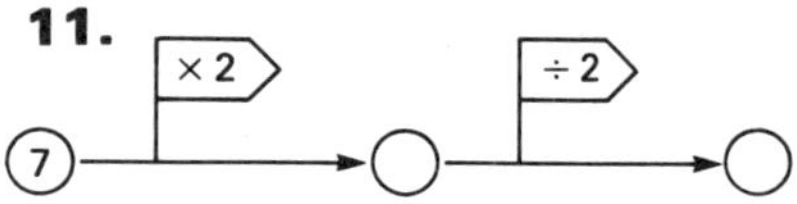

12.

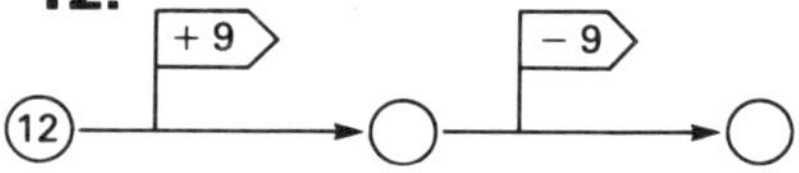

Exercise 3 M

Copy and complete:

1. (*a*)

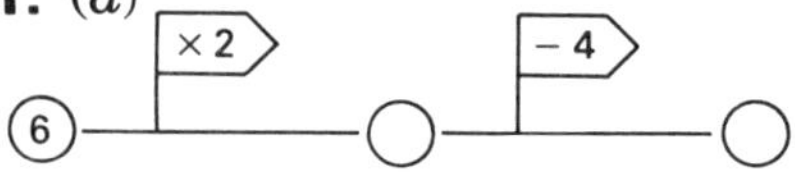

(*b*)

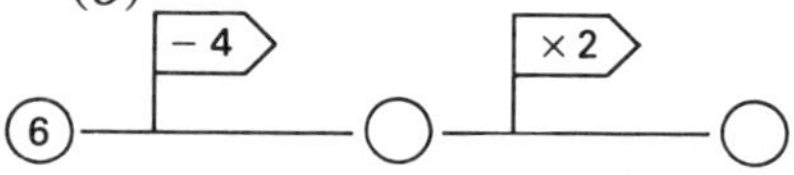

2. (*a*)

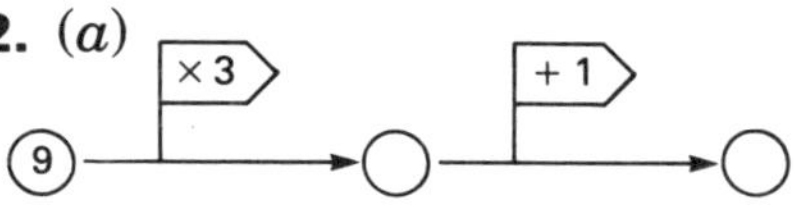

(*b*)

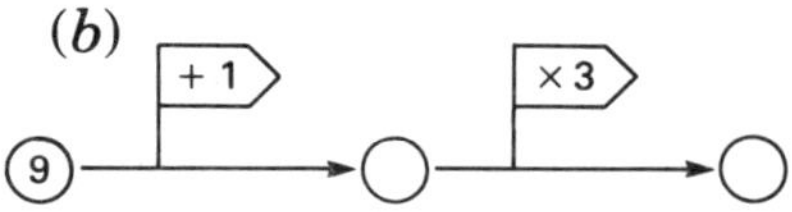

3. (*a*)

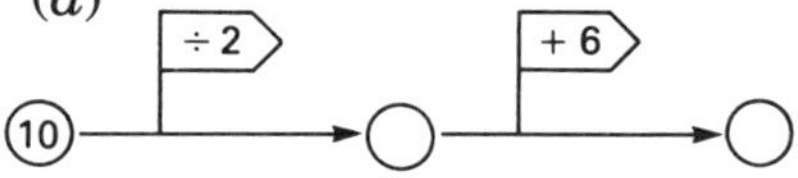

(*b*)

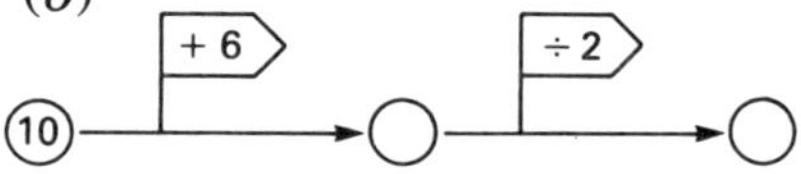

4. (*a*)

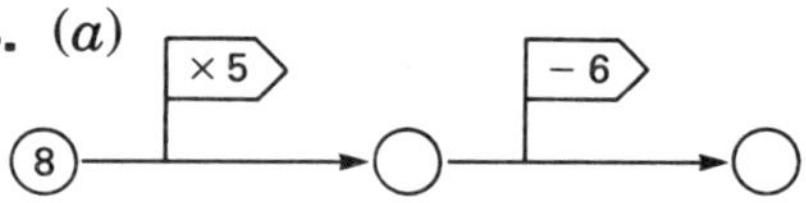

(*b*)

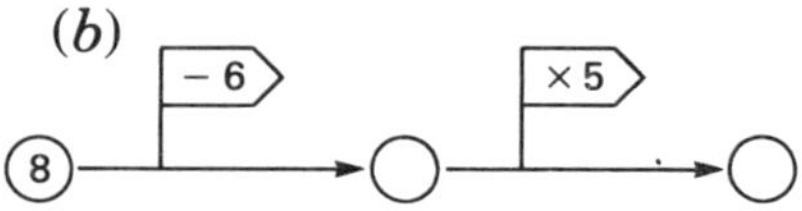

5. (*a*)

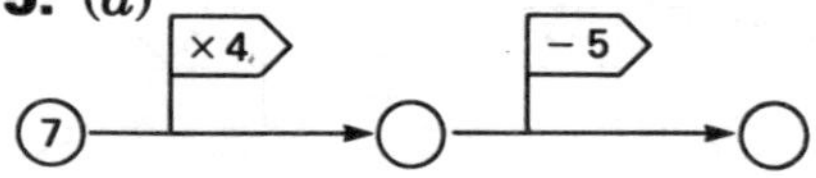

(*b*)

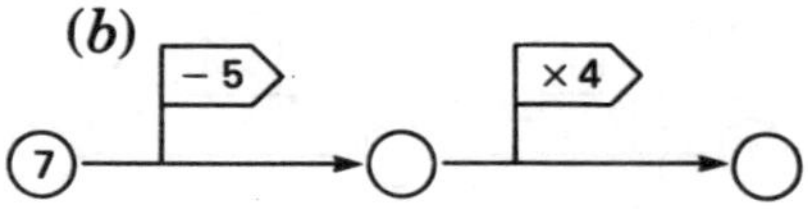

6. (*a*)

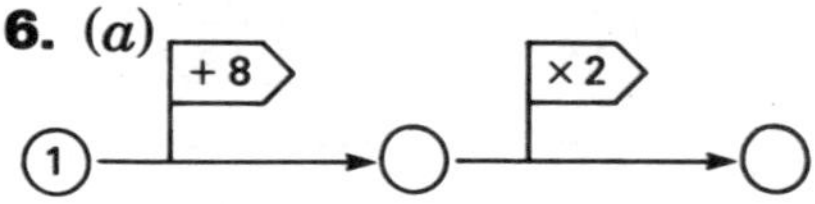

(*b*)

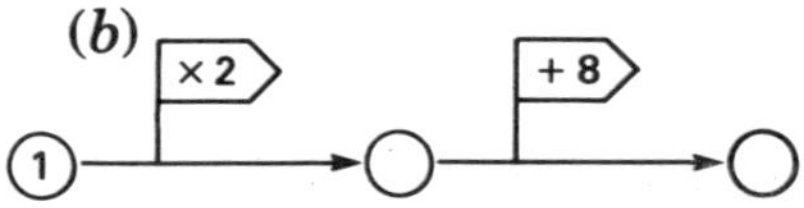

7. (*a*)

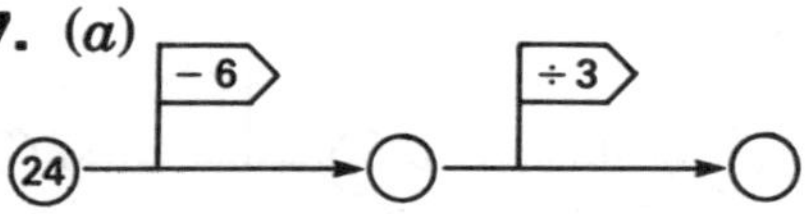

(*b*)

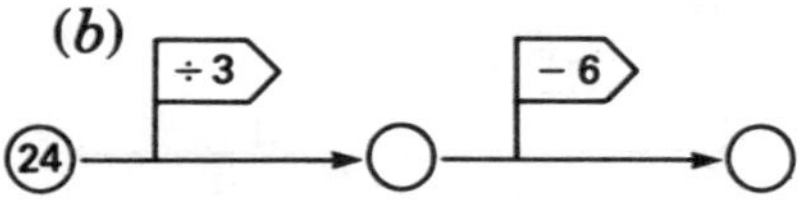

8. (*a*)

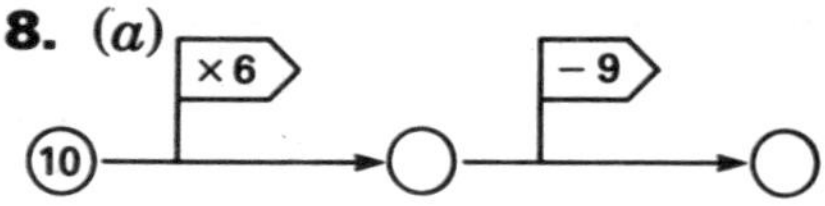

(*b*)

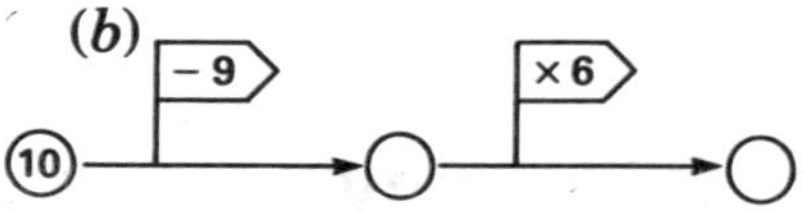

Exercise 4 **M**

Copy and complete:

1. (*a*)

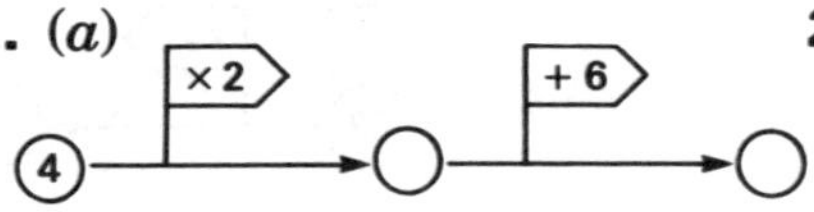

(*b*)

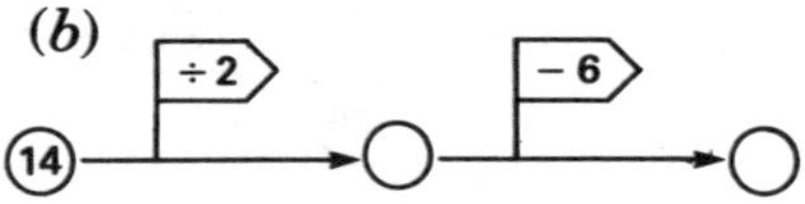

(*c*)

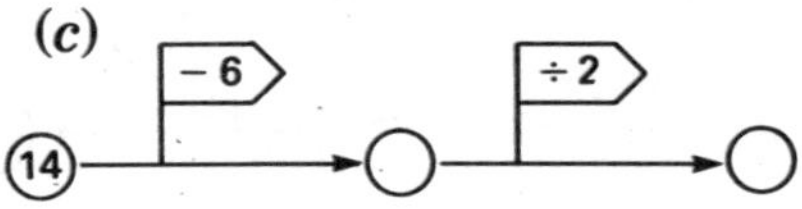

2. (*a*)

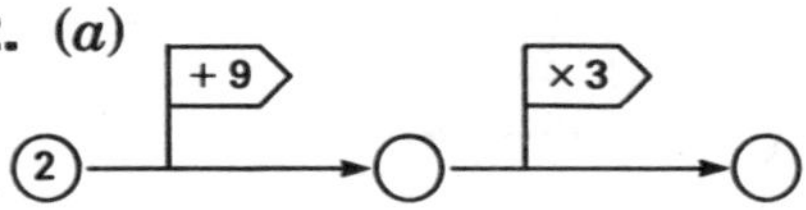

(*b*)

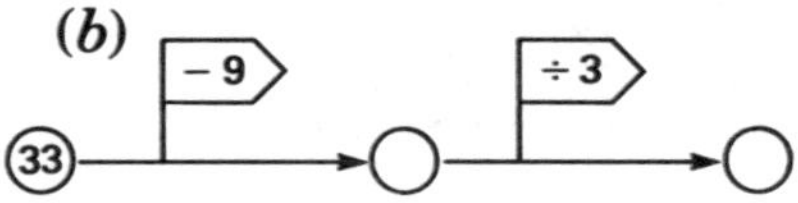

(*c*)

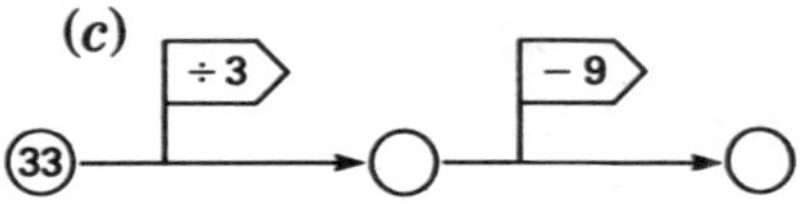

3. (*a*)

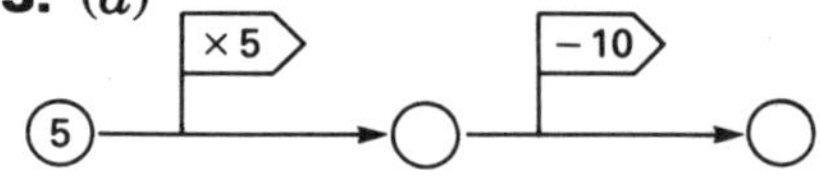

(*b*)

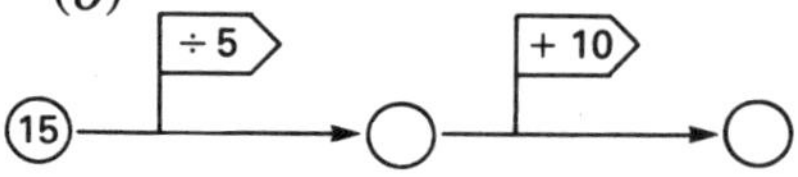

(*c*)

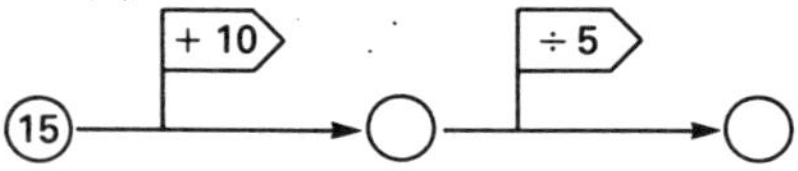

4. (*a*)

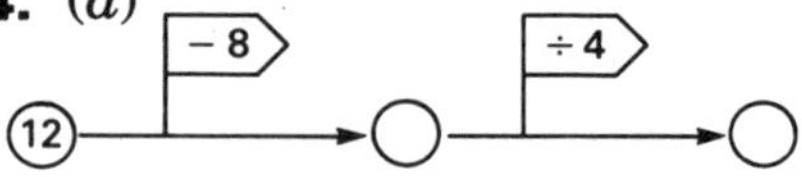

(*b*)

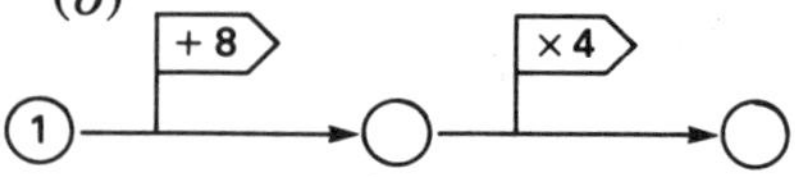

(*c*)

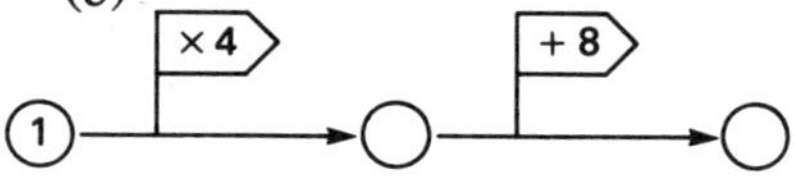

Function Keys on a Calculator

Exercise 5 M

Copy and complete the following. Use a calculator.

A

1.

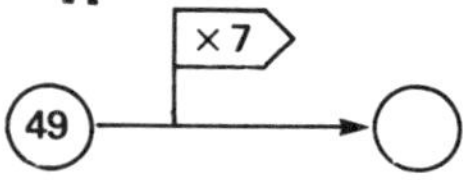

2.

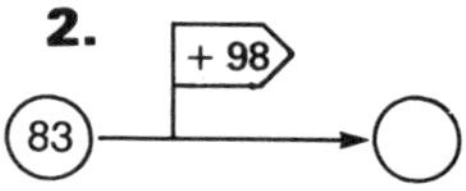

3.

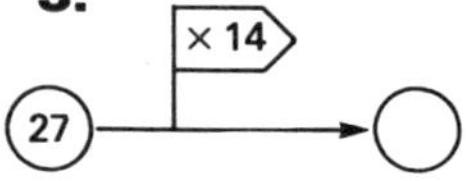

4.

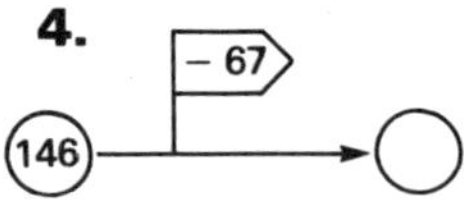

5.

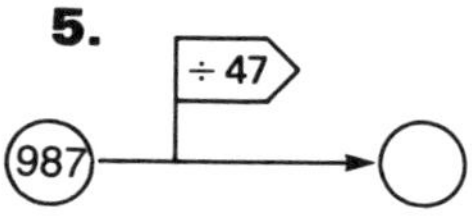

6.

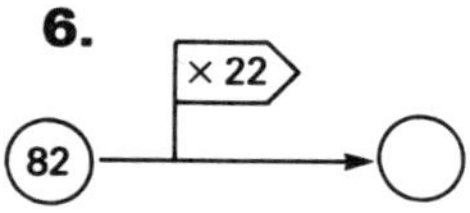

7.

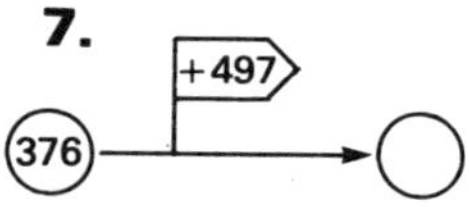

8.

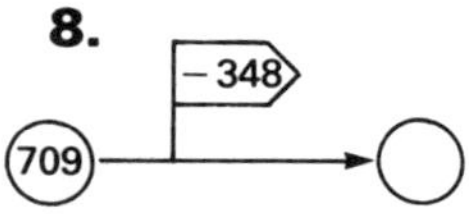

B

1. 871 → [÷ 13] → ○ → [+ 33] → ○

2. 18 → [× 26] → ○ → [−185] → ○

3. 361 → [−109] → ○ → [÷ 21] → ○

4. 28 → [+ 17] → ○ → [× 9] → ○

5. ○ → [× 11] → ○ → [+145] → 772

Exercise 6 M

Copy and complete the following. Use the $\boxed{x^2}$ and $\boxed{\sqrt{\ }}$ keys on a calculator.

1. 24 → [x^2] → ○

2. 1.9 → [x^2] → ○

3. 784 → [$\sqrt{\ }$] → ○

4. 961 → [$\sqrt{\ }$] → ○

5. 27 → [x^2] → ○

6. 169 → [$\sqrt{\ }$] → ○

7. 841 → [$\sqrt{\ }$] → ○

8. 16 → [x^2] → ○

16 Two-state Systems

Exercise 1 Railway Sidings

The drawing shows some railway sidings.
Every 100 m there is a set of points at which the line splits into two lines.

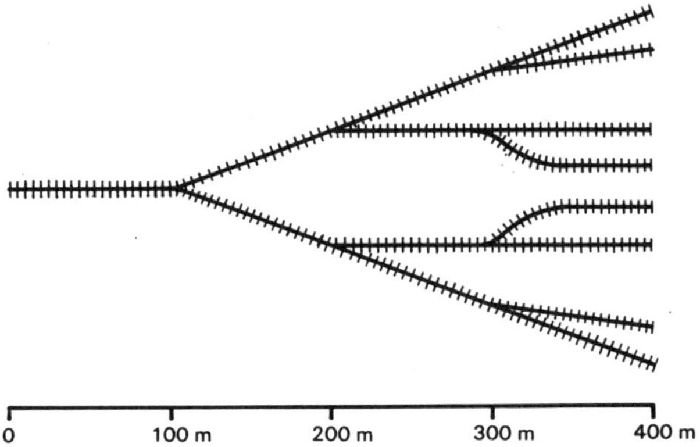

Copy and complete these sentences:

1. At 100 m the line splits into [?] lines.

2. At 200 m the line splits into [?] lines.

3. At 300 m the line splits into [?] lines.

4. At 400 m the line splits into [?] lines.

5. At 500 m the line splits into [?] lines.

6. At 600 m the line splits into [?] lines.

Exercise 2 A Family Tree

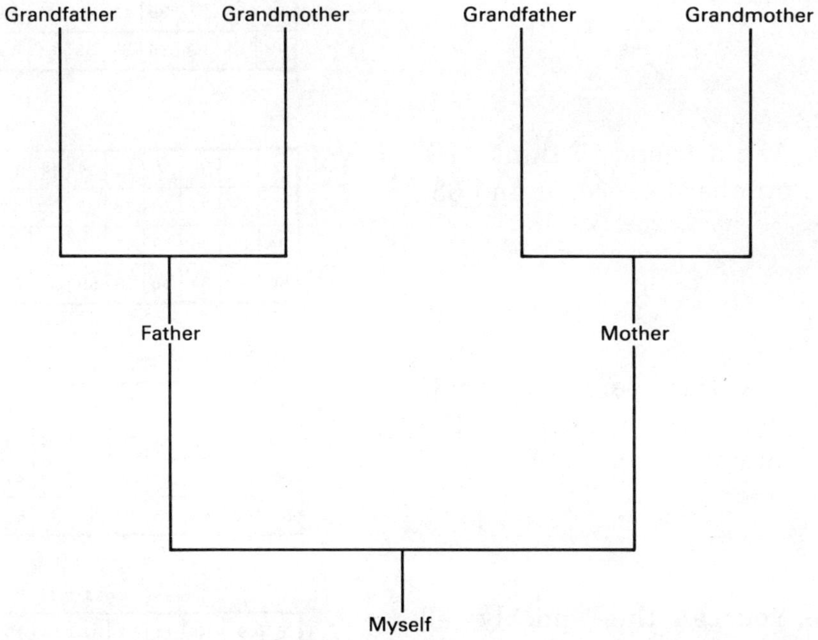

Copy the family tree shown above and continue it for two more steps (as far as great great grandfathers and great great grandmothers).

Copy and complete these sentences:

1. There are ? parents.

2. There are ? grandparents.

3. There are ? great grandparents.

4. There are ? great great grandparents.

5. There are ? great great great grandparents.

Exercise 3 Magic Cards M

Either use the given set of six magic cards or make your own set.

1. Ask a friend to think of a number between 1 and 63.

2. Ask that person to show you which cards the number is on.

3. You can then quickly tell your friend the number he or she thought of.

   To do this, add the numbers in the top left-hand corners of the cards your friend shows you. This total is the required number.

1	3	5	7	9	11	13	15
17	19	21	23	25	27	29	31
33	35	37	39	41	43	45	47
49	51	53	55	57	59	61	63

2	3	6	7	10	11	14	15
18	19	22	23	26	27	30	31
34	35	38	39	42	43	46	47
50	51	54	55	58	59	62	63

4	5	6	7	12	13	14	15
20	21	22	23	28	29	30	31
36	37	38	39	44	45	46	47
52	53	54	55	60	61	62	63

8	9	10	11	12	13	14	15
24	25	26	27	28	29	30	31
40	41	42	43	44	45	46	47
56	57	58	59	60	61	62	63

16	17	18	19	20	21	22	23
24	25	26	27	28	29	30	31
48	49	50	51	52	53	54	55
56	57	58	59	60	61	62	63

32	33	34	35	36	37	38	39
40	41	42	43	44	45	46	47
48	49	50	51	52	53	54	55
56	57	58	59	60	61	62	63

Exercise 4 Punched Cards

Make a set of 15 cards like the cards shown below. (The holes should be in the same position on each card.) Cut out slots as shown.

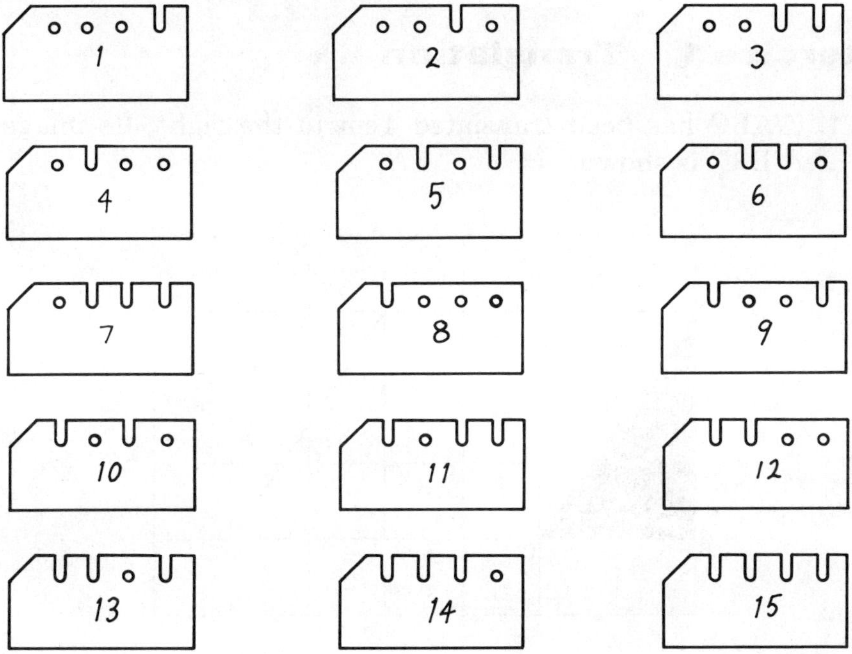

Number the cards from 1 to 15 as shown.
Place them together with the cut-off corner at the left, then shuffle them. Keeping the cut-off corner to the left, push a needle through the right-hand holes in the cards. Lift the needle. (The cards with slots in this position will be left behind.) Place the set of cards picked up with the needle at the front of the pack. Now push the needle through the second hole from the right. The cards lifted out should again be placed at the front.

Repeat this for the third hole from the right and then the fourth hole (the left-hand hole.)

The cards should now be sorted into order from 1 to 15.

17 Transformation Geometry

Exercise 1 Translation

1. $\triangle ABC$ has been translated 4 cm to the right. Its image A′B′C′ is shown.

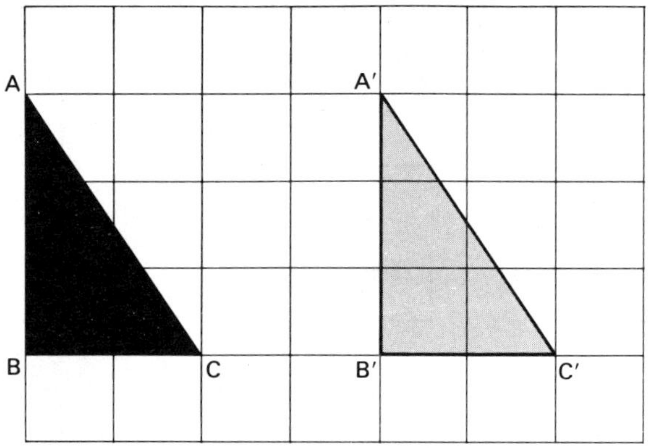

Make a copy of $\triangle ABC$ on 1 cm squared paper. Draw its image after a translation 5 cm downwards.

2. Make another copy of $\triangle ABC$.
Draw its image after a translation 3 cm to the left.

3. Part of a pattern is shown below. It has been obtained by translating a triangle. Copy and complete the pattern.

Reflection

Exercise 2 M

A The diagram shows a triangle that has been reflected in the given mirror line.

Copy the following shapes and for each one, draw its reflection in the given mirror line:

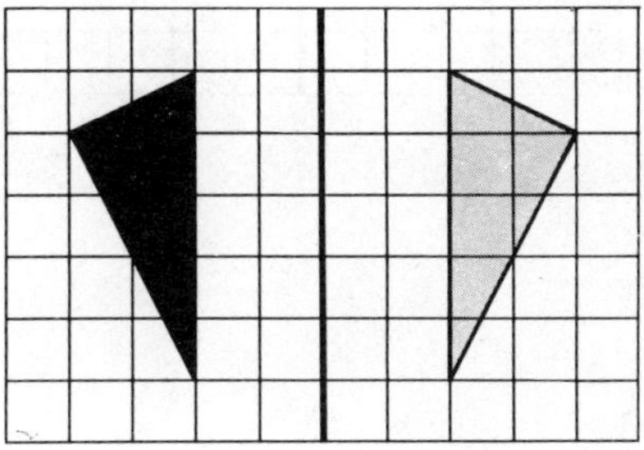

1.

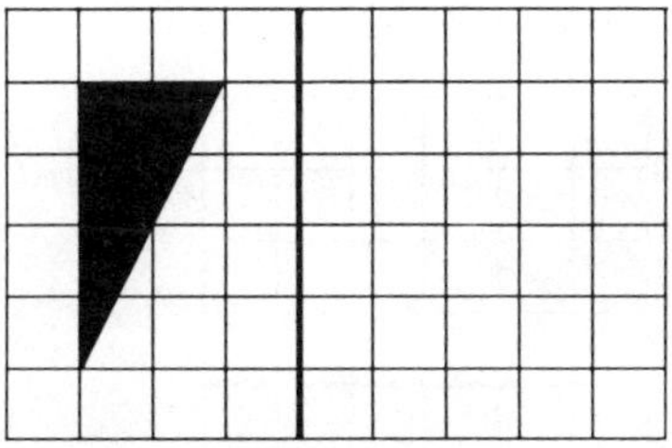

2.

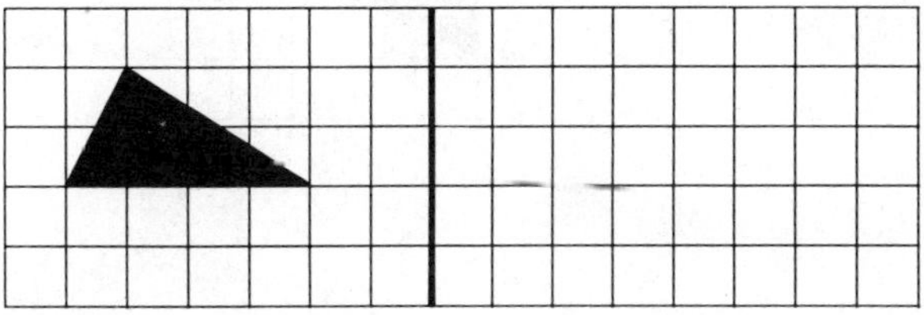

3.

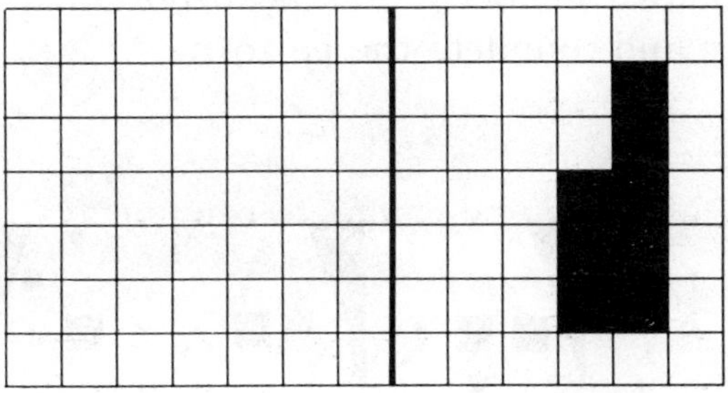

4.

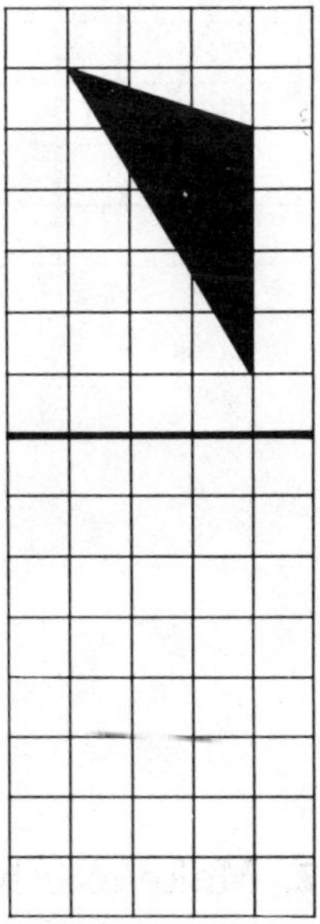

5.

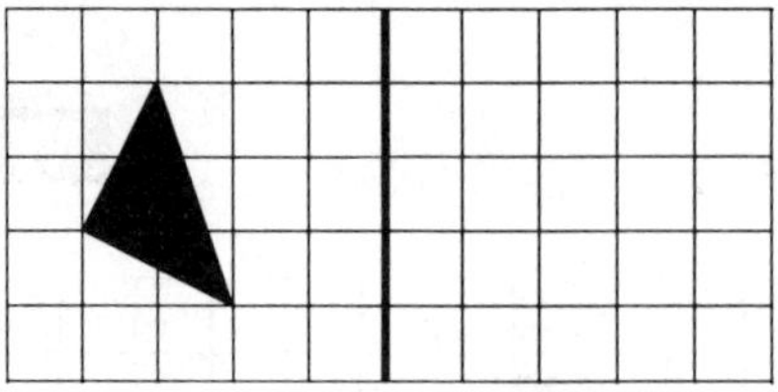

7.

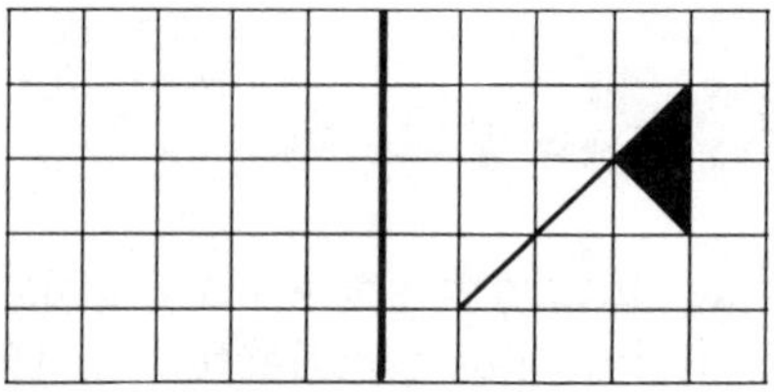

6.

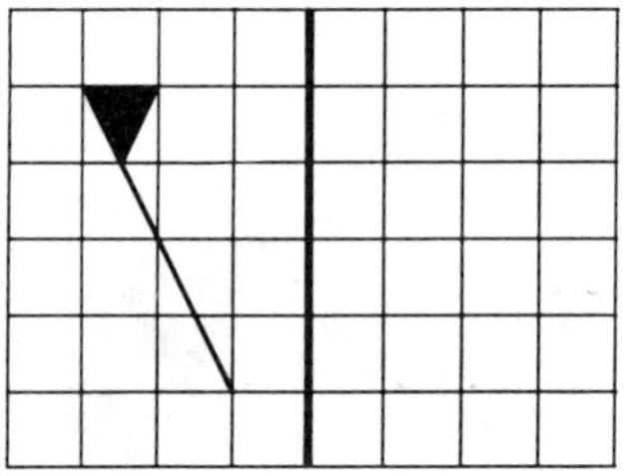

8.

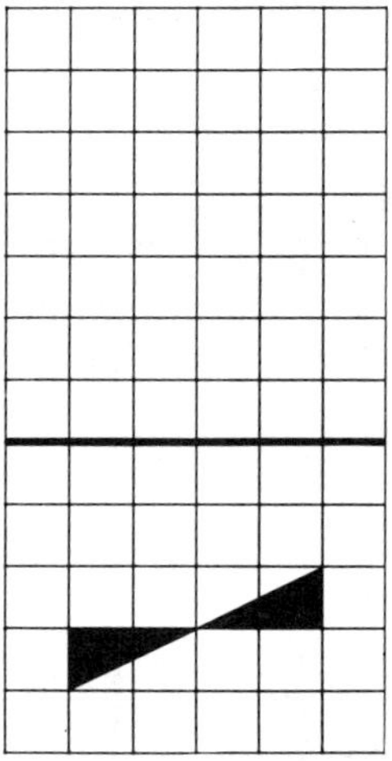

B Part of a pattern is shown below. It has been obtained by reflecting shapes. Copy and complete the pattern.

Exercise 3

Place a pencil, as shown, in front of an upright plane mirror.

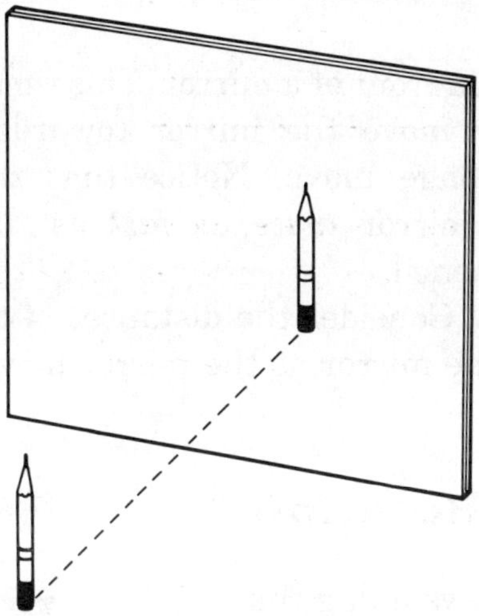

1. (*a*) If the pencil is 16 cm from the mirror, how far is its image from the mirror?
   (*b*) If the pencil is 12 cm from the mirror, how far is its image from the mirror?

2. (*a*) If the pencil is moved closer to the mirror, what happens to its image?
   (*b*) If the pencil is moved further away from the mirror, what happens to its image?

3. (*a*) If the pencil is moved 6 cm closer to the mirror, how far does its image move?
   (*b*) If the pencil is moved 2 cm further away from the mirror, how far does its image move?

4. If a pencil 16 cm from a mirror is moved 4 cm closer to the mirror:
   (*a*) How far did its image move?
   (*b*) How far from the mirror is the pencil now?
   (*c*) How far from the mirror is its image now?

5. Hold a pencil in front of a mirror. Keep the mirror still. Steadily move the pencil towards the mirror. Does the image move faster, slower or at the same speed as the pencil itself?

6. Hold a pencil in front of a mirror. This time, keep the pencil still. Steadily move the mirror towards the pencil and watch the image move. Notice that the image moves towards the mirror *twice as fast* as the mirror moves towards the pencil.
Explain why. (Consider the distances of the pencil and its image from the mirror as the mirror moves.)

Exercise 4 Rotation **M**

The diagram shows a flag that has been rotated through a $\frac{1}{4}$ turn clockwise about point O.

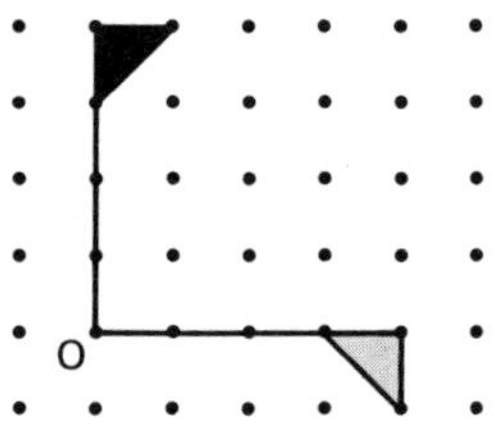

Copy the following flags. Draw the image of each flag after a rotation about O as instructed.

1. $\frac{3}{4}$ turn clockwise:

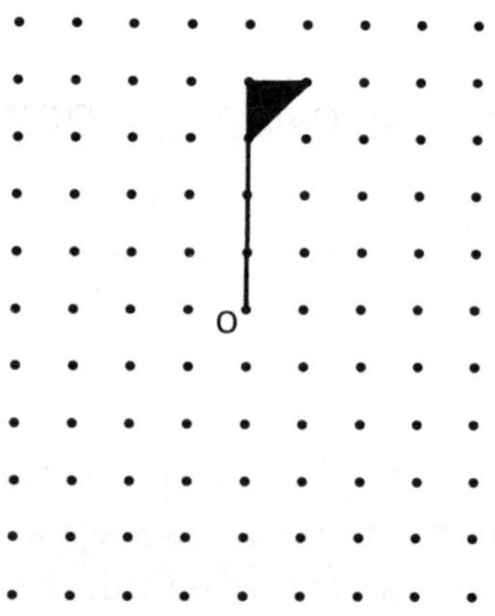

2. $\frac{1}{2}$ turn anticlockwise:

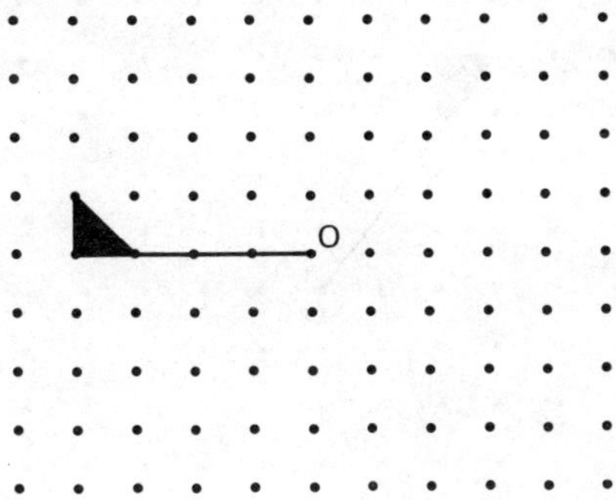

3. $\frac{1}{4}$ turn anticlockwise:

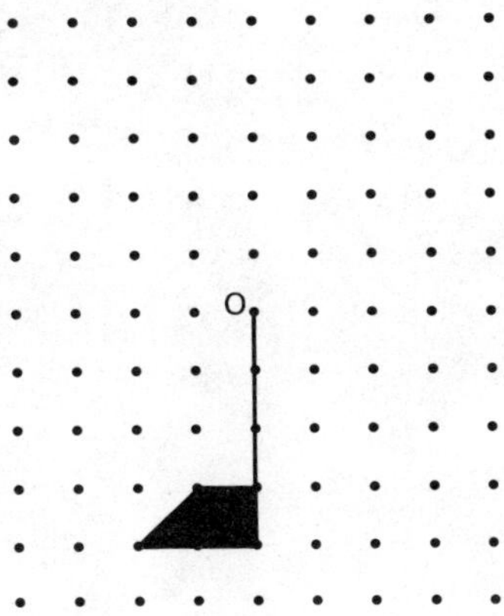

4. $\frac{3}{4}$ turn clockwise:

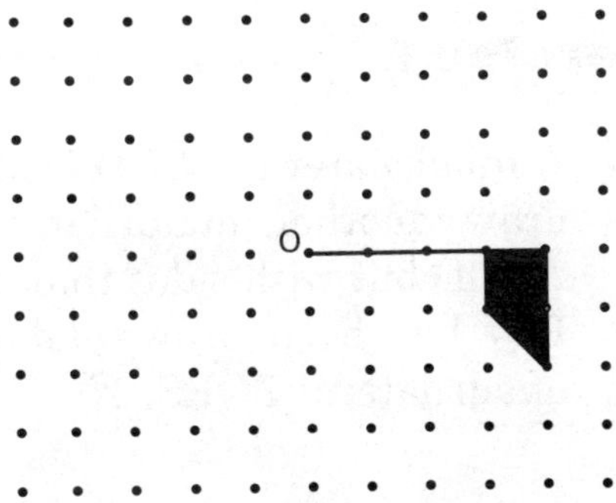

5. $\frac{1}{4}$ turn clockwise:

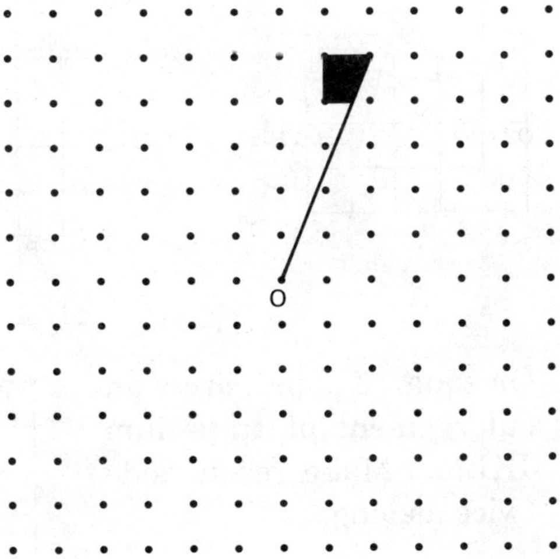

6. $\frac{3}{4}$ turn anticlockwise:

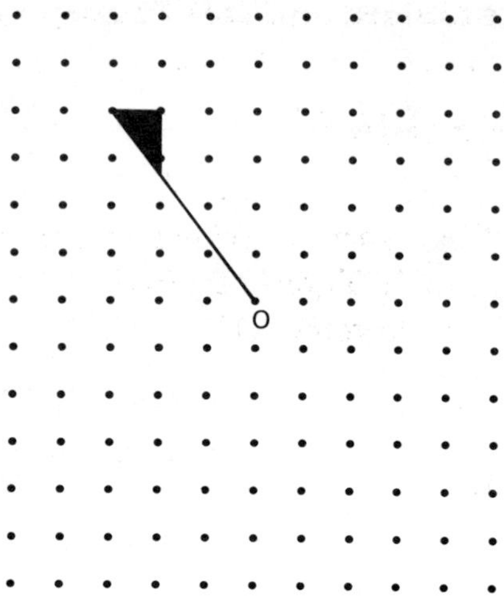

Exercise 5 Enlargements **M**

1. A quadrilateral ABCD is shown below. On squared paper, draw another quadrilateral PQRS the same shape as ABCD but with sides that are twice as long. One side (side PQ) has been drawn for you. Copy PQ, then complete quadrilateral PQRS.

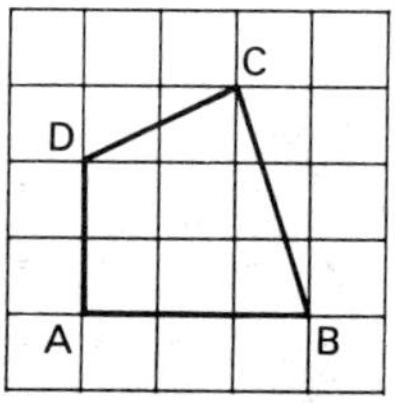

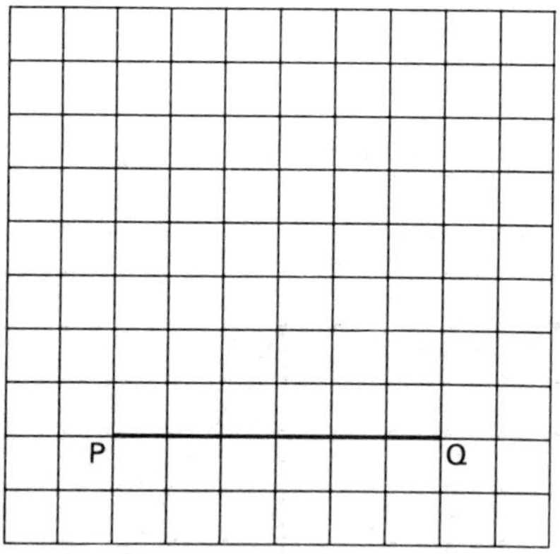

2. On squared paper, draw an enlargement of trapezium JKLM. Make each side twice as long.

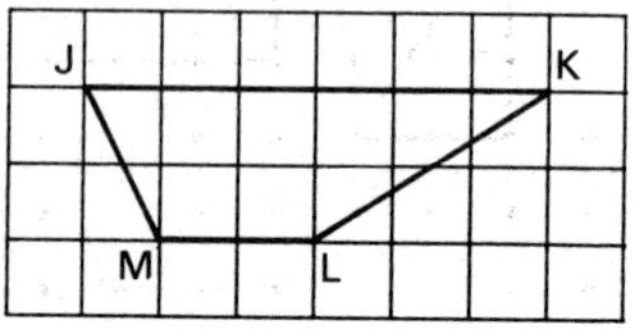

Miscellaneous Transformations

Exercise 6

What sort of transformation has taken place in mapping triangle T_1 onto triangle T_2 (a translation, reflection, rotation or enlargement)?

1.

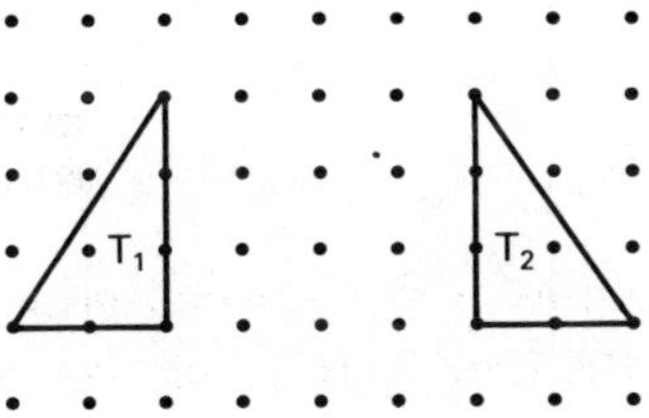

3.

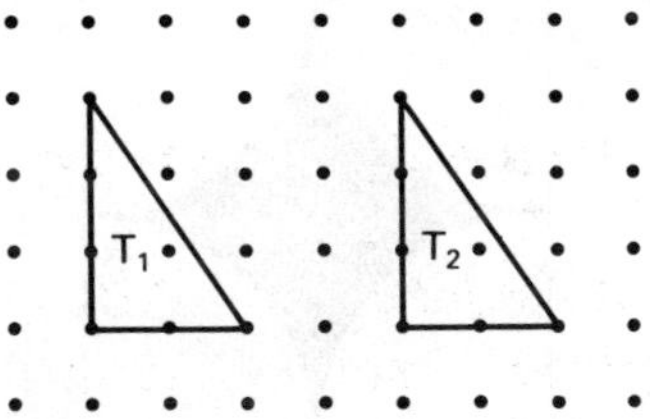

2.

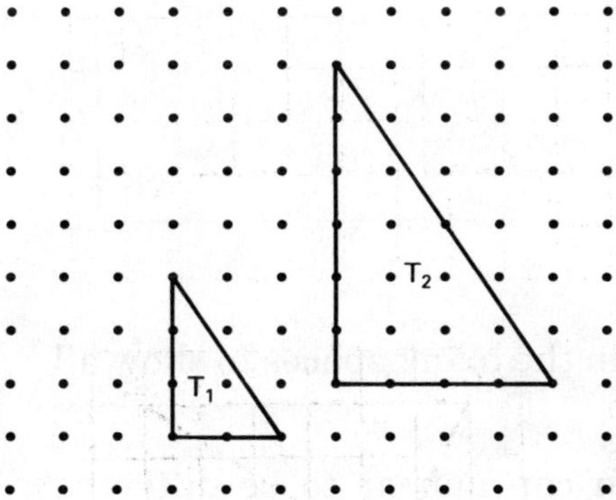

4.

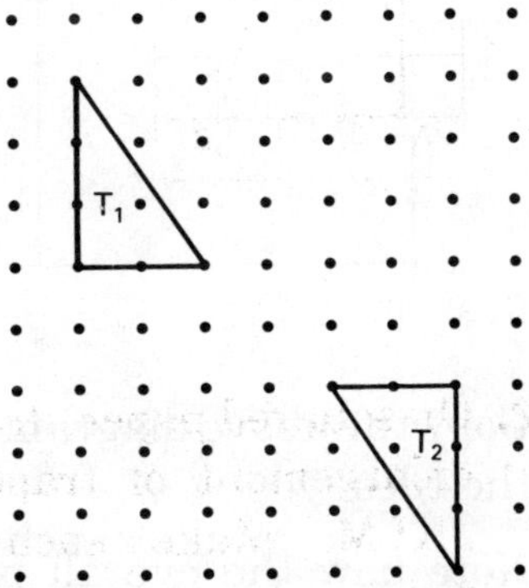

Wallpaper designs and the patterns made by wall tiles show the ideas of transformation geometry. The patterns are based on translations, reflections and rotations.

Exercise 7 M

1. The diagram shows one tile and spaces for 8 more tiles of the same design:

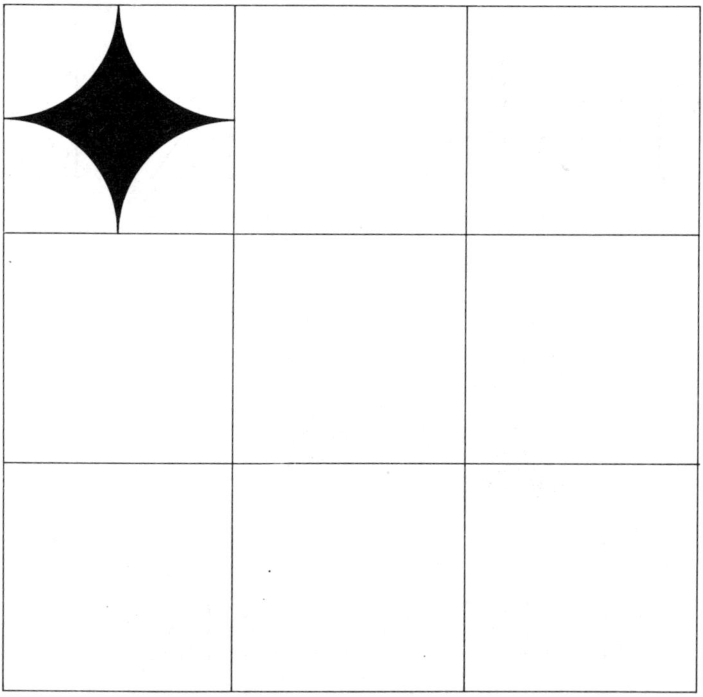

Copy the above diagram. Fill in the blank spaces to show all the tiles.

Note how the overall pattern can appear to be different when several tiles are put together.

2. (*a*) The diagram shows two tiles, and spaces for seven more tiles of the same design.
Copy the diagram.
Fill in the blank spaces to show all the tiles.

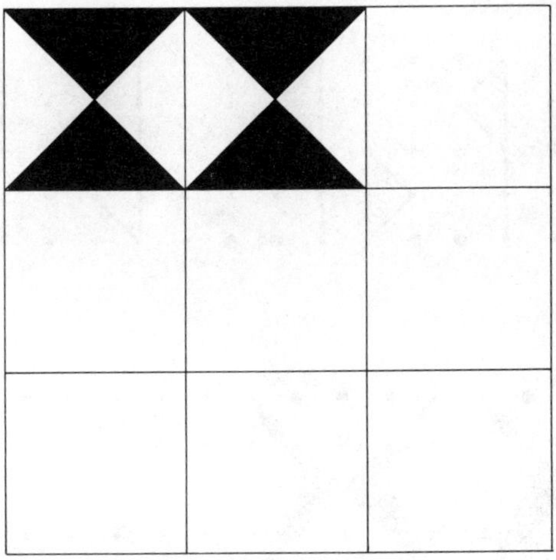

(*b*) This diagram shows the same two tiles as in part (*a*).
They now start to form a different pattern.
Copy the diagram.
Fill in the blank spaces to show all the tiles, and to form the different pattern.

3. Design your own tile.
Show the pattern formed by your tile by drawing a diagram. Draw at least nine tiles.

Exercise 8 **M**

Copy and complete the given patterns:

1.

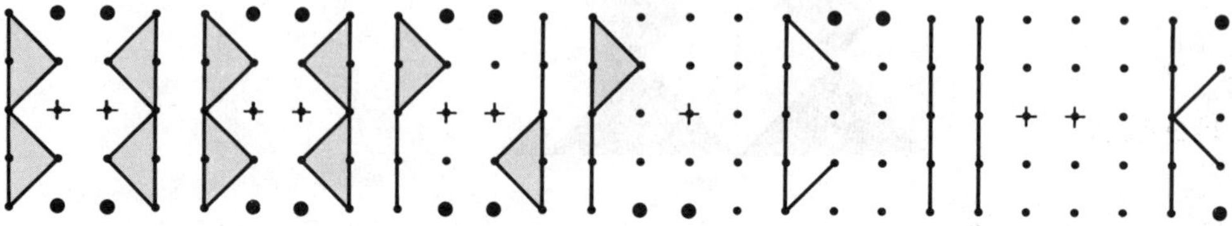

2.

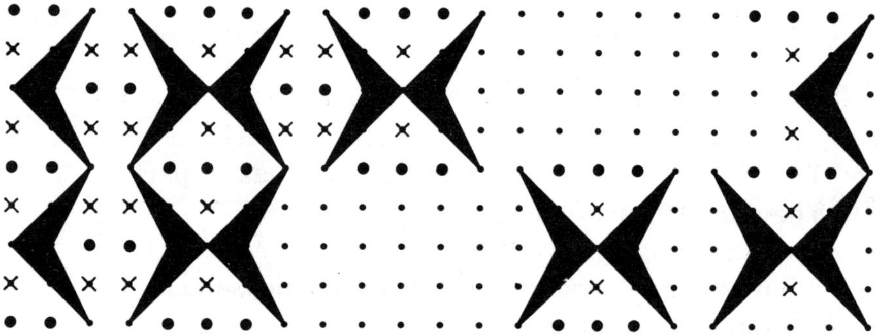

18 Ratio and Proportion

Ratio

Exercise 1

1. Draw a straight line that is three times the length of the given line:

2. Draw a straight line that is half the length of this line:

3. Draw a rectangle with sides that are twice as long as the sides of the given rectangle.

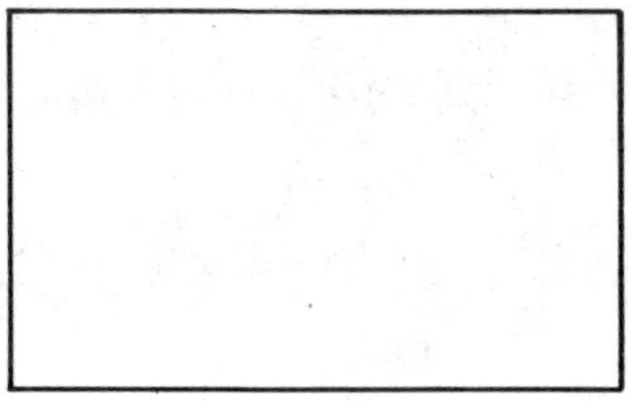

4. Draw a right-angled triangle having sides that are three times as long as the sides of the given triangle.

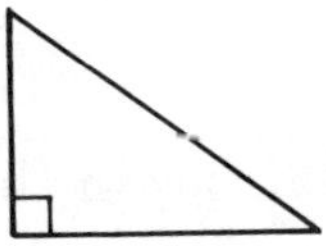

Exercise 2

A By how many times is the first quantity bigger than the second?

1. £20, £4
2. £60, £10
3. 40 p, 8 p
4. £1.50, 75 p

5. 21 m, 7 m
6. 24 ft, 4 ft
7. 30 km, 6 km
8. 28 lb, 7 lb
9. 400 g, 50 g
10. 70 ℓ, 5 ℓ
11. 60 mm, 15 mm
12. 2 min, 30 s
13. 4 weeks, 2 days
14. 4 cm, 20 mm
15. 2 m, 25 cm
16. 3 km, 200 m

B What fraction of the second quantity is the first?

1. £5, £20
2. 4 p, 36 p
3. 6 ft, 12 ft
4. 8 m, 24 m
5. 2 miles, 8 miles
6. 9 ℓ, 27 ℓ
7. 20 kg, 100 kg
8. 7 cm, 56 cm
9. 15 mm, 45 mm
10. 75 p, £1
11. 8 km, 12 km
12. 120 g, 160 g

If Martin has £4 and spends £3, he has spent $\frac{3}{4}$ of his money.
We can write:

The ratio of the money spent to the money started with is:

3 to 4
or 3 : 4
or $\frac{3}{4}$

Exercise 3

Write these ratios in their simplest form:

e.g. 1 6 : 8 = $\underline{\underline{3:4}}$ *e.g. 2* 6 to 8 = $\underline{\underline{\text{3 to 4}}}$

1. 2 : 4
2. 3 : 9
3. 2 to 6
4. 6 : 2
5. 6 to 10
6. 4 to 10
7. 8 : 12
8. 8 : 4
9. 10 : 2
10. 12 : 4
11. 12 : 6
12. 12 : 8
13. 12 : 9
14. 10 : 25
15. 16 to 20
16. 20 to 16
17. 24 : 4
18. 9 to 21
19. 14 to 21
20. 18 : 27
21. 20 to 25
22. 36 : 24
23. 40 : 16
24. 35 : 30

Exercise 4

Write the following as ratios in their simplest form:

1. £6 to £3
2. 4 m to 6 m
3. 9 cm to 6 cm
4. 12 km to 3 km
5. 4 ft to 16 ft
6. 10 kg to 6 kg
7. 15 ℓ to 5 ℓ
8. 6 gal to 15 gal
9. 30 m to 10 m
10. 50 mm to 30 mm
11. 24 h to 8 h
12. 15 lb to 21 lb
13. 15 min to 40 min
14. 21 cm to 12 cm
15. 40 s to 25 s
16. 24 mm to 30 mm
17. 240 g to 200 g
18. 50 ℓ to 35 ℓ
19. 6 mm to 2 cm
20. 3 m to 50 cm

Exercise 5

Give each answer in its simplest form:

1. There were 16 boys and 8 girls in a class. Find:
   (*a*) the ratio of the number of boys to the number of girls,
   (*b*) the ratio of the number of girls to the number of boys,
   (*c*) the ratio of the number of girls to the total number of pupils in the class,
   (*d*) the ratio of the total number of pupils in the class to the number of boys.

2. A shopkeeper bought a game for £8, then sold it for £10. Find:
   (*a*) the profit,
   (*b*) the ratio of the selling price to the cost price,
   (*c*) the ratio of the profit to the cost price.

3. A nurse mixed a drug with water. If 40 mℓ of the drug was used and 200 mℓ of water, find:
   (*a*) the ratio of the amount of drug to the amount of water,
   (*b*) the ratio of the amount of drug to the total amount of the mixture of water and drug.

4. Diana and Pete shared £270. If Diana got £180 and Pete £90, find:
   (*a*) the ratio of Diana's share to Pete's share,
   (*b*) the ratio of Pete's share to the total,
   (*c*) the ratio of the total to Diana's share.

Exercise 6

1. If the large gear wheel turns clockwise, which way does the small gear wheel turn?

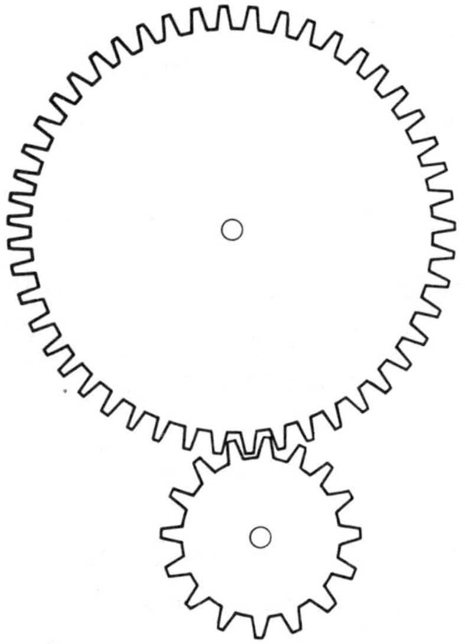

2. The large wheel has 48 teeth and the small wheel has 16 teeth.
   (*a*) If the large wheel makes one full turn, how many full turns does the small wheel turn?
   (*b*) Find the ratio of the number of teeth on the large gear wheel to the number of teeth on the small gear wheel.

3. If the large wheel has 48 teeth and the small wheel 24 teeth, find:
   (*a*) the number of full turns of the small wheel when the large wheel makes one full turn,
   (*b*) the ratio of the number of teeth on the large wheel to the number of teeth on the small wheel.

Rates

Exercise 7

A These statements are incomplete. Copy and complete them.

1. The car travelled quite fast at 70 miles.

2. The car's petrol consumption was 38 miles.

3. Mrs Downing's car uses too much petrol. It only does 6 km.

4. I. J. Kay earns £4.

5. We go to school for about 33 h.

B Select the correct answer:

1. Milk costs 24 p. Is that:
A. Cheap?
B. Expensive?
C. A fair price?
D. I don't know?

2. Milk costs 48 p. Is that:
A. A better value than in question 1?
B. A poorer value than in question 1?
C. The same value as in question 1?
D. I don't know?

The statements as written in Exercise 7 are meaningless. Each needs two quantities to make sense, but only one has been given.

To answer section B, we need to know how much milk we get for our money before we can decide whether or not it is good value. When both quantities are given we have what is called a *rate*. Miles per hour, pence per pint, etc., are all rates.

The second quantity in a rate is usually given as one unit, so we have:

the number of miles travelled in *one* hour (m.p.h.);
the cost of *one* pint (cost per pint);
the number of miles travelled on *one* gallon (m.p.g.);
the cost of *one* kilogram (cost per kilogram);

and so on.

If steak costs £6.95 per kilogram then its price per kilogram can also be written as £6.95/kg, which is still read as £6.95 per kilogram. Metres per second can be written m/s and the same applies to other rates.

Exercise 8

1. A train travels at a steady speed of 50 miles per hour. How far does it travel in:
(*a*) 2 h? (*b*) 3 h? (*c*) 5 h?

2. A bus travels at a steady speed of 70 kilometres per hour. How far does it travel in:
(*a*) 2 h? (*b*) 5 h? (*c*) 6 h?

3. If milk costs 25 p per pint, find the cost of:
(*a*) 2 pt (*b*) 3 pt (*c*) 8 pt

4. If milk costs 48 p per litre, find the cost of:
(*a*) 2 ℓ (*b*) 500 mℓ (*c*) 1.5 ℓ

5. A car gets 40 miles per gallon of petrol. How far does it travel on:
(*a*) 2 gal? (*b*) 5 gal? (*c*) 7 gal?

6. A shop gives a discount of 20 p in the pound. Find the discount on an article selling at:
(*a*) £4 (*b*) £10 (*c*) £15

7. If net curtaining costs £4.50 per metre, find the cost of:
(*a*) 2 m (*b*) 3 m (*c*) 5 m

8. Mrs Wise earned £4.20 per hour. Find her earnings for:
(*a*) 4 h work (*b*) 10 h work (*c*) 30 h work

Exercise 9

1. A car used 2 gal of petrol in travelling 70 miles.
How many miles per gallon did it get?

2. A car travelled at a steady speed of 80 miles per hour.
How long did it take to travel 160 miles?

3. Mr Hart earned £5 per hour.
How many hours did he need to work to earn £100?

4. Mrs Vickers earned £80 for 20 h work.
How much did she earn per hour?

5. Mr Bond got 54 French francs for £6 at a bank.
What was the exchange rate in francs per pound?

6. A cricketer scored 35 runs in 7 overs.
Find the batting rate in runs per over.

7. Kelly paid £2.40 for a box of a dozen pencils.
What was the cost per pencil?

8. A car gets 14 km per litre of petrol. At that rate, how many litres are needed for a journey of:
(*a*) 28 km? (*b*) 140 km? (*c*) 210 km?

Proportion (Direct Proportion)

If you earn £3 per hour you should be paid

£6 for working 2 h
£9 for working 3 h
£12 for working 4 h

and so on.

Your earnings depend on the number of hours worked.

More hours ⟶ more pay
Less hours ⟶ less pay

Hours worked and earnings are said to be in *direct proportion.*

The table shows that the two quantities are in direct proportion:

Hours worked	1	2	3	4	5	6	7	8
Earnings (£)	3	6	9	12	15	18	21	24

$$\frac{1}{3} = \frac{2}{6} = \frac{3}{9} = \frac{4}{12} = \frac{5}{15} = \frac{6}{18} = \frac{7}{21} = \frac{8}{24}$$

Note If you double the hours worked you double the earnings; in fact, whatever number the hours worked are multiplied by, the earnings are multiplied by the same number.

Exercise 10

A For each of the following, write whether it is 'TRUE' or 'FALSE':

1. If 2 cartons of fruit juice cost £1 then 4 cartons of the same sort, size and make cost £2

2. Henry VIII had 6 wives so Henry IV had 3 wives.

3. A 15-year-old boy takes size 8 in shoes. When he is twice as old he will take size 16.

4. Two friends living near each other travel to school together. If Belinda travelled on her own, the journey would take 25 min. When she travels with Fariyal, the journey must take 50 min.

5. A car travelling at a steady speed travels 60 km in 1 h, so it will travel 120 km in 2 h.

6. Amy, aged 13 years, is 1.57 m tall. When she is 26 years old, she will be 3.14 m tall.

7. In the high jump Joe, aged 15 years, can jump 1.4 m. At the age of 60 years he will be able to jump 5.6 m.

8. On his own, Tony takes 3 min to sing a song. If 10 people sing the same song together, it will take them 30 min.

B Which of the following are in direct proportion and which are not?

1. The amount of water in a bath and the time the tap runs (at a steady rate).

2. The age of a person and how heavy that person is.

3. The number of eggs boiled and the time taken.

4. Distance travelled by a mower while mowing and the number of turns of the roller on the mower.

5. The distance a person can throw a javelin and that person's age.

Exercise 11 M

The tables give sets of numbers that are in direct proportion. Copy and complete them:

1.

Number of items bought	1	2	3	4	5	8	10	12
Cost (£)	6	12			30			

2.

Area of carpet (m^2)	1	2	3	5	8	10	12	20
Cost (£)	7			35				

3.

Distance travelled (km)	30			120			210	
Time taken (h)	1	2	3	4	5	6	7	8

4.

Distance travelled (km)	12		36		60			120
Petrol used (ℓ)	1	2	3	4	5	7	9	

Exercise 12

1. If it cost 90 p to post one parcel, what would it cost to post two more parcels that are identical to the first one?

2. If 2 stamps cost 40 p what would 6 of the same sort of stamp cost?

3. If 2 ℓ of orange juice fills 6 identical glasses, how many of the glasses will 4 ℓ fill?

4. At the daily rate, a hotel charges £60 for a 2 day stay. At the same rate what would be the charge for 4 days?

5. A ticket for a 6 mile journey costs 42 p. What should it cost, at the same rate, for a 12 mile journey?

6. If 15 m of material will make 3 pairs of curtains, how many pairs of curtains of the same size as the others can be made from 30 m?

7. Using the same amount of milk in each drink, if I use 120 mℓ of milk in 4 cups of coffee, how much milk do I need for 12 cups?

8. If 4 three-pin plugs cost £3.36 what would 8 cost?

9. If 5 miles is the same distance as 8 km, how many kilometres is:
(*a*) 10 miles? (*b*) 25 miles?

10. If 2 cm on a map represents 5 km:
(*a*) What distance is represented by 4 cm?
(*b*) What distance is represented by 14 cm?
(*c*) What length represents a distance of 15 km?
(*d*) What length represents a distance of 50 km?

Exercise 13

1. If 450 g of self-raising flour is used to make 2 cherry cakes of the same size:
(*a*) How many grams of flour are needed to make 4 cherry cakes?
(*b*) How many kilograms of flour are needed to make 8 cherry cakes?
(*c*) How many kilograms of flour are needed to make 6 cherry cakes?

2. If 100 g of butter and 250 g of self-raising flour are used to make 30 gingerbread biscuits, how much butter and how much flour would be used to make only 15 biscuits?

3. 75 g of brown sugar is needed to make 20 chocolate cookies. How much brown sugar is needed to make:
(*a*) 40 cookies? (*b*) 60 cookies?

4. The ingredients to make 24 nut cookies are:

225 g plain flour
pinch of salt
75 g butter or margarine
grated rind of $\frac{1}{2}$ lemon
1 egg, separated
45 mℓ milk
100 g walnuts

Copy the list of ingredients but give the quantities needed to make 48 nut cookies.

5. The ingredients to make 3 kg ($6\frac{1}{2}$ lb) of beetroot chutney are:

900 g (2 lb) raw beet, shredded or grated
450 g (1 lb) onions, skinned and chopped
700 g ($1\frac{1}{2}$ lb) apples, peeled and chopped
450 g (1 lb) seedless raisins
1ℓ ($1\frac{3}{4}$ pt) malt vinegar
900 g (2 lb) sugar
30 mℓ (2 level tbsps) ground ginger

Copy the list of ingredients but give the quantities needed to make:
(*a*) 6 kg (13 lb) of chutney, (*b*) 1.5 kg ($3\frac{1}{4}$ lb) of chutney.

19 Flow Charts

It is important to do things in the correct order:

e.g. To open the window and *then* to look around outside is correct,

but to look around outside *then* to open the window is not!

Exercise 1

Here are the chapter headings for a story. Put them in a sensible order.

Captured
The return
Planning the journey
The dark island
The first sail
Setting out
Escape

Exercise 2

Write the following statements in the correct order:

1. *Posting a letter*:
Stick the stamp on.
Post the letter.
Take the letter to the post office.
Buy a stamp.

2. *Taking a sweater out of a drawer*:
Look for the sweater.
Decide which drawer the sweater is in.
Take out the sweater.
Close the drawer.
Open the drawer.

3. *Going to the cinema*:
Sit down.
Buy a ticket.
Enter the building.
Go to the correct screen (studio).
Find an empty seat.

4. *Making a cup of tea*:
Remove the teabag.
Pour on the water.
Boil the kettle.
Drink the tea.
Add the milk.
Put a teabag in the cup.

5. *To draw a circle*:
Put more pressure on the point of the needle than on the pencil.
Set the pair of compasses to the size of the radius.
Sweep the pencil in one direction only.
Place the point of the needle at the required centre.

Exercise 3

The words in each sentence have been given in the wrong order. Write each sentence correctly.

1. is two than nine bigger
2. two to two four equals added
3. a circle half is of a semicircle
4. factor is 4 a 12 of
5. is 5 of multiple 20 a
6. make four whole a quarters
7. four square a has sides equal
8. faces a six has cube
9. angle 90° is an of a called right angle

10. sides pentagon five has a

11. symmetry rectangle a of lines has bilateral two

12. diameter circumference circle the a the of about is times three

In carrying out a task, a flow chart (or flow diagram) can be used to help you to sort out the order of doing things. Here are two flow charts: the first shows how to use a cassette recorder and the second shows how to find the perimeter of a rectangle.

How to Use a Cassette Recorder

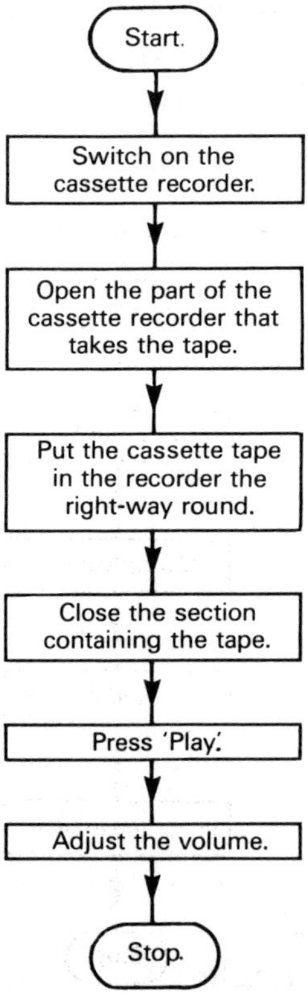

Finding the Perimeter of a Rectangle

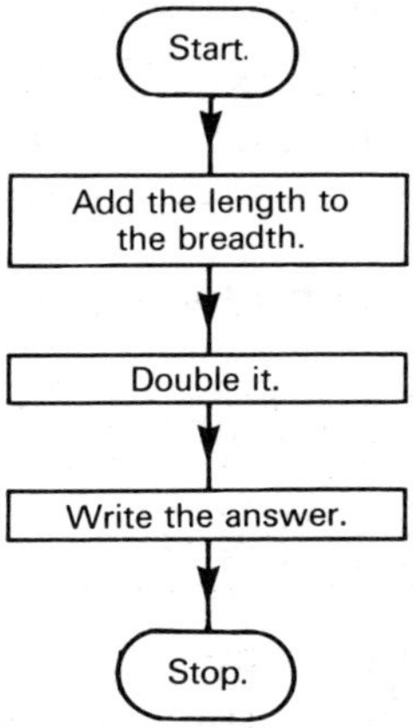

The flow charts clearly show an order (simply follow the arrows). Note the shapes of the boxes used. For *Start* and *Stop*, a 'running-track' shape is used. *Instructions* are written in rectangular boxes.

Exercise 4

1. Use the flow chart called 'Finding the Perimeter of a Rectangle' on page 234 to find the perimeters of the following rectangles:

(*a*) A rectangle of length 9 cm and breadth 7 cm
(*b*) A rectangle of length 6.5 cm and breadth 3.5 cm
(*c*) A rectangle of length 8.7 cm and breadth 4.9 cm

2. Answer the following from the given flow chart:

(*a*) When the number 4 is input find the answer that is output.
(*b*) When the input is 7 find the output.
(*c*) Find the output for an input of 9.
(*d*) Find the output for an input of $1\frac{1}{2}$.
(*e*) Find the output for an input of 2.5.
(*f*) Find the output for an input of 4.2.

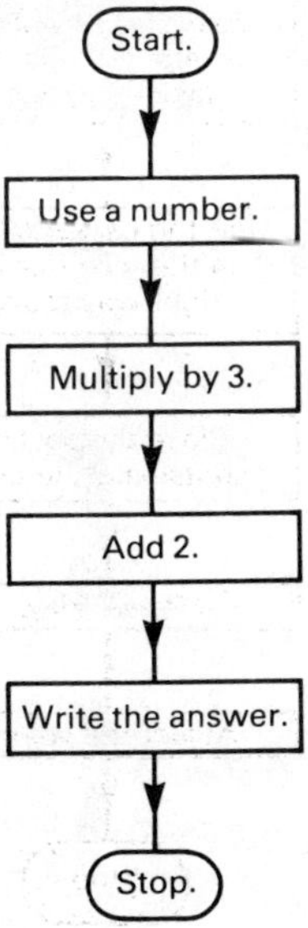

3. Use the flow chart to find the output for the following inputs:

(*a*) 3
(*b*) 10
(*c*) 15
(*d*) 23
(*e*) 2.5
(*f*) 3.9

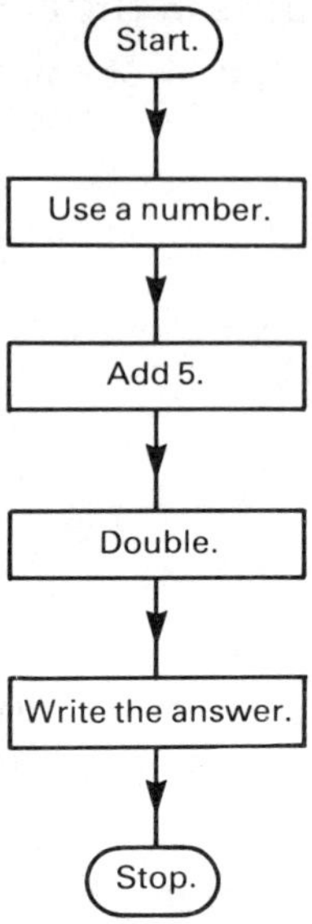

4. Use the flow chart to find the output for the following inputs:

(*a*) 5
(*b*) 10
(*c*) 14
(*d*) 25
(*e*) 8.5
(*f*) 4.3

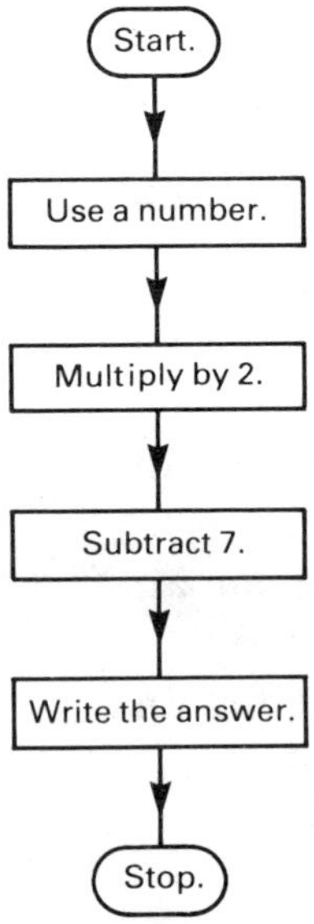

5. Use the flow chart to find the output for the following inputs:
(*a*) 3
(*b*) 6
(*c*) 8
(*d*) 10
(*e*) 16
(*f*) 27

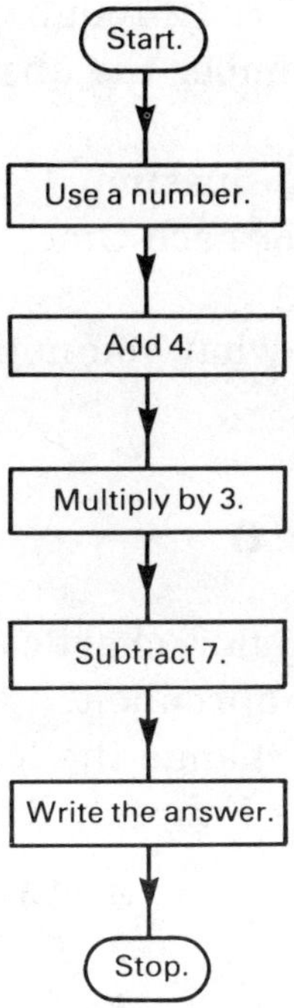

Exercise 5

1. (*a*) Work through flow chart A using any number you wish.

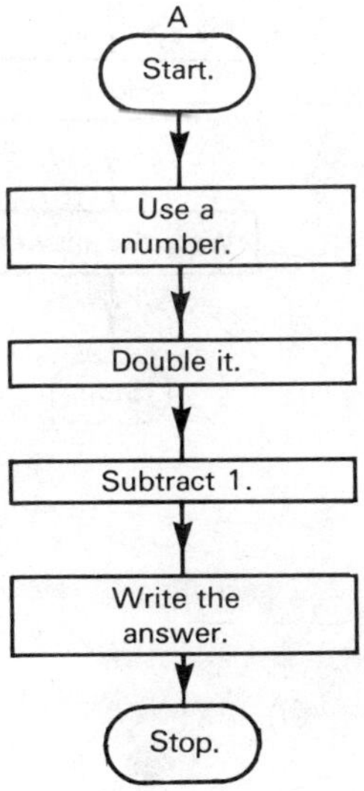

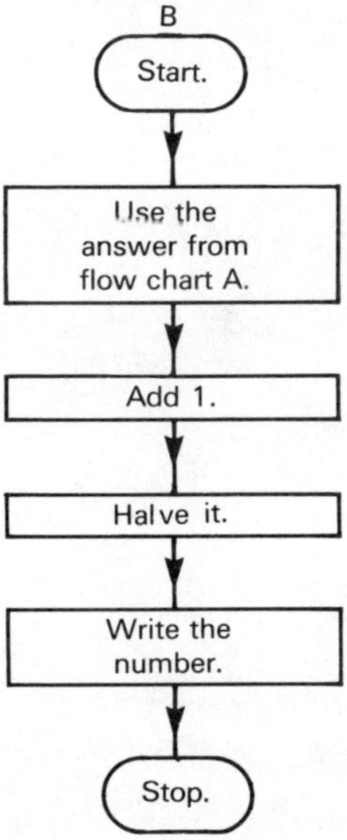

(*b*) Work through flow chart B using as your starting number the answer obtained in part (*a*).

2. Repeat question 1 several times using a different starting number each time.

3. Write what you notice.

Exercise 6

The given flow chart can be used to change degrees Celsius to degrees Fahrenheit.

Use it to change the following to degrees Fahrenheit:

1. 20 °C
2. 40 °C
3. 0 °C
4. 100 °C
5. 25 °C
6. 15 °C
7. 50 °C
8. 45 °C
9. 60 °C
10. 35 °C

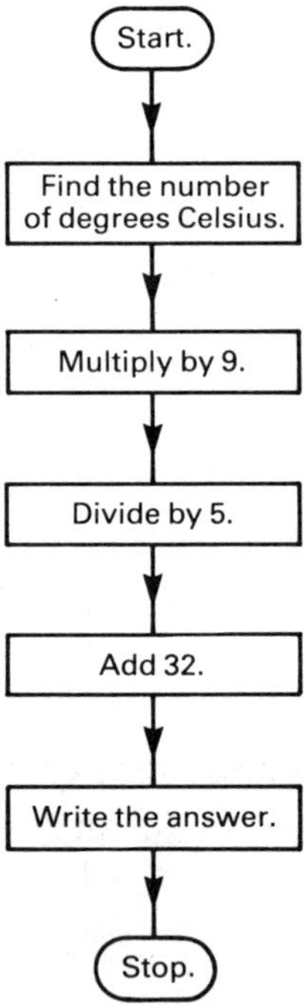

Revision Exercises XIV to XIX

Revision Exercise XIV

1. Copy the given shape on to squared paper. Complete it so that it has a horizontal and a vertical line of symmetry as shown by the broken lines.

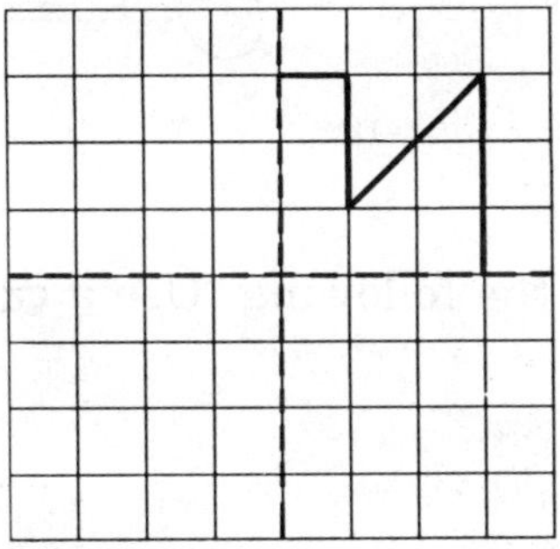

2. (*a*) Draw a triangle that has exactly one line of symmetry.
   (*b*) What sort of triangle has exactly one line of symmetry?

3. For each shape write its number of lines of bilateral symmetry (if any) and its order of rotational symmetry:

(*a*)

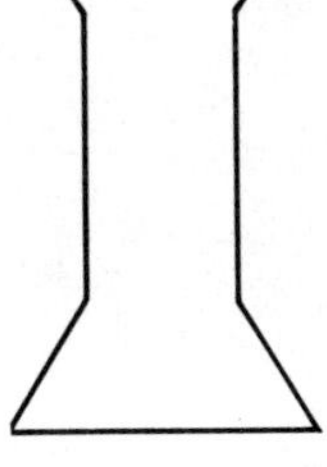

(*b*)

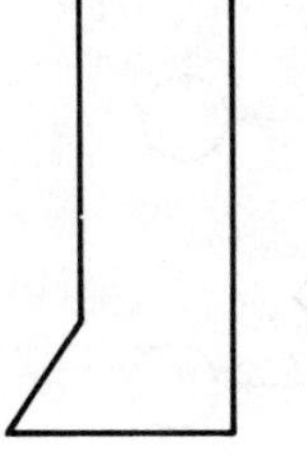

Revision Exercise XV

A Copy and complete:

1.

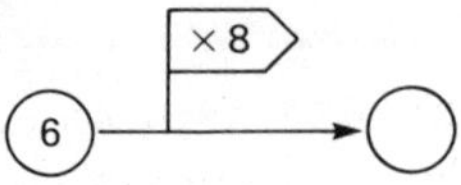

2.

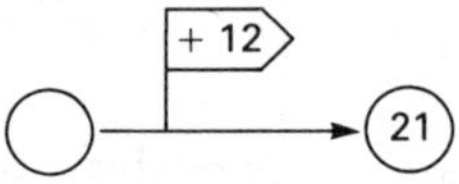

3.

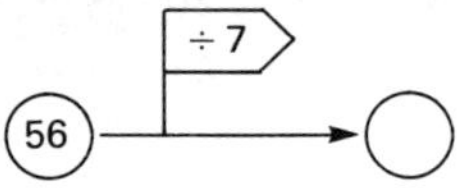

4.

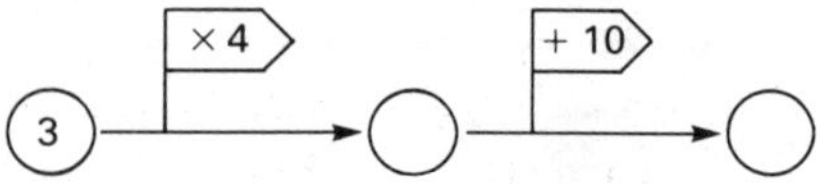

5.

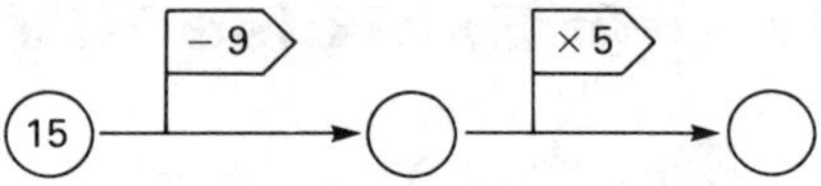

6.

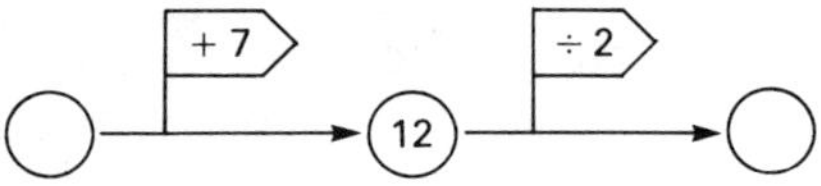

B Copy and complete the following. Use a calculator.

1.

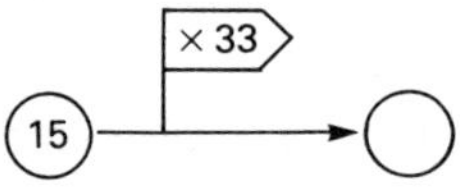

2.

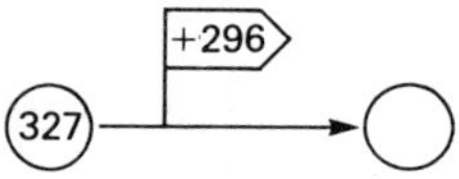

3.

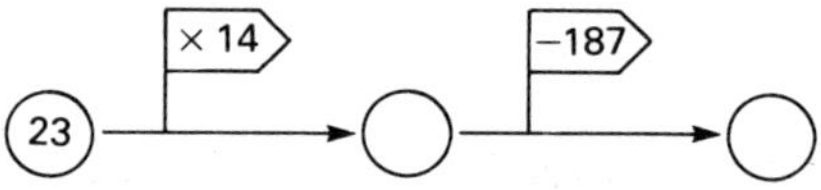

4.

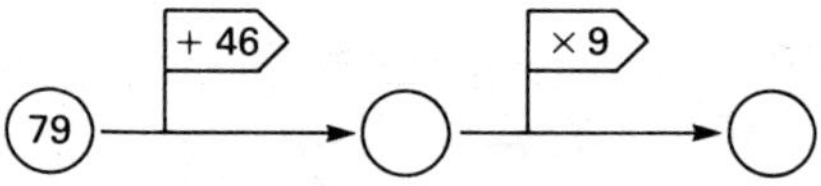

5.

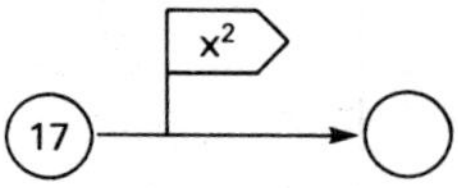

6.

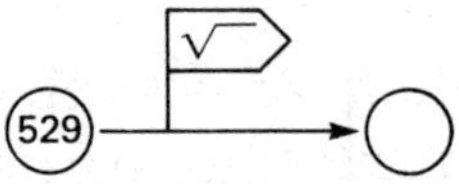

Revision Exercise XVI **M**

1. Copy the table.

2. Use a large sheet of paper. Before it has been folded, the paper is one sheet thick (this is shown in the table).

Number of folds	Number of sheets thick
0	1
1	2
2	
3	
4	
5	
6	
7	

Fold the sheet of paper in half as shown. It is now 2 sheets thick. This has been entered in the table:
1 fold, 2 sheets thick.

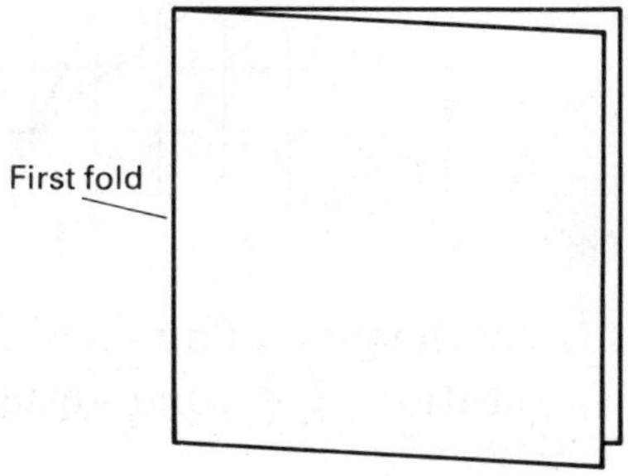

Fold the paper in half again (as shown). How many sheets thick is it? Enter the answer into your table next to the 2 folds.

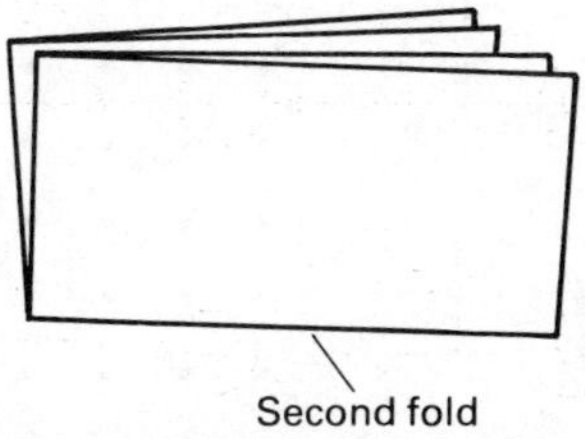

Continue folding the piece of paper in half. After each fold, write in your table how many sheets thick the paper is.

Revision Exercise XVII

M

1. $\triangle ABC$ has been drawn on 5 mm squared paper. Copy the triangle. Draw its image after a translation 4 cm to the left.

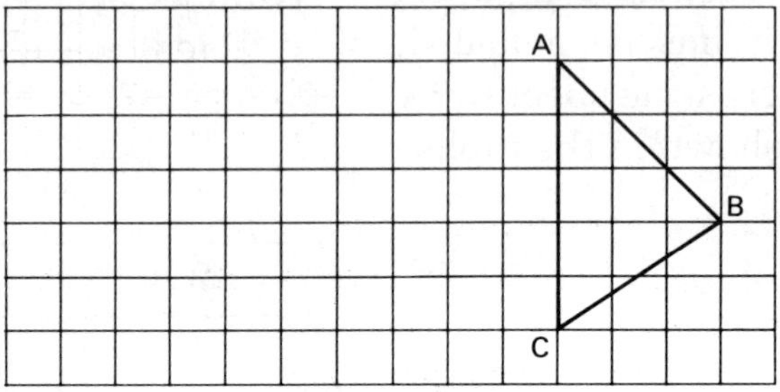

2. Copy $\triangle DEF$. Draw its image after a reflection in the given mirror line:

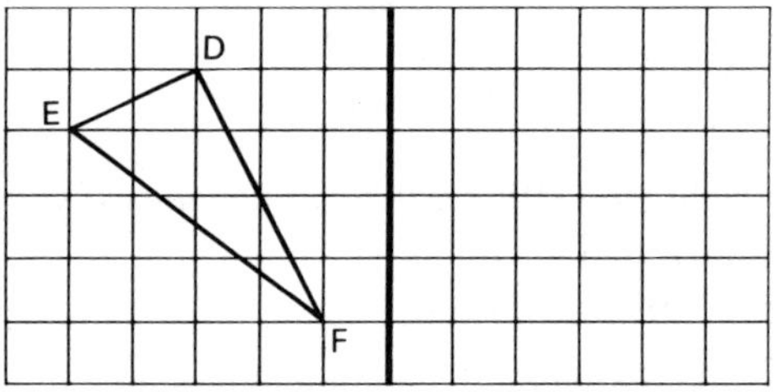

3. Copy the given flag on to dotty paper. Draw its image after a rotation of $\frac{3}{4}$ turn anticlockwise about O.

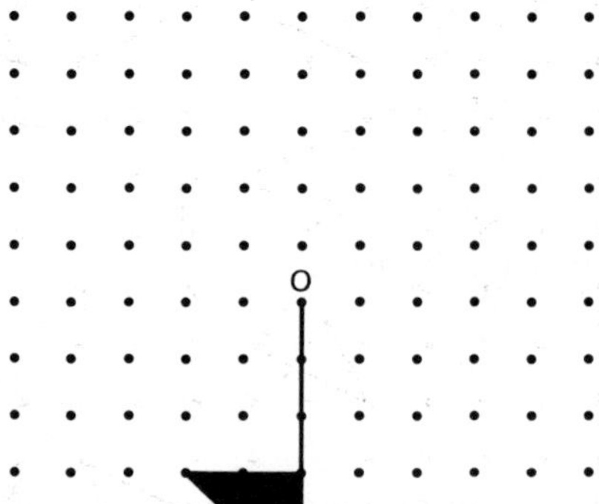

4. On squared paper, draw an enlargement of quadrilateral PQRS. Make each side twice as long.

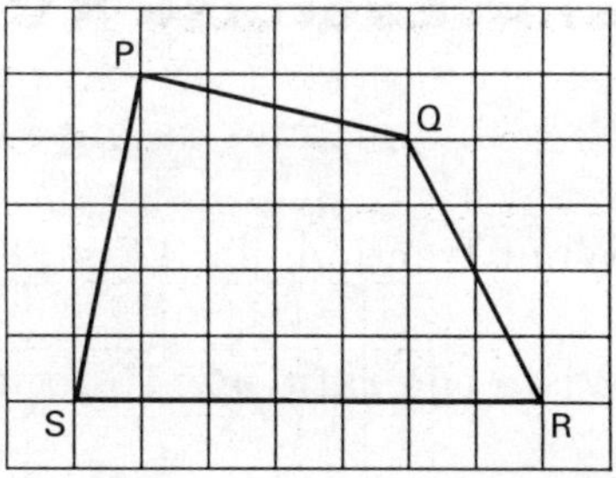

5. The diagram shows one tile and spaces for 8 more tiles of the same design. Copy the diagram. Fill in the blank spaces to show all 9 tiles.

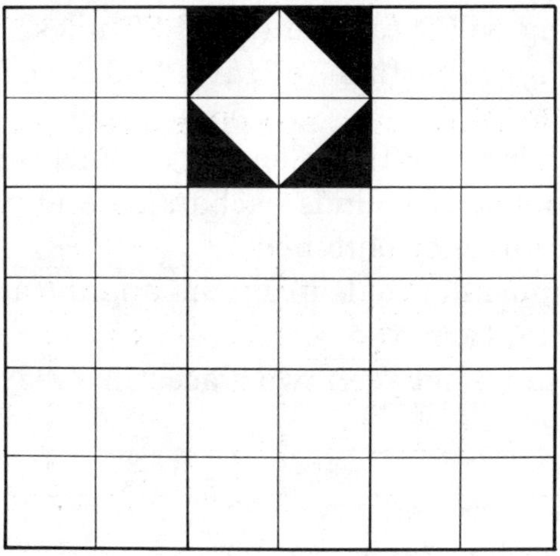

6. Copy and complete the pattern:

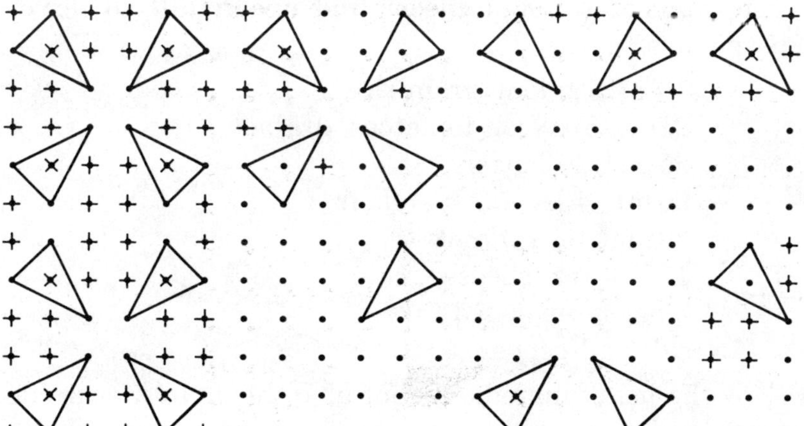

Revision Exercise XVIII

1. How many times as big as 9 km is 36 km?

2. What fraction of £32 is £8?

3. Write the ratio 12 : 16 in its simplest form.

4. Write 28 m to 16 m as a ratio in its simplest form.

5. A car travels at a steady speed of 40 miles per hour.
(*a*) Find the distance it will travel in 4 h.
(*b*) How long will it take to travel 80 miles?
(*c*) How long will it take to travel 200 miles?

6. Write whether or not the following are in direct proportion:
(*a*) The number of pounds exchanged and the amount of foreign currency obtained.
(*b*) The number of people in a room and the amount of heat needed to warm the room.
(*c*) The distance between two places on a map and the true distance.

7. At a steady speed Alan walked 12 km in 2 h. At the same steady speed, how far would he walk in 6 h?

8. The ingredients for a cheese and rice soufflé to serve 4 people are:

50 g (2 oz) long grain rice
226 g (8 oz) can tomatoes, drained
25 g (1 oz) butter
60 mℓ (4 level tbsps) flour
300 mℓ ($\frac{1}{2}$ pt) milk
175 g (6 oz) cheese, grated
3 eggs, separated
salt and pepper

Copy the list of ingredients but give the quantities needed for 8 people.

Revision Exercise XIX

1. Use the flow chart to find the output for the given inputs:

(*a*) 2
(*b*) 4
(*c*) 7
(*d*) 10
(*e*) 3.5
(*f*) 2.6

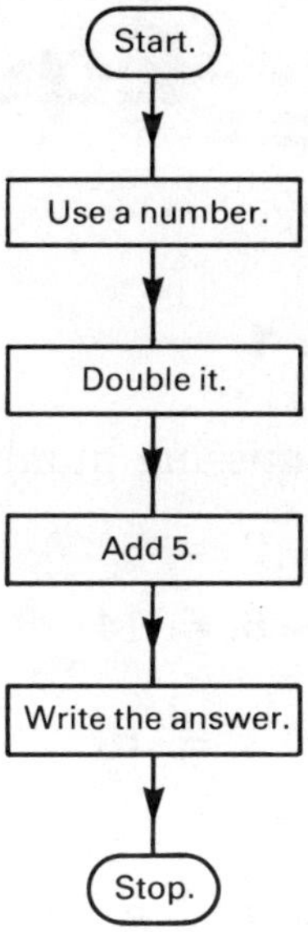

2. Use the flow chart to find the output for the given inputs:

(*a*) 2
(*b*) 3
(*c*) 5
(*d*) 10
(*e*) 3.5
(*f*) 4.1

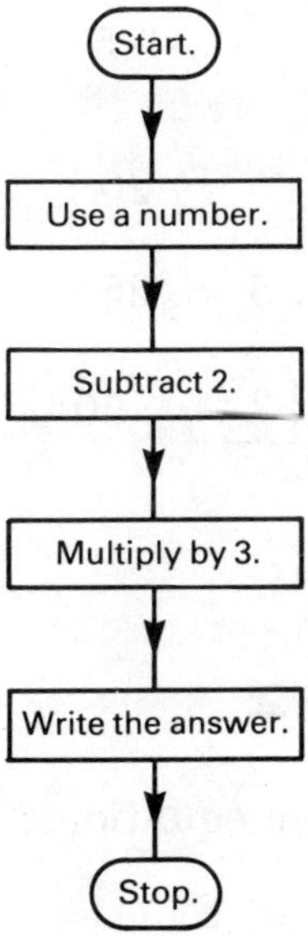

20 Simple Equations

Exercise 1

Find the missing numbers to make the statements correct:

1. $\boxed{?} + 2 = 8$
2. $\boxed{?} + 6 = 14$
3. $\boxed{?} + 8 = 15$
4. $13 + \boxed{?} = 25$
5. $\boxed{?} - 3 = 8$
6. $\boxed{?} - 6 = 11$
7. $12 - \boxed{?} = 5$
8. $2 \times \boxed{?} = 20$
9. $\boxed{?} \times 5 = 25$
10. $10 \times \boxed{?} = 50$
11. $\boxed{?} \div 2 = 4$
12. $\boxed{?} - 8 = 2$
13. $\dfrac{\boxed{?}}{5} = 4$
14. $\boxed{?} - 7 = 10$
15. $\boxed{?} + 15 = 21$
16. $\boxed{?} \times 9 = 45$
17. $\dfrac{\boxed{?}}{2} = 9$
18. $\dfrac{\boxed{?}}{6} = 3$
19. $7 + \boxed{?} = 26$
20. $\dfrac{\boxed{?}}{3} = 12$

Exercise 2

Solve these equations:

1. $x + 3 = 8$
2. $c + 6 = 12$
3. $m - 5 = 9$
4. $t - 8 = 3$
5. $5u = 50$
6. $2k = 26$
7. $p \div 2 = 5$
8. $\dfrac{e}{3} = 10$

9. $y + 8 = 16$
10. $9w = 18$
11. $q - 4 = 12$
12. $v - 16 = 16$
13. $s + 16 = 16$
14. $4x = 28$
15. $\frac{u}{3} = 11$
16. $m \div 5 = 6$
17. $\frac{t}{9} = 3$
18. $w + 11 = 22$
19. $n - 11 = 22$
20. $15 + y = 24$
21. $d - 10 = 13$
22. $\frac{h}{2} = 12$
23. $19 - f = 15$
24. $14 + g = 25$
25. $10g = 70$
26. $a - 15 = 6$
27. $17 + z = 17$
28. $\frac{n}{7} = 8$
29. $l + 24 = 39$
30. $b - 19 = 21$
31. $h - 23 = 1$
32. $f + 9 = 30$
33. $c - 13 = 16$
34. $7 + z = 51$
35. $\frac{d}{4} = 7$
36. $a + 22 = 31$
37. $8p = 40$
38. $l \div 9 = 4$
39. $b - 17 = 23$
40. $v + 34 = 49$

Exercise 3

Answer the following:

1. I add 5 to a number and get 12. Find the number.
2. When 8 is added to a number, the answer is 20. Find the number.
3. 7 is subtracted from a number. The answer is 11. Find the number.
4. A number is doubled. If the answer is 18, find the number.
5. When a number is divided by 8 the answer is 3. Find the number.
6. A number is multiplied by 6. The answer is 30. Find the number.

7. Hayley thinks of a number, then adds 12. If the answer is 23, find the number.

8. Andy thinks of a number. He multiplies it by 9 and gets the answer 72. Find the number.

9. Moti thinks of a number. He subtracts 8 and gets the answer 13. Find the number.

10. Lisa thinks of a number. She adds 13 and gets the answer 22. Find the number.

Exercise 4

1. I think of a number, double it, then add 1. The answer is 9. Find the number.

2. I think of a number, double it, then add 3. The answer is 7. Find the number.

3. Carl thinks of a number, doubles it and subtracts 3. The answer is 5. Find the number.

4. Tony thinks of a number, doubles it and adds 5. The answer is 11. What number did Tony think of?

5. Holly thinks of a number, adds 3, then doubles the result. The final answer is 12. Find the number.

6. Gina thinks of a number, subtracts 5, then multiplies the result by 3. The final answer is 6. Find the number.

7. Judy thinks of a number. She multiplies it by 3, then adds 4. The final answer is 19. Find the number.

8. Gary thinks of a number. He multiplies it by 2, then subtracts 6. The final answer is 4. Find the number.

21 Formulae

Exercise 1

1. The diameter of a circle is double the radius. Find the diameter of a circle when the radius is:
(*a*) 7 cm (*b*) 18 cm (*c*) 1.6 cm

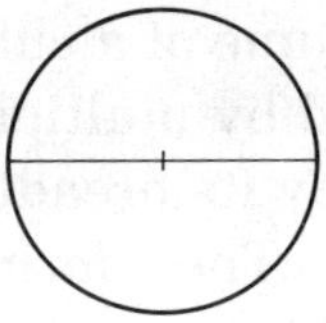

2. The circumference of a circle can be estimated by multiplying the diameter by 3.
Estimate the circumference when the diameter measures:
(*a*) 8 cm (*b*) 26 mm (*c*) 1.2 m

3. The perimeter of a regular hexagon can be found by multiplying the length of a side by 6. Find the perimeter of a hexagon with sides measuring:
(*a*) 7 cm (*b*) 13 cm (*c*) 2.8 m

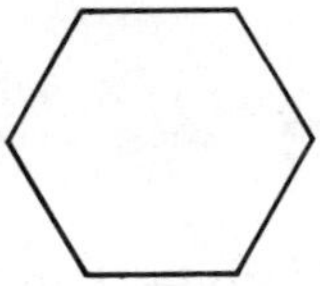

4. (*a*) To change kilometres into miles, multiply by 5 then divide by 8. Change 24 km into miles.
(*b*) Would the formula in part (*a*) still work if we divided first? Try it. The new formula is 'divide by 8, then multiply by 5'.

5. To change miles into kilometres, divide by 5 then multiply by 8. Change 20 miles into kilometres.

6. Another formula that can be used to change miles into kilometres is 'multiply by 1.6'.
   (*a*) Change 9 miles into kilometres.
   (*b*) Change 3.5 miles into kilometres.

7. Crown roast of lamb needs to be roasted in an oven at gas mark 4 (180 °C/356 °F) allowing 30 min per lb plus 30 min. For how long should 3 lb of lamb stay in the oven?

8. The volume of a cuboid can be found by multiplying its length by its breadth by its height. The formula is $V = lbh$.
   Use the formula to find the volume of a cuboid in which the length = 6 cm, the breadth = 5 cm and the height = 3 cm.

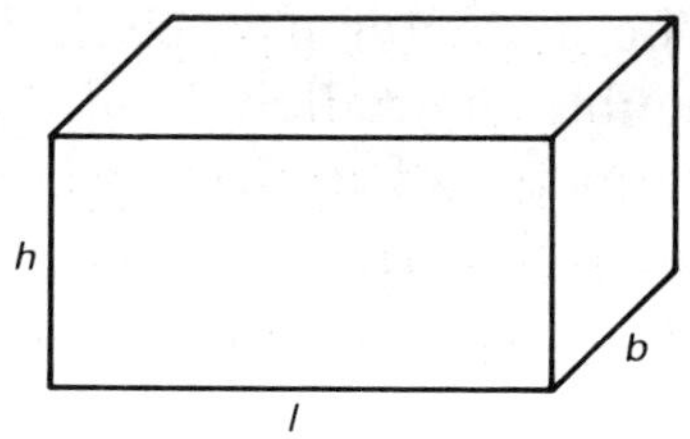

Exercise 2

1. $W = 3n$ gives the wages earned (in pounds) by someone working for n hours at £3 an hour. Calculate the earnings of someone who works for:
   (*a*) 5 h, (*b*) 16 h, (*c*) 35 h.

2. The formula $d = 7w$ gives the number of days in w weeks. Use it to find the number of days in:
   (*a*) 3 weeks, (*b*) 4 weeks, (*c*) 10 weeks.

3. The formula $a = 180° - b$ gives angle a in terms of angle b.

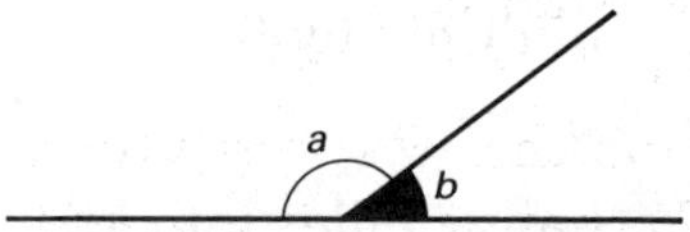

   Find angle a when angle $b = 42°$.

4. Net pay = gross pay − deductions.
Find the net pay if the gross pay is £120 and the deductions come to £25.

5. Sale price = normal price − discount.
A shirt costing £11.99 is sold at a discount. Find the sale price if the discount is £3.

6. HP price = deposit + balance.
(*a*) Find the HP price when the deposit is £65 and the balance is £125.
(*b*) Find the balance when the HP price is £118 and a deposit of £34 is paid.
(*c*) If the HP price is £92, what deposit is needed to leave a balance of £57?

7. HP price = loan + interest.
(*a*) Find the HP price on a loan of £35 if the interest charge is £10.50.
(*b*) Find the interest charge if the total HP price on a loan of £112 is £133.60.

8. The total cost of buying some items is given by the formula:

Total cost = number of items × cost of one item.

(*a*) Find the total cost of buying 4 items at £7 each.
(*b*) Find the total cost of buying 6 items at £19 each.
(*c*) Find the total cost of buying 3 items at £8.75 each.

Exercise 3

1. If 2 resistors, r_1 and r_2, are connected in series, the total resistance in the circuit, R, is the sum of the 2 resistances.
So $R = r_1 + r_2$.
Find the total resistance, R ohms, if $r_1 = 2.5$ and $r_2 = 3.5$.

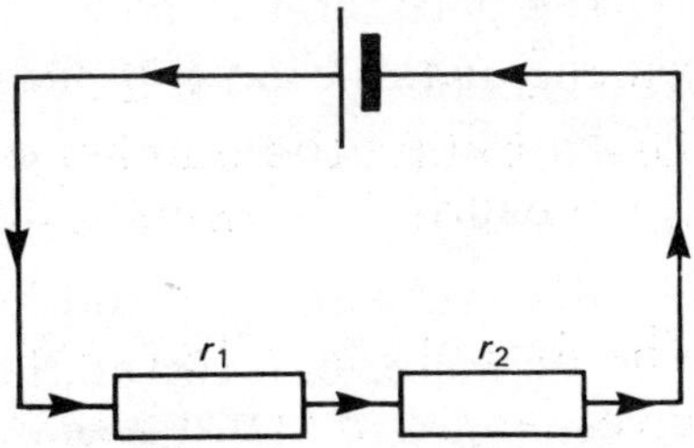

2. The area of a parallelogram can be found by multiplying its base by its perpendicular height.
We can write $A = bh$.

Calculate the area of a house drive that is in the shape of a parallelogram. Its dimensions are given in the sketch.

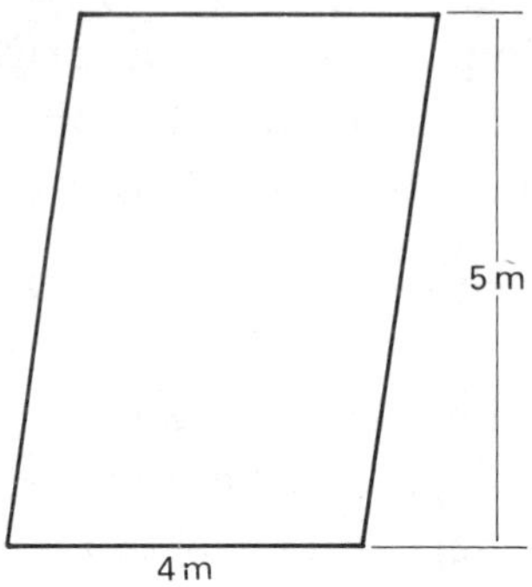

3. $A = \frac{1}{2}bh$ gives the area of a triangle with base b and height h.
Find the area of a triangle with base $b = 12$ cm and height $h = 7$ cm.

4. The perimeter of isosceles triangle ABC is given by the formula, $P = 2m + b$. Find the perimeter of a triangle in which $m = 7$ and $b = 4$.

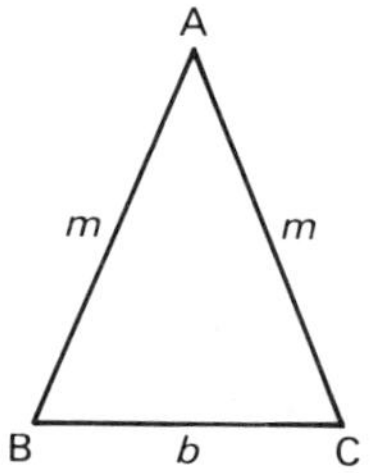

5. (*a*) $P = 2l + 2b$ gives the perimeter of a rectangle. Find P if length $l = 8$ cm and breadth $b = 7$ cm.

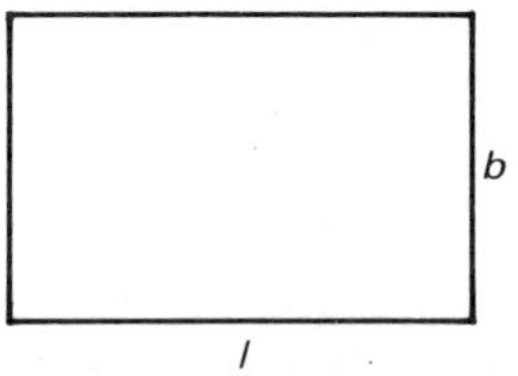

(*b*) $P = 2(l + b)$ is another formula that gives the perimeter of a rectangle. Use it to find P when $l = 8$ and $b = 7$. Check your answer with the answer to part (*a*).

(*c*) Use $P = 2(l + b)$ to find P when $l = 12$ and $b = 8$.

6. The volume of a cuboid can be found from the formula, $V = Ah$, where A is the area of the base and h the perpendicular height.
Find the volume of a cuboid whose base has an area of $20\,\text{cm}^2$ and whose perpendicular height is $3\,\text{cm}$.

7. $A = \frac{1}{2}Dd$ gives the area of a rhombus when the length of its diagonals are D and d. Find A when long diagonal, $D = 8\,\text{cm}$ and short diagonal, $d = 5\,\text{cm}$.

8. $I = \dfrac{Prn}{100}$ can be used to calculate interest.
P is the amount invested (called the principal), r is the rate of interest (a percentage), n is the number of years the money is invested for and I is the interest. Find the interest, I (in pounds) where the principal, $P = £400$, the rate, $r = 7\%$, and the number of years, $n = 2$.

9. $A = l^2$ gives the area of a square with side l units.
Find the area of a square with side:
(*a*) 7 cm (*b*) 2.5 cm

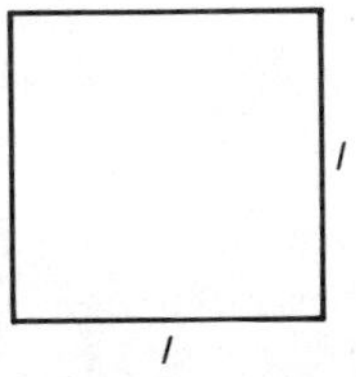

10. $V = l^3$ gives the volume of a cube with edge l units. Find the volume of a cube if its edge measures 2 cm.

22 Co-ordinates and Graphs

Co-ordinates

Exercise 1

1. (*a*) Copy the given pair of axes (use 1 cm squared paper *or* graph paper).

(*b*) A is the point (1, 5). Plot A.

(*c*) Plot the point B(4, 1).

(*d*) Measure AB.

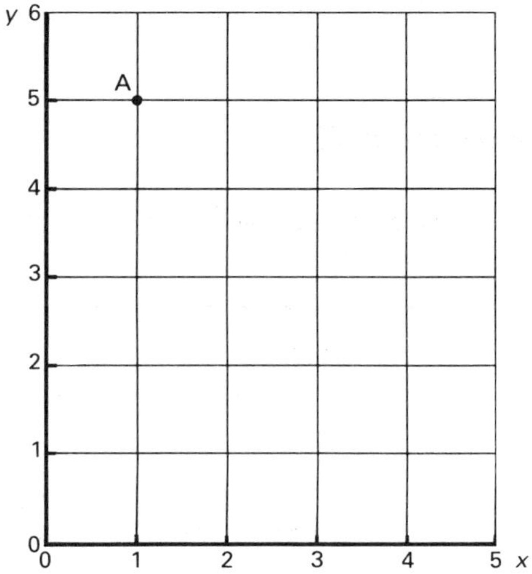

2. (*a*) Draw another pair of axes as shown above.

(*b*) Plot the points C(0, 4) and D(4, 6). Join CD with a straight line.

(*c*) Label the mid-point of CD as M. What are the co-ordinates of M?

(*d*) Plot the points E(0, 3) and F(4, 1). Join EF with a straight line.

(*e*) Label the mid-point of EF as N. What are the co-ordinates of N?

(*f*) Join MN with a straight line. How long is MN?

3. (*a*) Draw a pair of axes as shown.

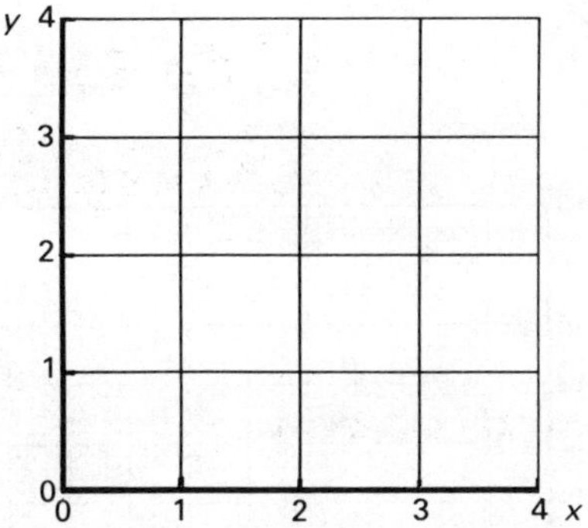

(*b*) Plot the points G(0, 3) and I(4, 1), then join them with a straight line.

(*c*) Plot the points H(3, 4) and J(1, 0) and join them with a straight line.

(*d*) What are the co-ordinates of the point of intersection of lines GI and HJ?

(*e*) At what angle do the lines GI and HJ cross?

(*f*) Join G to H to I to J to G. What shape is GHIJ?

Exercise 2

1. (*a*) Draw a pair of axes as shown.

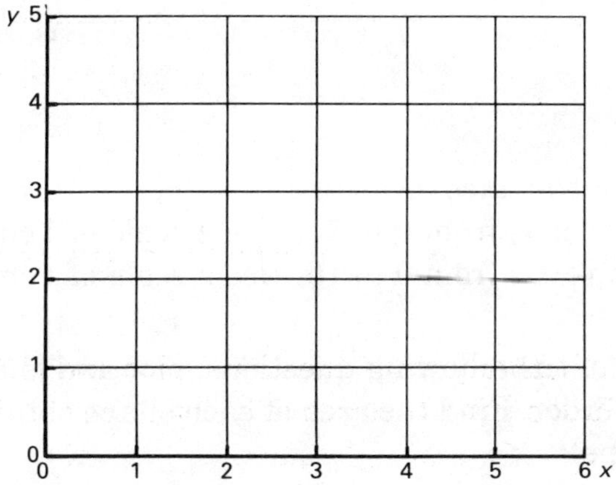

(*b*) Plot the points P(0, 1), Q(0, 4) and R(5, 4).

(*c*) Join PQ and QR.

(*d*) Find and plot a point S such that PQRS is a rectangle. Complete the rectangle.

(*e*) What are the co-ordinates of S?

(*f*) Find the area of rectangle PQRS.

2.

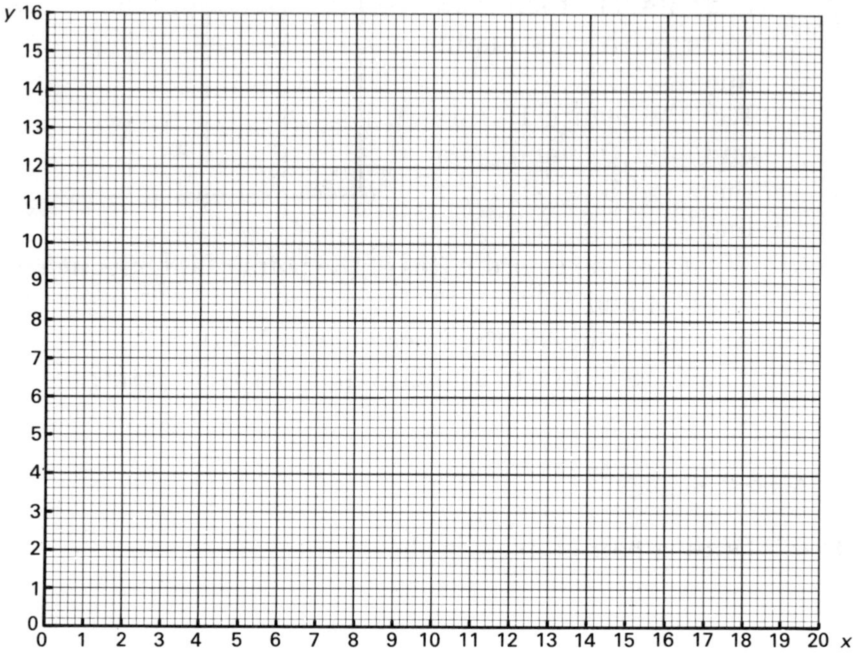

Draw a pair of axes as shown.
Label the x-axis from 0 to 20. Use a scale of 1 cm to 1 unit.
Label the y-axis from 0 to 16. Use a scale of 1 cm to 1 unit.

For each of the following questions, plot and join the given points in order. Find the area of each shape obtained (name each shape):

(*a*) (2, 2), (2, 5), (6, 5), (6, 2), (2, 2)
(*b*) (8, 1), (14, 1), (14,6), (8, 1)
(*c*) (5, 11), (11, 7), (3, 7), (5, 11)
(*d*) (19, 12), (20, 15), (14, 15), (13, 12), (19, 12)
(*e*) (2, 14), (12, 14), (9, 11), (2, 14)

Exercise 3

Use the same pair of axes throughout this exercise. Draw the axes as shown below but use a scale of 1 cm to 1 unit.

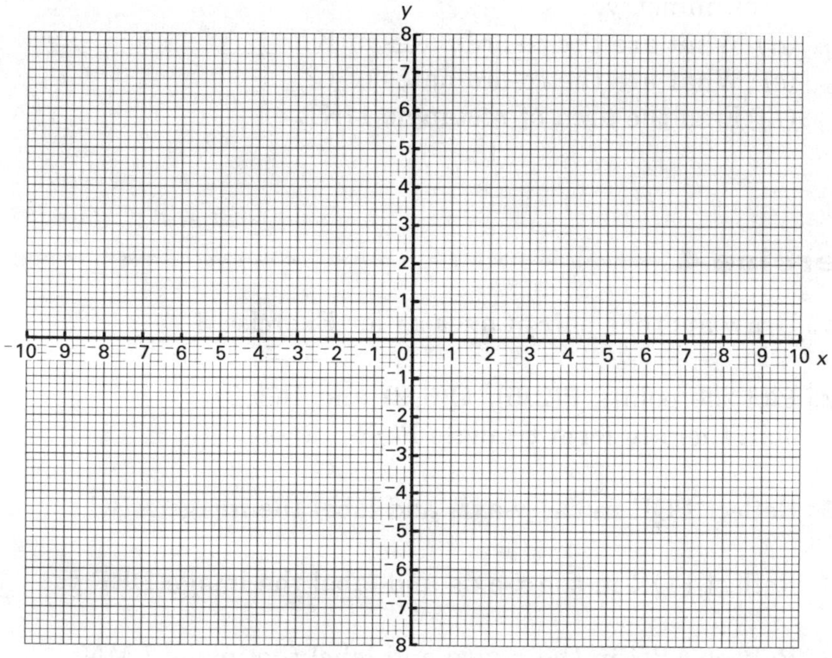

1. (*a*) Plot the points A(2, 5), B(3, 7) and C(7, 5).
(*b*) Draw triangle ABC.
(*c*) If AC is the line of symmetry for shape ABCD, find the point D and draw ABCD.
(*d*) What are the co-ordinates of point D?
(*e*) What is the name of shape ABCD?

2. (*a*) Plot the points E($^-2$, $^-2$), F(1, 1) and G(4, $^-2$).
(*b*) Find a point H such that EFGH is a square. Draw the square.
(*c*) What are the co-ordinates of point H?
(*d*) Draw the lines of symmetry of square EFGH.
(*e*) Does one of the lines of symmetry of EFGH, if produced, pass through the point ($^-2$, 1)?

3. (*a*) Plot the points $P(^-8, 1)$, $I(^-8, 5)$, $J(^-5, 5)$ and $Q(^-5, 1)$. Join them in that order.
(*b*) Join P to Q using a broken line.
(*c*) Complete rectangle IJKL such that PQ is a line of symmetry.
(*d*) What are the co-ordinates of K?
(*e*) What are the co-ordinates of L?
(*f*) Find the area of rectangle IJKL.

Exercise 4

1. Draw another pair of axes as for Exercise 3.

2. Plot the points P(2, 5), Q(6, 2) and R(1, 2). Join them to form triangle PQR.

3. Reflect PQR in the y-axis and label the image ABC.

4. Reflect PQR in the x-axis and label the image DEF.

5. Reflect ABC in the x-axis and label the image LMN.

Drawing Graphs from Given Data

Exercise 5

1. A bath contains some water. More water is poured into the bath at a steady rate until the bath is full. The following table shows the number of litres of water in the bath at each minute:

Time (min)	0	1	2	3	4	5	6	7
Quantity of water (ℓ)	20	31	42	53	64	75	86	97

(*a*) Draw a pair of axes as shown. Time should range from 0 min to 7 min (use a scale of 2 cm to 1 min). The quantity of water should range from 0 ℓ to 100 ℓ (use a scale of 1 cm to 5 ℓ).

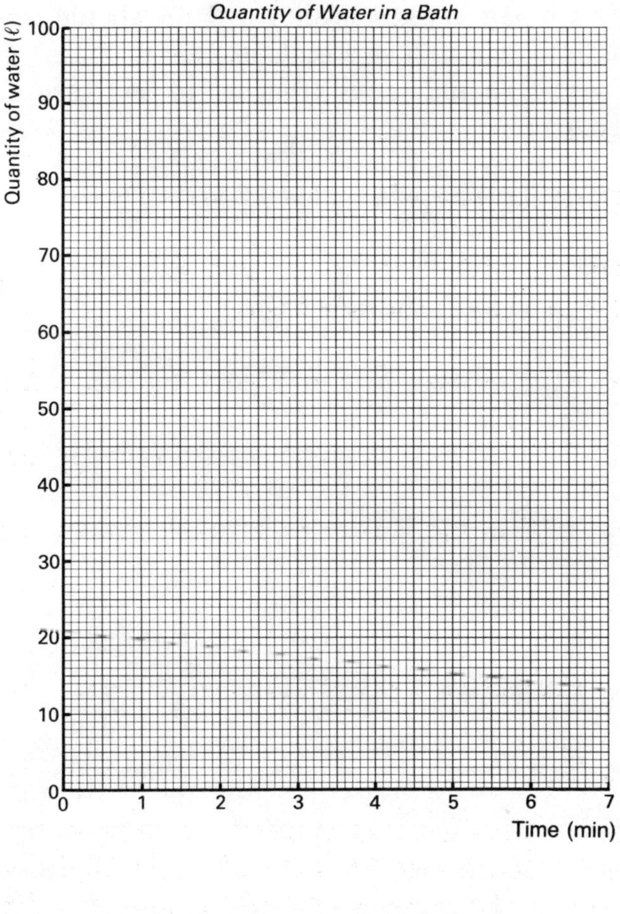

(*b*) Draw a graph from the given table.

Use the graph to help you to find:

(*c*) How much water there was in the bath to start with.

(*d*) How much water there was in the bath after $3\frac{1}{2}$ min.

2. A bath is being emptied at the rate of 8 ℓ/min. The following table shows the number of litres of water in the bath at each minute.

Time (min)	1	2	3	4	5	6	7	8	9
Quantity of water (ℓ)	68	60	52	44	36	28	20	12	4

(*a*) Draw a pair of axes as shown. Time should range from 0 min to 10 min (use a scale of 2 cm to 1 min). Quantity of water should range from 0 ℓ to 80 ℓ (use a scale of 1 cm to 5 ℓ).

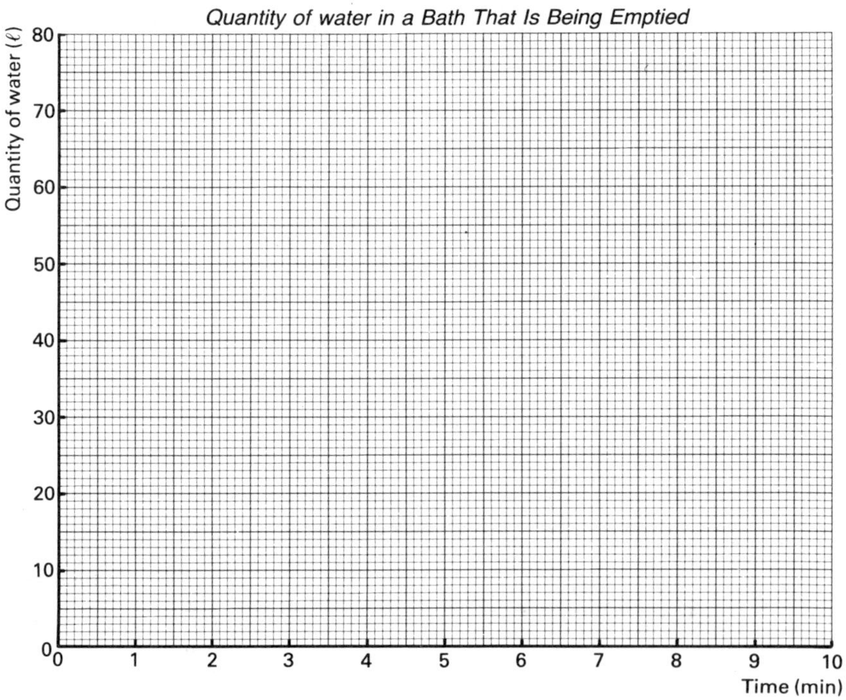

(*b*) Draw a graph using the given table.

Use the graph to help you to find:

(*c*) How much water there was in the bath to start with.

(*d*) How much water remains in the bath after $5\frac{1}{2}$ min.

(*e*) How long it takes to empty the bath.

3. The table shows the height of a stone measured every second as it falls from the top of a cliff.

Time of fall (s)	0	1	2	3	4
Height of stone (m)	80	75	60	35	0

(*a*) Draw a pair of axes as shown. Time should range from 0 to 4 s (use a scale of 5 cm to 1 s). Height should range from 0 to 80 m (use a scale of 1 cm to 5 m).

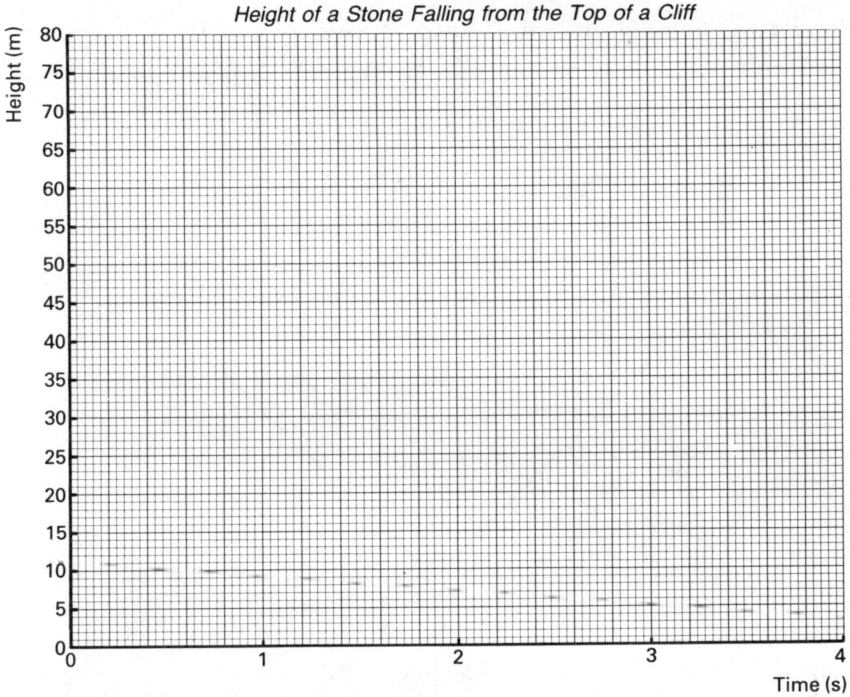

(*b*) Draw a graph using the given table. (Make a smooth curve.)

Use the graph to help you to find:

(*c*) The height of the cliff.

(*d*) The number of seconds the stone takes to hit the ground.

(*e*) The height of the stone after $1\frac{1}{2}$ seconds.

(*f*) The height of the stone after 2.4 seconds.

23 Scale Drawings

Exercise 1 M

Copy the given diagram. The direction of North is marked. Complete your copy to show S, E, W, NE, SW, NW and SE.

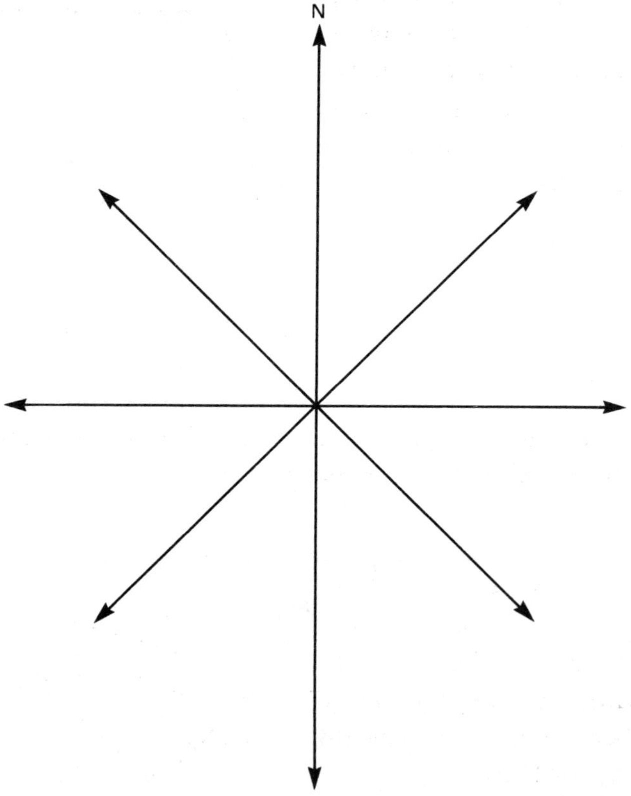

Exercise 2

1. A ship sailed 3 km due east followed by 4 km due south. The scale drawing shows the journey. It has been drawn to a scale of 1 cm to 1 km.
By measuring, find the distance the ship would have sailed if it had sailed a direct route.

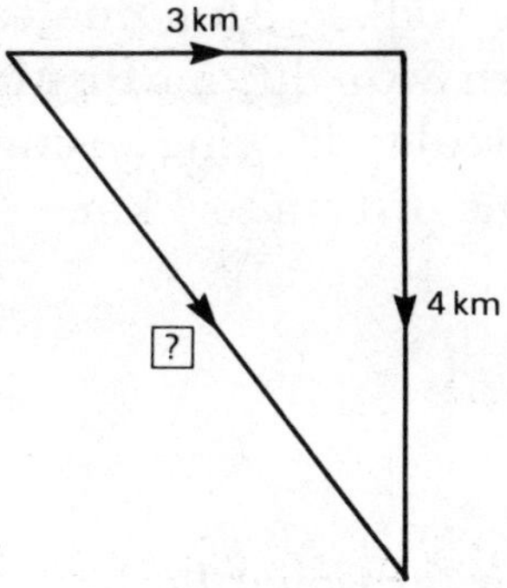

2. A hiker walked 7 km north-east, then 4 km due west. The scale drawing shows the route. It has been drawn to a scale of 1 cm to 1 km.

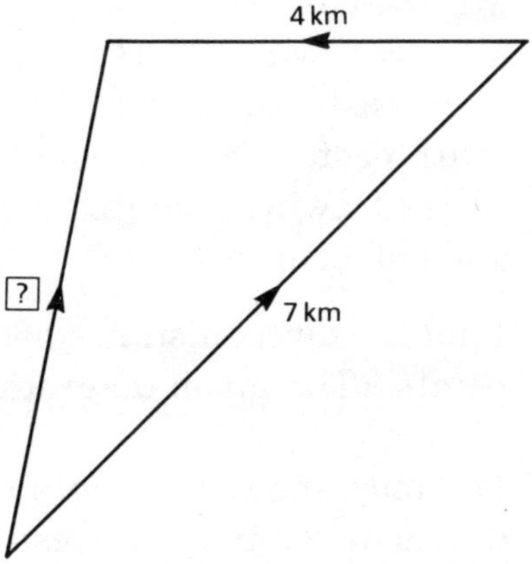

By measuring, find the distance the hiker would have walked by following a direct route.

3. Paula walked 8 km south-west, then 4 km north-west. The scale drawing of her journey has been drawn to a scale of 1 cm to 2 km.
By measuring, find the distance Paula would have walked by following a direct route.

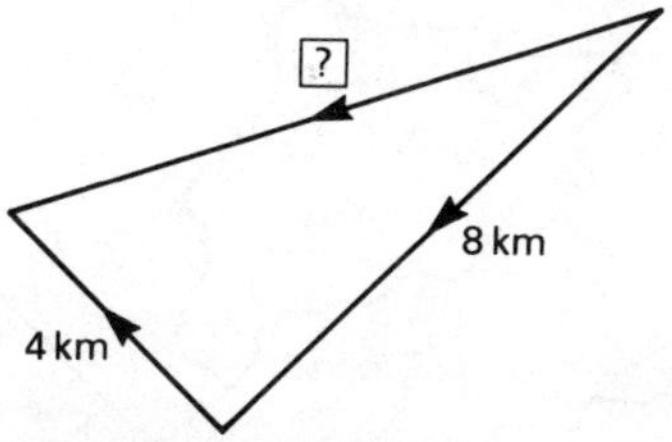

Exercise 3

1. Joe walked 6 km due east, then 8 km due north. Make a scale drawing using a scale of 1 cm to 1 km.

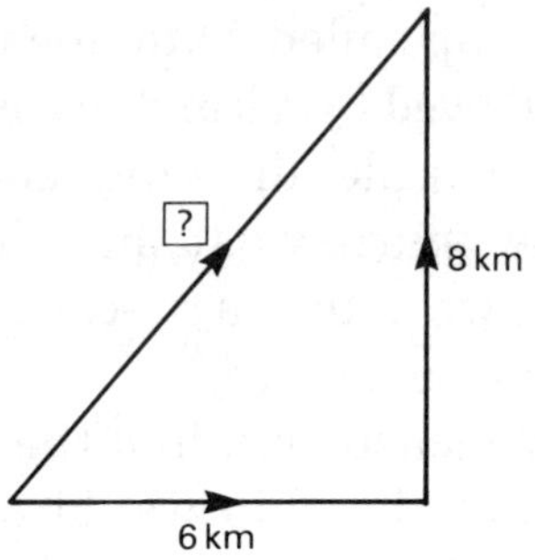

How far would Joe have walked if he had walked by a direct route? (The given diagram has not been drawn to scale.)

2. Mr Walker went hiking. He walked 6 km due east then 5 km south-east. Make a scale drawing using a scale of 1 cm to 1 km.

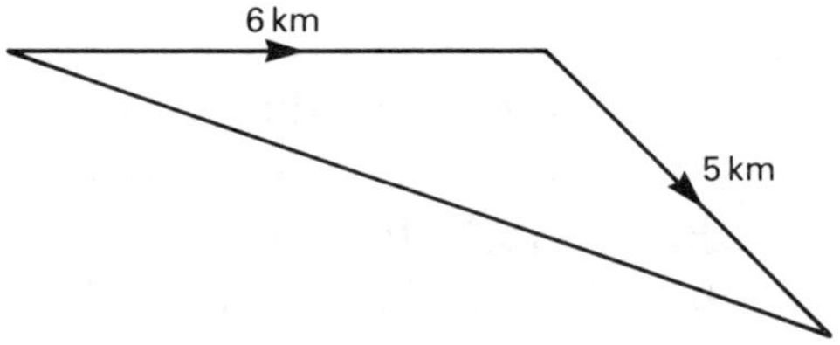

Find the direct distance between the starting and finishing points. (The given diagram has not been drawn to scale.)

3. The map shows the route followed by a ship.
It sailed 68 km south-west, then sailed 35 km due west.

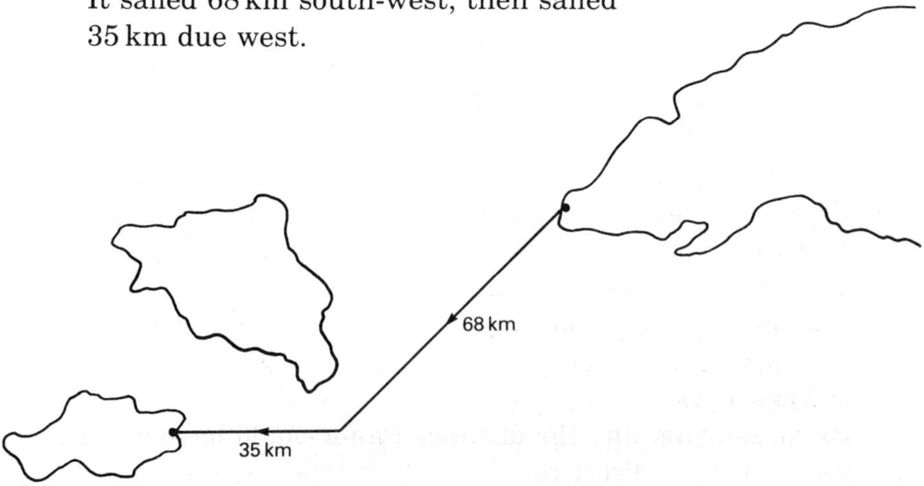

(*a*) How far did the ship sail altogether?
(*b*) A helicopter set off from the same port as the ship and flew a direct route to the same finishing port. Make a scale drawing. Use a scale of 1 cm to 10 km (i.e. 1 mm to 1 km). From your scale drawing, find the distance flown by the helicopter.

Exercise 4

1. A rectangular garden is 15 m long and 10 m wide. Make a scale drawing. Use a scale of 1 cm to 2 m. From your scale drawing, find the length of one of the diagonals of the garden.

2. A firm erects garages. Before a garage is erected the corner positions must be marked to show its final position. The two diagonals must then be measured to check accuracy.
Make a scale drawing to show the corner positions of a garage measuring 6 m by 3 m. Use a scale of 1 cm to 1 m. What should the diagonals measure?

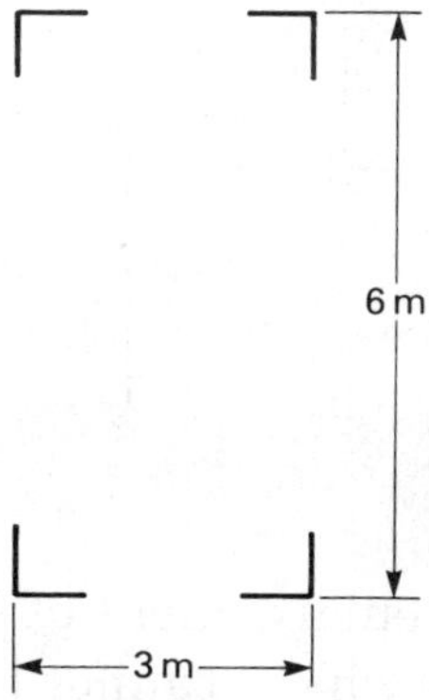

3. A ladder leans against a wall and makes an angle of 60° with the ground. The foot of the ladder is 2.5 m from the wall. Make a scale drawing. Use a scale of 2 cm to 1 m (1 cm to $\frac{1}{2}$ m). Find the length of the ladder.

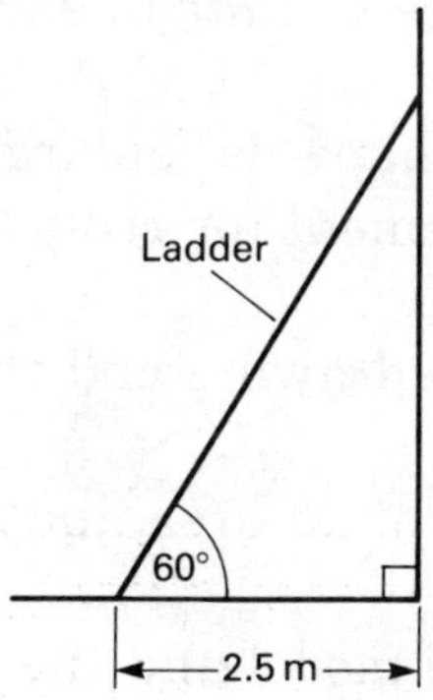

24 Similarity

Exercise 1

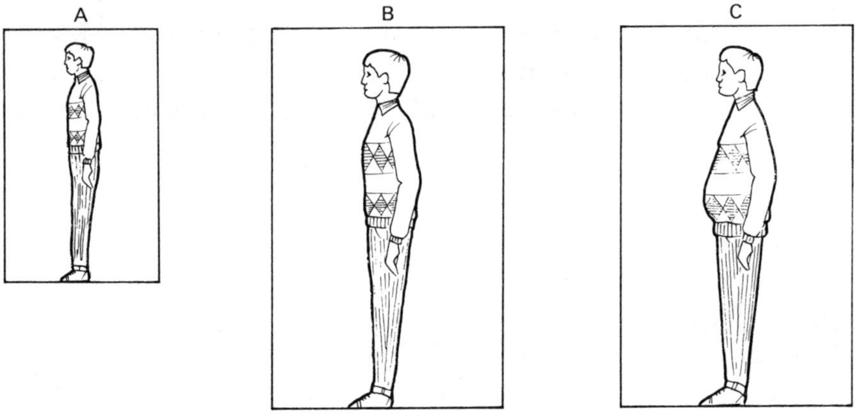

B is an enlargement of A.
Explain why C cannot be an enlargement of A.

Figures or solids that have the same shape but have different sizes are *similar* to each other.

If a shape is enlarged (or reduced) the shape and its enlargement (or reduction) are similar.

A scale drawing and the original are similar.

A map of an area and the real area are similar.

A scale model and the original are similar.

Exercise 2

For each question, write which two of the three given shapes are similar:

1. A B C

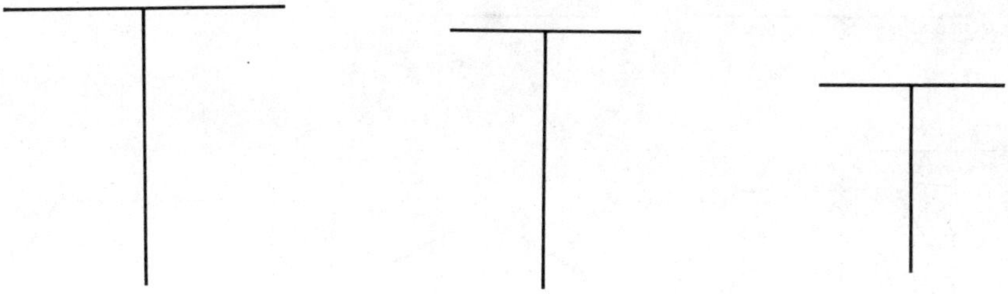

2. A B C

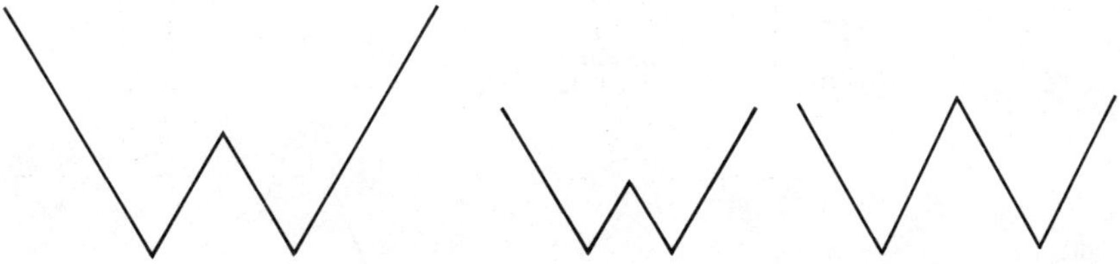

3. A B C

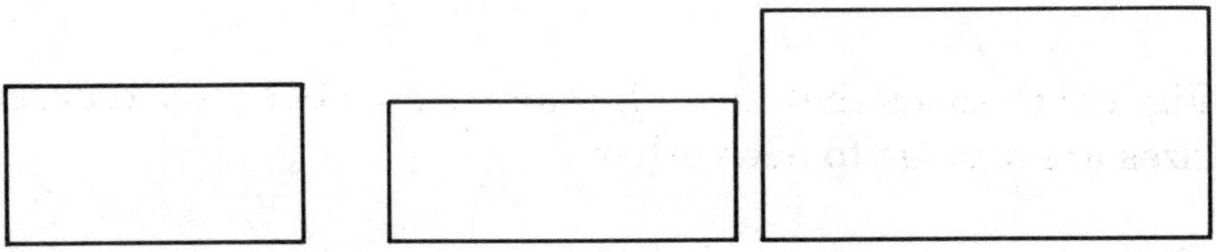

4. A B C

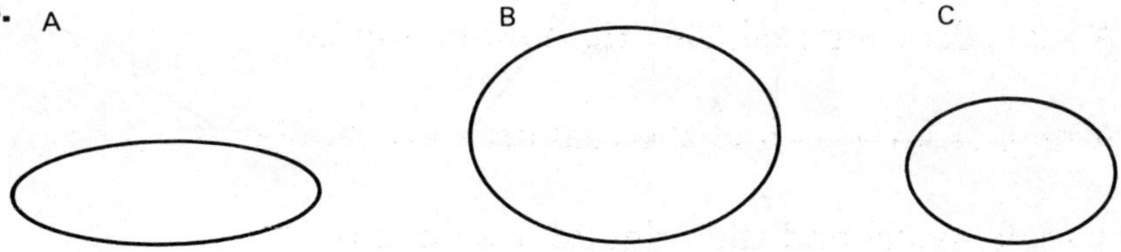

5.

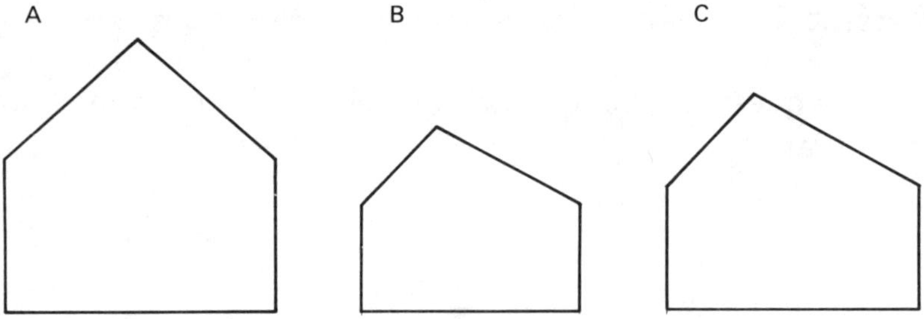

6.

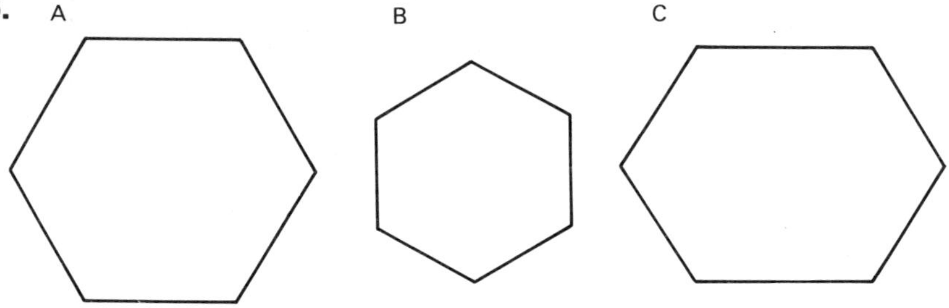

7.

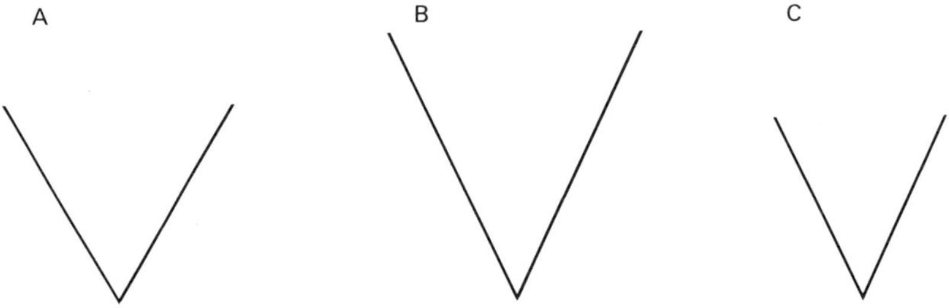

8.

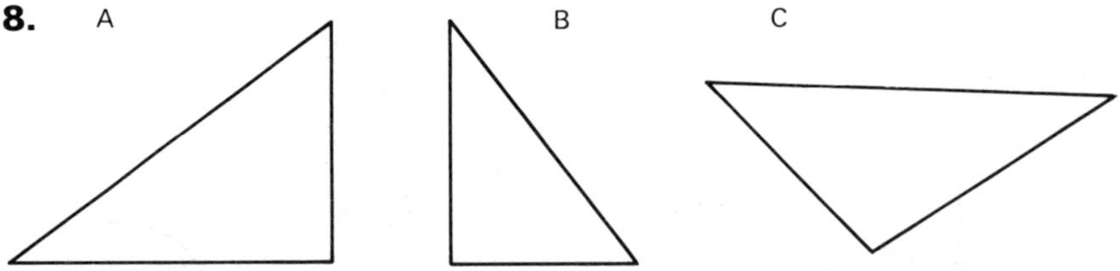

25 Pythagoras

Exercise 1 M

Copy the given diagram on to squared paper. It shows a right-angled triangle with squares on each side.

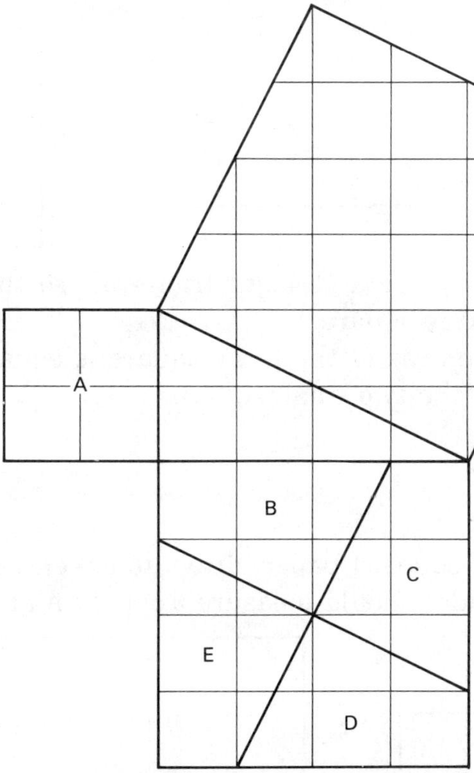

Cut out the small square (labelled A).
Cut out the middle-sized square, then cut it into 4 quadrilaterals as shown (the quadrilaterals are labelled B, C, D and E).

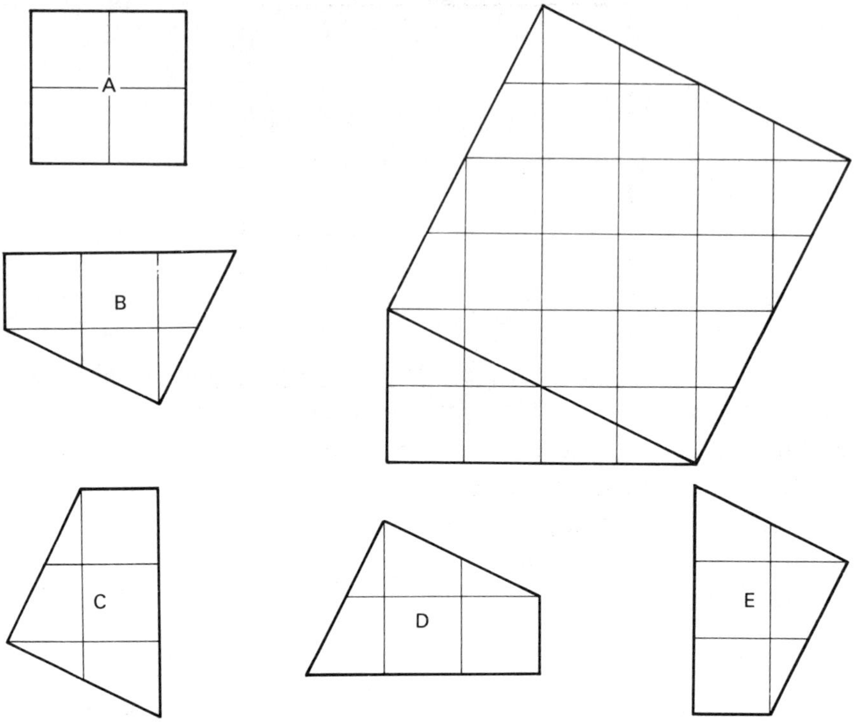

Position the small square and 4 quadrilaterals so that they exactly cover the large square.
This shows that the area of the large square is equal to the sum of the areas of the two small squares.

Exercise 2

1. Using a piece of squared paper, draw and then cut out 3 squares. Their sides should measure 3 cm, 4 cm and 5 cm.

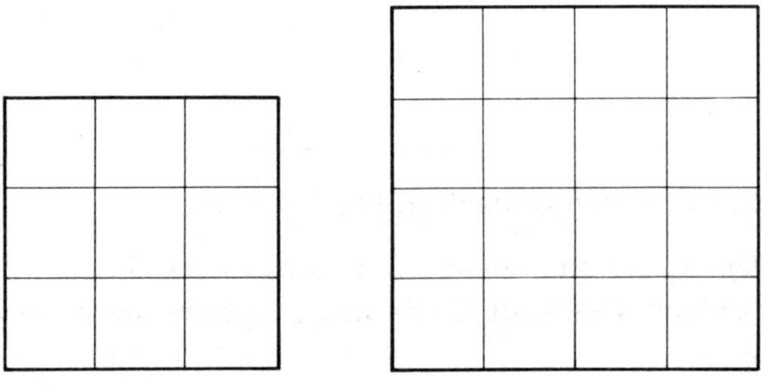

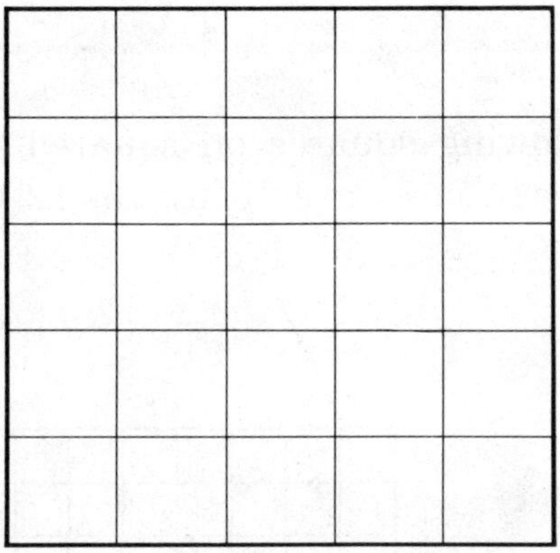

2. On a piece of plain paper, draw and then cut out a right-angled triangle with sides measuring 3 cm, 4 cm and 5 cm.

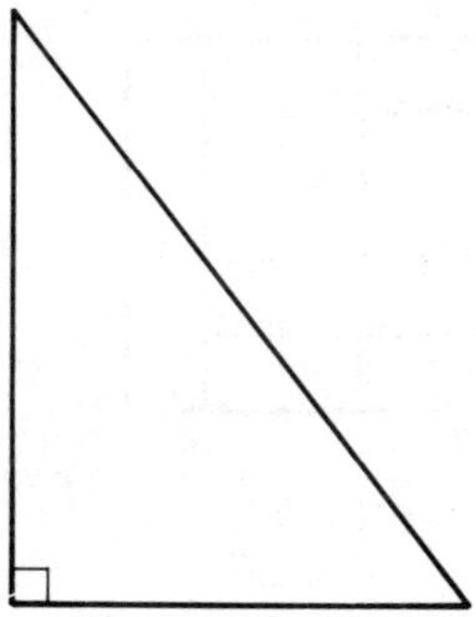

3. Stick the triangle into your exercise book, then stick the three squares in turn along each side of the triangle. The position of one of the squares, the 4 by 4, is shown below:

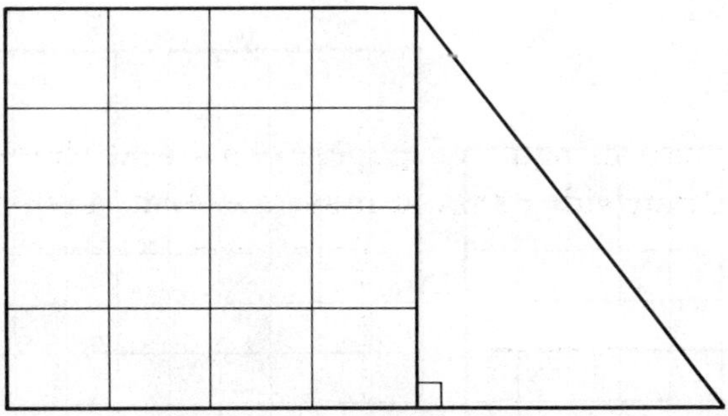

4. What is the area of each of the 3 squares?

5. Add the areas of the two small squares and compare your answer with the area of the large square. Write what you notice.

Exercise 3 **M**

1. Draw the following squares on squared paper, then cut them out. The squares are 13 by 13, 12 by 12, 10 by 10 and 5 by 5.

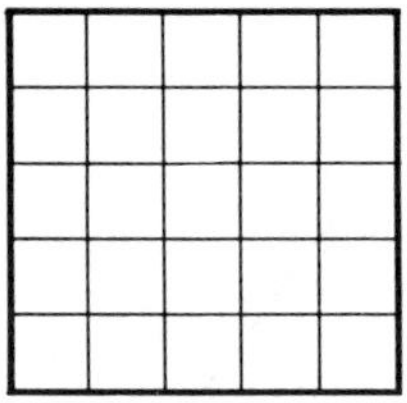

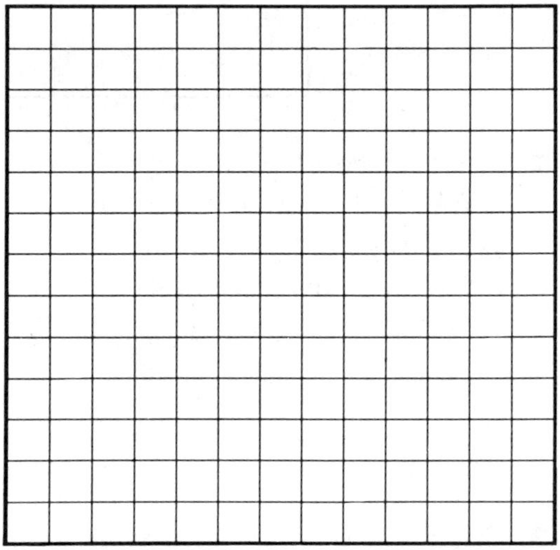

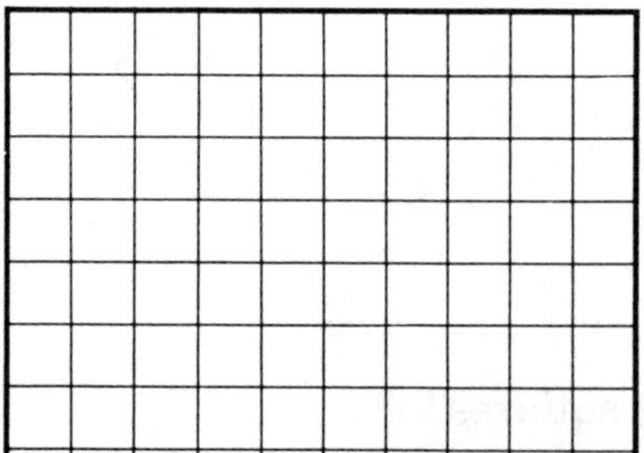

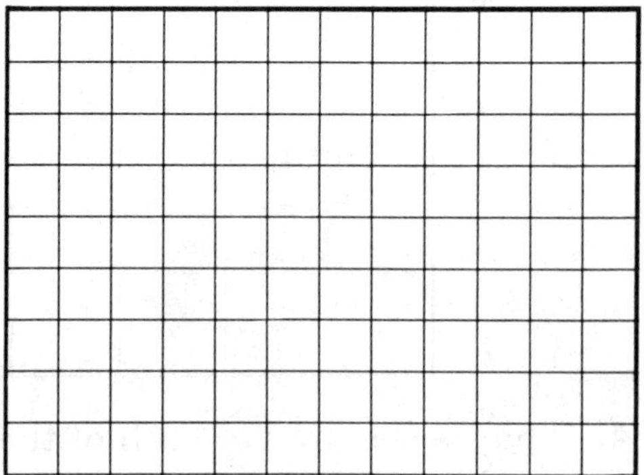

2. By fitting three of the four squares together as shown in the sketch, find which three squares fit together to make a right-angled triangle in the middle.

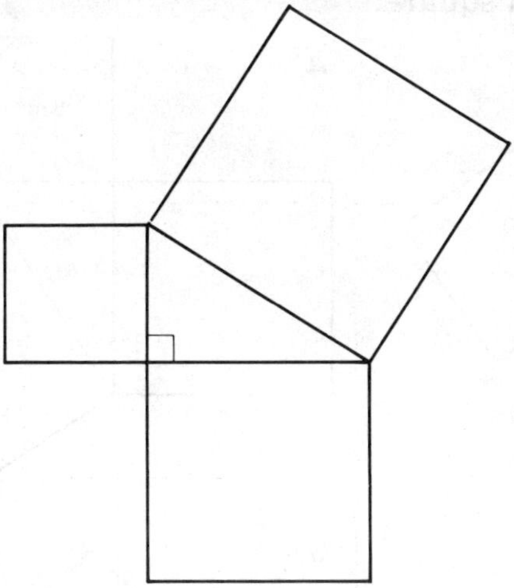

3. The area of the 13 by 13 square is 169 square units (since $13 \times 13 = 169$).

(*a*) Find the area of each of the 3 squares that fit together to make a right-angled triangle.

(*b*) Add together the areas of the two smaller squares of part (*a*) and compare the total area with the area of the large square. Write what you notice.

Exercises 1, 2 and 3 all show that for a right-angled triangle the area of the square on the long side is equal to the sum of the areas of the squares on the other two sides. This is called the theorem of Pythagoras.*

In the given sketch,

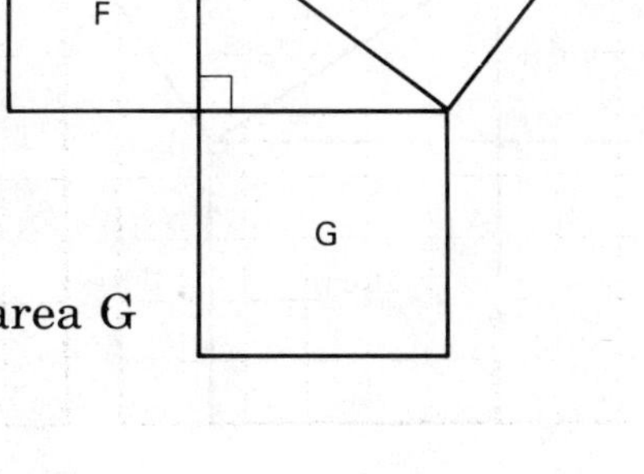

$$\text{Area H} = \text{area F} + \text{area G}$$

*See the appendix, p.296

Exercise 4

For each question, the areas of two squares are given. Find the area of the third square:

1.

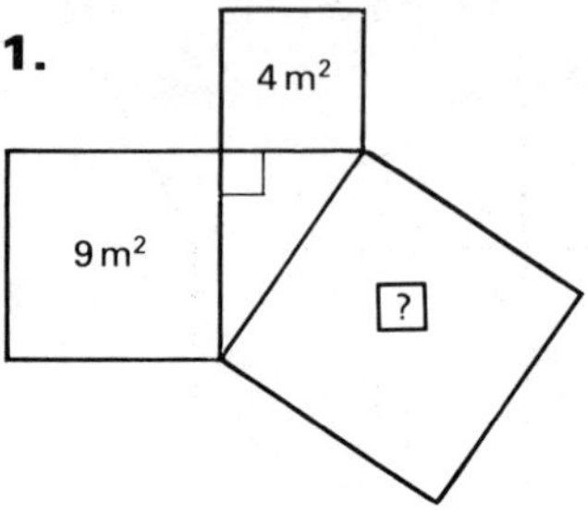

2.

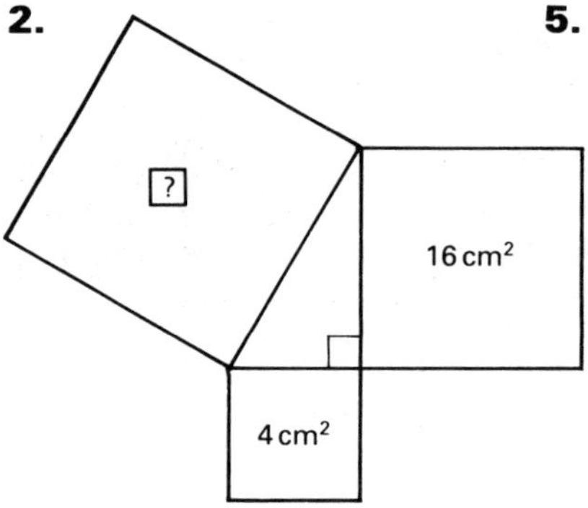

3.

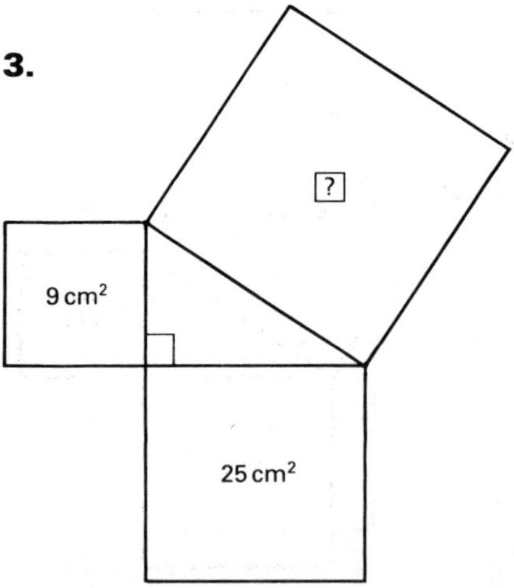

4.

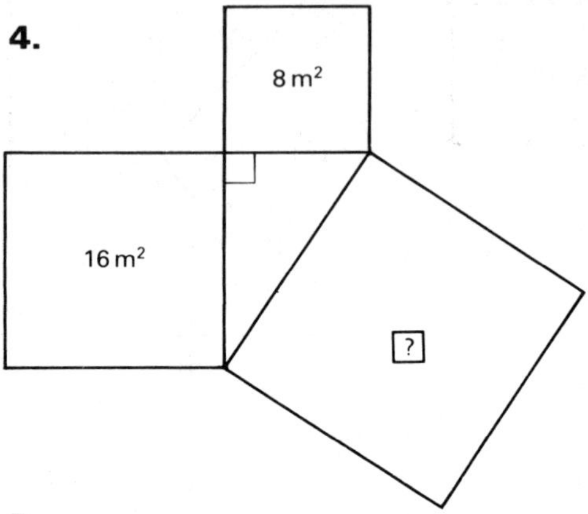

5.

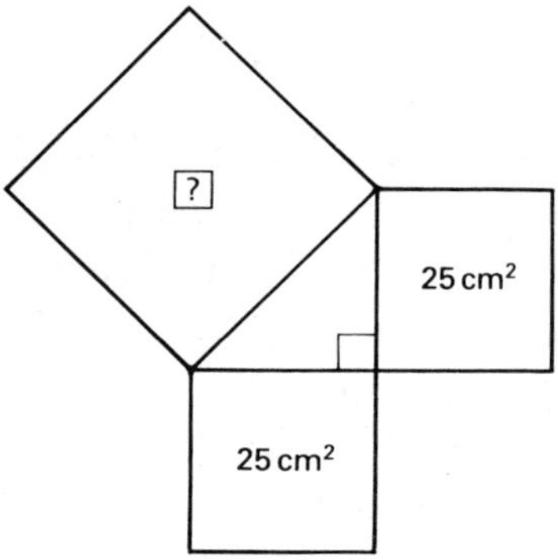

6.

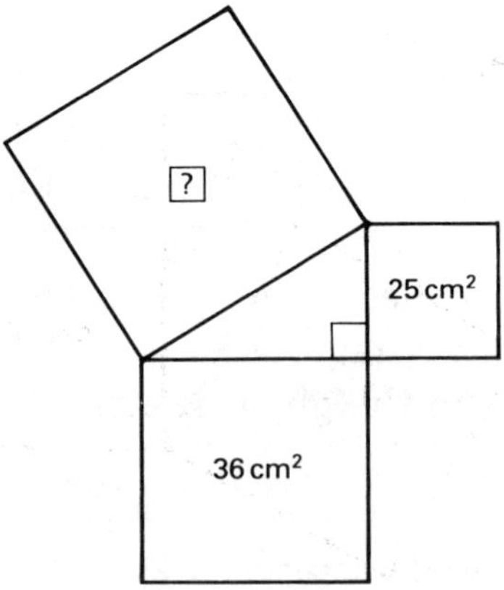

7.

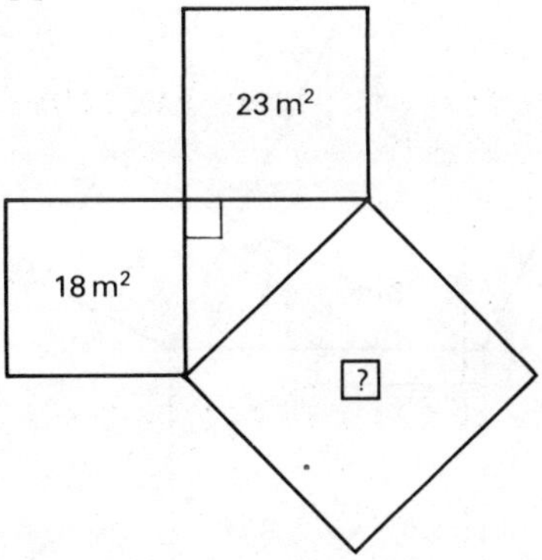
23 m²
18 m²
?

10.

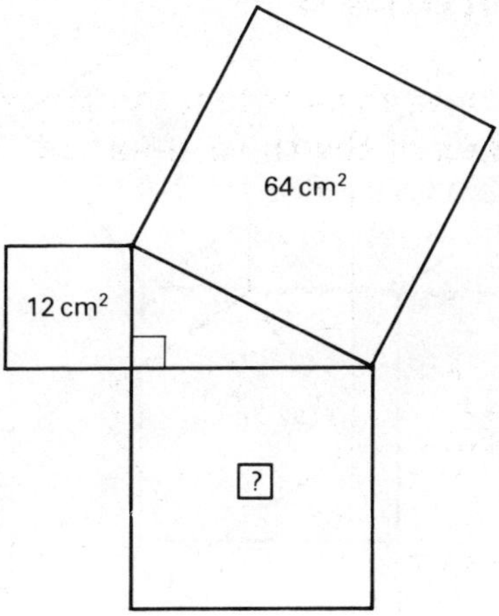
64 cm²
12 cm²
?

8.

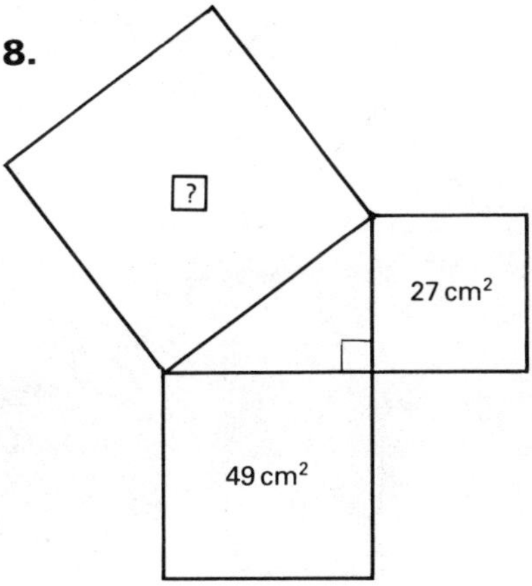
?
27 cm²
49 cm²

11.

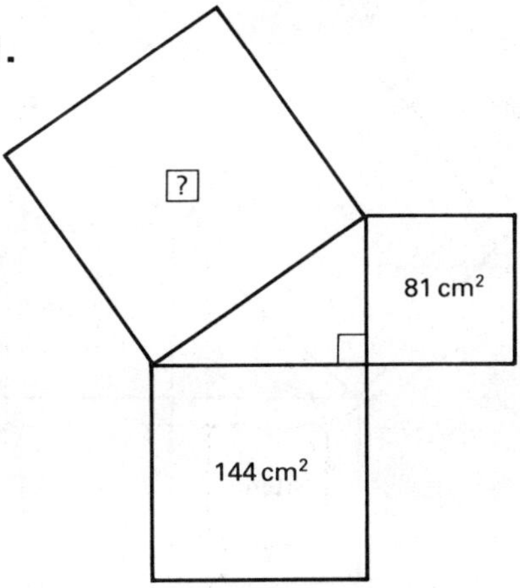
?
81 cm²
144 cm²

9.

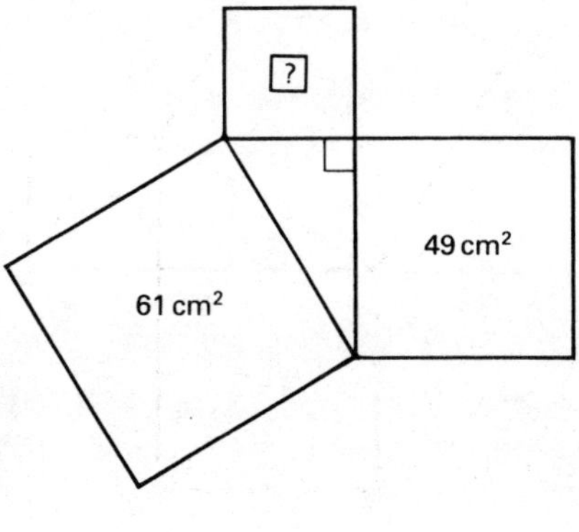
?
49 cm²
61 cm²

12.

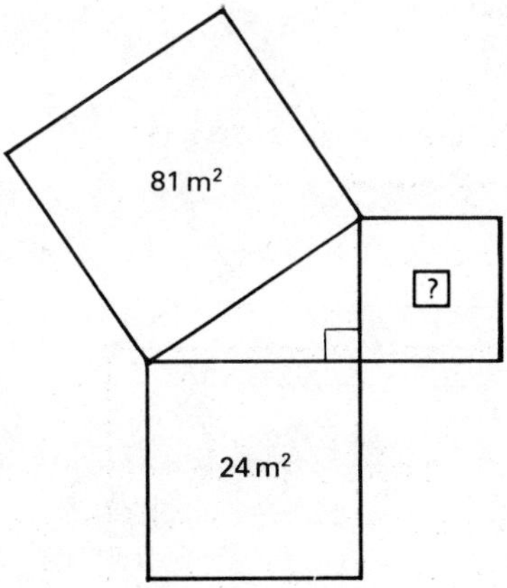
81 m²
?
24 m²

13.

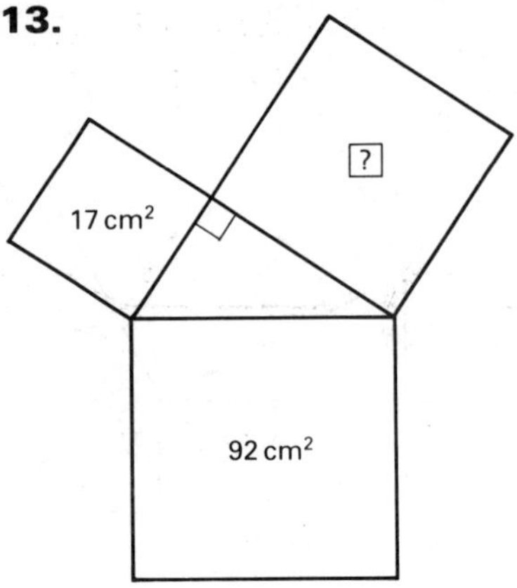

15.

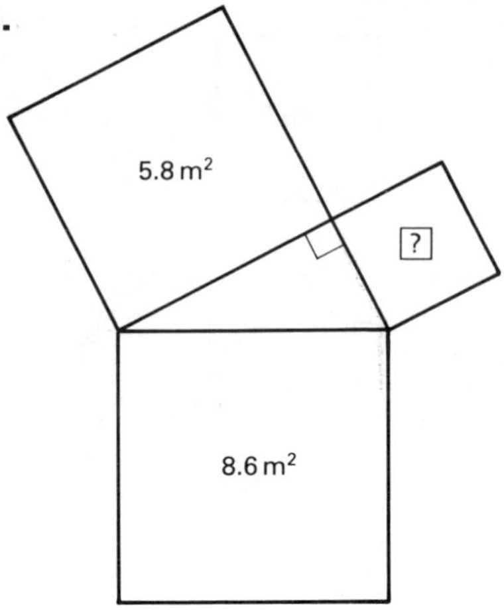

14.

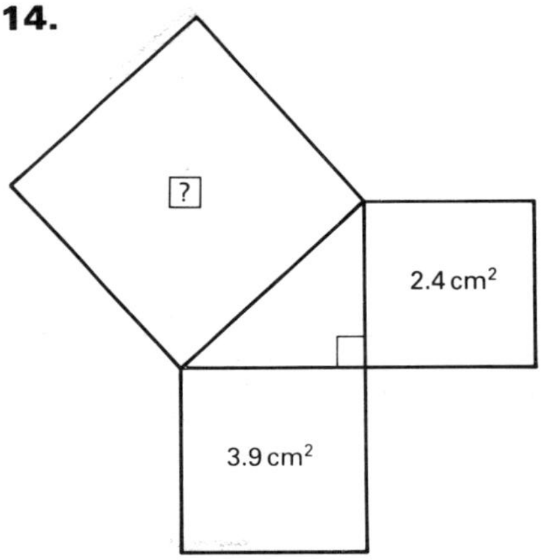

26 Statistics

Bar Charts

Exercise 1

1. In a survey, the following statements were made to a number of people and it was noted whether or not they agreed.
The bar chart shows the results of the survey:

A 'Boys should not do cookery.'
B 'Girls should not do woodwork or metalwork.'
C 'Only girls should show visitors around schools.'
D 'Only boys should carry books or move furniture in schools.'
E 'Only boys should play football.'

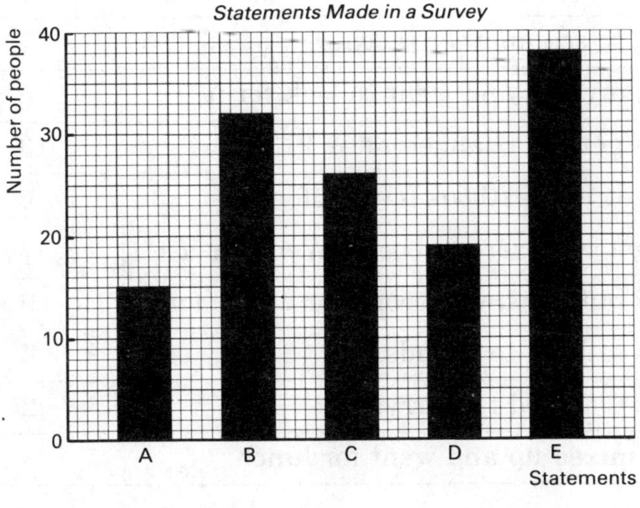

Use the graph to help you with these questions:

(*a*) How many people think that boys should not do cookery?
(*b*) How many people think that only girls should show visitors around school?
(*c*) How many people think that only boys should play football?
(*d*) How many people think that girls should not carry books or move furniture in school?

2. (*a*) Carry out the same survey as in question 1 for pupils in your class.
(*b*) Draw a bar chart to show the results of the survey for pupils in your class.

Exercise 2 M

1. The table shows the reasons given by several pupils for being late for lessons:

Reason	Number of pupils
I went to the wrong room.	7
I had to see another teacher.	5
I lost my bag and went to look for it.	8
I was getting changed after PE.	13
I forgot my textbook. (Went for it.)	6
I forgot my exercise book. (Went for it.)	7
I was putting away books for a teacher.	9
I was talking to friends.	4
I was too tired to hurry.	2
I got mixed up and went for lunch.	3

Answer the following questions from the given table:

(*a*) How many pupils were late because of a lost bag?
(*b*) Which was the most popular reason given?
(*c*) Which was the least popular reason given?
(*d*) How many pupils were late because they forgot a book?
(*e*) How many pupils went to the wrong place?

2. The 3rd year pupils in a school were given an end of year treat for working so hard. Their choices are shown in the table below:

Choice	Frequency	Choice	Frequency
Barbecue	22	Disco	59
Cinema	24	Ice skating	30
Coach trip	9	Safari park	12

Draw a bar chart to show the choices. Make each column 1 cm wide.
For the frequency, use a scale of 1 cm to 5 pupils.

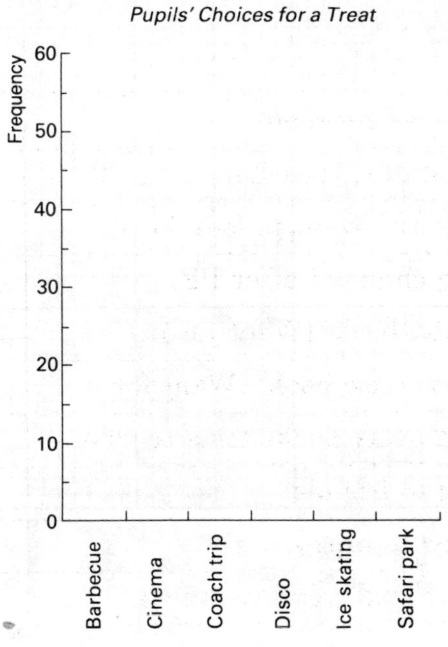

3. 40 people in the village of Firtree were asked the name of the road on which they lived. The given table and graph together show the results of the survey. Copy and complete them.

Name of street	Number of people	Name of street	Number of people
Ash Lane		Lime Street	
Beech Avenue	5	Oak Road	3
Birch Way	1	Sycamore Drive	9
Elm Close		Yew Crescent	

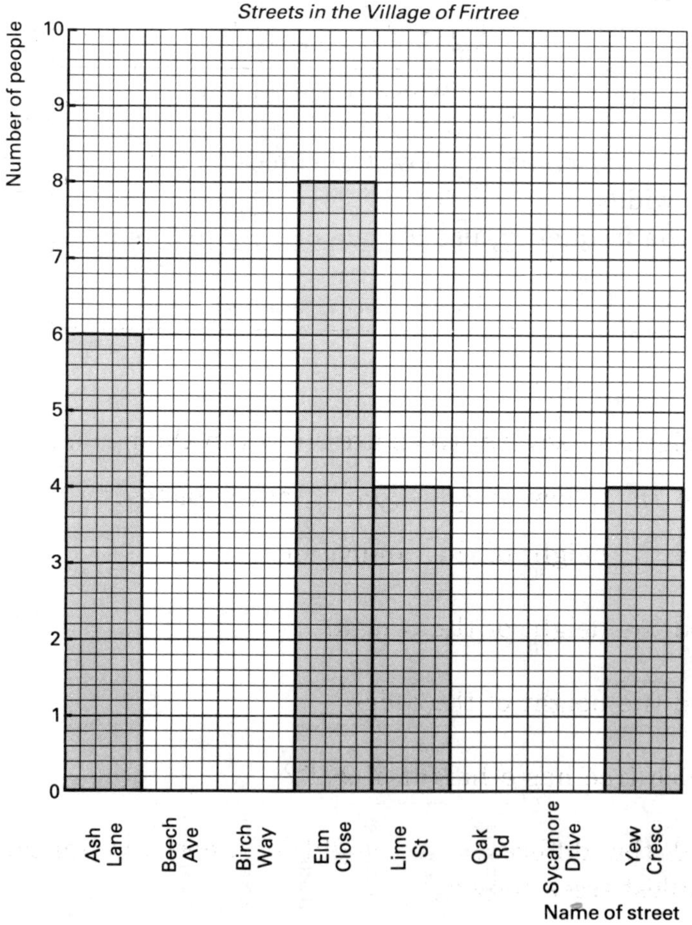

Exercise 3 Pictograms

Heights of Some Trees

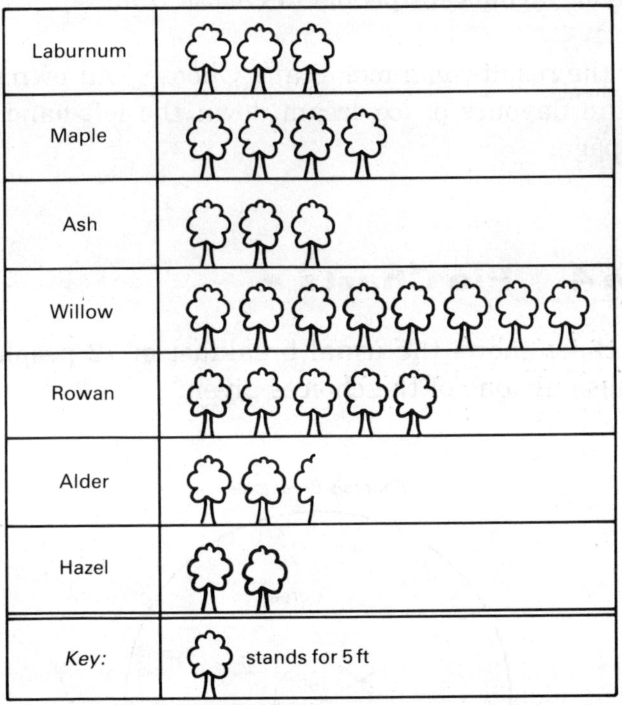

A The heights of several trees are shown in the pictogram.

Answer the following questions from the pictogram:

1. Find the height of the laburnum.

2. Find the height of the rowan.

3. Find the height of the alder.

4. Which tree has a height of 20 ft?

5. Find the difference in height between the tallest and smallest trees shown.

B **1.** Find which flavour of ice-cream pupils in your class prefer. They should each choose one flavour. Choose from vanilla, strawberry, chocolate, caramel, none (or list your own choice of flavours for people to choose from).

2. Show the results on a pictogram. Choose your own symbol. List the flavours of ice-cream down the left-hand side of your page.

Exercise 4 Pie Charts

A The pie chart shows the usual breakfast of 72 people. Each person has only one of the choices given.

People's Breakfasts

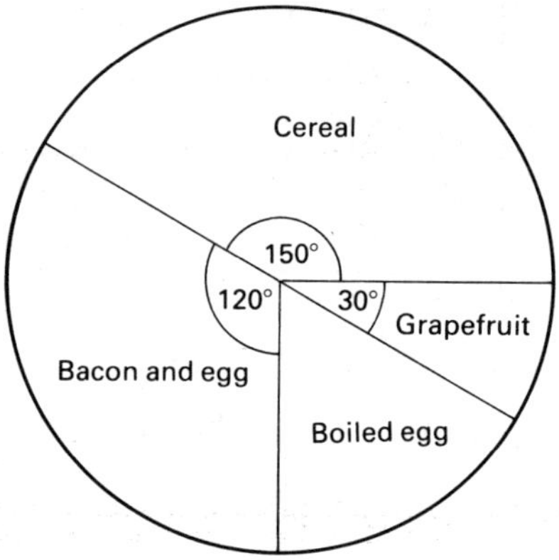

1. How many people have bacon and egg?

2. (*a*) How many degrees should the boiled egg sector have?
(*b*) How many people have a boiled egg?

3. How many people have grapefruit?

4. How many people have a cereal?

B Draw a pie chart showing the usual breakfast of 180 people if 90 people have a cereal, 45 people have bacon and egg, 30 people have a boiled egg and 15 people have grapefruit.

Remember 180 people take up 360°
so 90 people take up [?]°
45 people take up [?]°
15 people take up [?]°
and 30 people take up [?]°

Exercise 5 Jagged-Line Graph

The following graph shows the number of people who stayed at a guest house during a certain week:

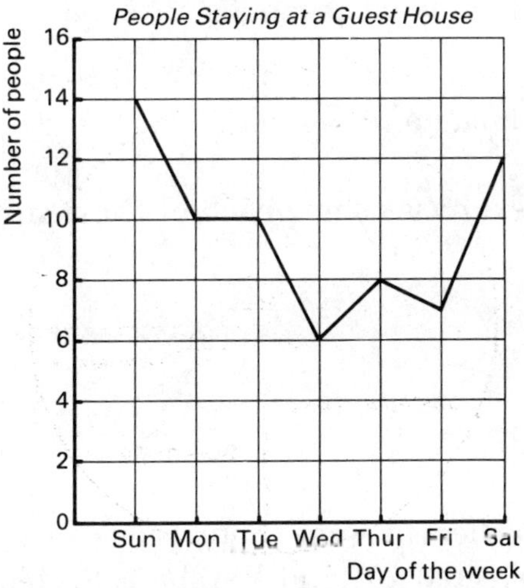

1. How many people were at the guest house on Monday?

2. How many people were at the guest house on Thursday?

3. On which day were there twice as many people as on Wednesday?

4. On which day were there half as many people as on Sunday?

5. On which two days did the most people stay at the guest house. Give a possible reason why those two days were the most popular.

Averages

The average speed was 42 miles per hour.
The cricket team averaged 4 runs per over.
Karen's average mark was 48%.
The team averaged 2 goals per match.
The car averaged 38 miles per gallon.
The bus averaged 50 km per hour.
His average score was 60.

The word 'average' is often used in everyday life, and so it is important to know what it means.

There are three different averages that are usually used. They are the *mean*, *mode* and *median*.

The following exercises show each of these in turn.

Mean

Exercise 6

A **1.** Three girls collected some money for charity. Val collected £8, Vicki collected £1, and Versha collected £3. Find the total amount collected by the 3 girls.

2. Five boys collected some money for charity. Dan collected £2, Doug £1, Dominic £4, Dinos £2 and Dennis £6. Find the total amount collected by the 5 boys.

3. Who collected the more money, the boys or the girls?

4. Did the boys do better than the girls at collecting money? Give your reasoning.

B The 3 girls collected £12 altogether. If each had collected £4 they would have reached the same total. For the girls, £4 is the average (or mean) amount of money collected.

Note

$$\text{Mean amount of money} = \frac{\text{Total amount of money}}{\text{Number of amounts of money}}$$

1. Work out the mean amount of money collected by the 5 boys in part A, question 2.

2. Compare the mean amount collected by the boys and the girls. Did the boys do better than the girls at collecting money?

> The *mean* of a set of numbers is the total of those numbers divided by how many numbers there are in that set.

Exercise 7

A Calculate the mean:

1. 3, 7, 2
2. 2, 4, 1, 5
3. 6, 8
4. 1, 2, 6
5. 8, 10
6. 9, 11
7. 1, 9, 11
8. 2, 4, 6
9. 1, 2, 8, 9
10. 5, 6, 7
11. 10, 2, 7, 9
12. 1, 2, 3, 1, 3
13. 5, 9, 10
14. 24, 19, 23
15. 31, 73, 46, 66

B **1.** In 4 matches, a football team scored 3, 1, 0 and 4 goals. What was the mean number of goals scored?

2. In 2 tests, Linda obtained 46% and 58%. What was her mean percentage mark?

3. A hockey team scored 21 goals in 7 matches. Find the mean number of goals per match.

4. There were 27 hours of sunshine in 3 days. Find the mean daily number of hours of sunshine.

5. Glen saved £4, £2, £3, £7 and £4 in 5 weeks. Find his mean weekly savings.

6. Mavis spent the following sums of money each week: 69 p, 48 p, 78 p and 85 p. Calculate the mean amount spent per week.

7. 6 lengths measured 4 cm, 2 cm, 9 cm, 5 cm, 8 cm and 14 cm. Calculate the mean length.

8. A cricketer scored 215 runs in 5 innings. Calculate the mean number of runs per innings.

Mode

In any number of items, the item that occurs the most is called the *mode*.

Exercise 8

Find the mode:

1. 2, 3, 3, 5
2. 4, 9, 2, 2, 1, 2, 7
3. 3, 4, 4, 1, 1, 1, 8, 2
4. 5, 9, 3, 3, 9, 9, 6, 8
5. 5, 0, 4, 5, 5, 4, 5
6. 3, 7, 1, 3, 7, 3, 2, 7, 7
7. 4, 4, 2, 0, 1, 0, 0, 5, 0
8. 2, 2, 1, 2, 4, 3, 8, 8, 8, 8
9. 9, 6, 9, 4, 6, 4, 2, 0, 6, 1
10. 6, 0, 4, 0, 4, 0, 4, 2, 4
11. a, b, a, c, c, d, b, a
12. a, e, i, e, o, u, u, i, e
13. 15, 13, 19, 15, 14, 16, 18, 17
14. 9, 8, 2, 6, 6, 8, 4, 9, 9, 1
15. 3, 0, 2, 5, 5, 1, 6, 5, 1, 1, 5
16. 2, 4, 2, 4, 4, 2, 2, 4, 2
17. 7, 0, 2, 0, 7, 0, 2, 0, 7
18. 6, 1, 1, 1, 6, 4, 6, 4, 4, 6
19. 24, 29, 35, 29, 37, 24, 29, 31
20. 56, 43, 70, 70, 43, 56, 43, 90

Median

> If a set of numbers are put in order of size, the middle number is the *median*.

Exercise 9

Find the median:

1. 2, 3, 4, 5, 6
2. 1, 4, 5, 6, 9
3. 2, 8, 9
4. 5, 9, 10, 14, 18
5. 7, 21, 46
6. 2, 3, 3, 8, 9
7. 4, 7, 9, 10, 12, 13, 17
8. 2, 2, 9, 12, 15
9. 4, 7, 7, 7, 9, 9, 6
10. 3, 8, 9, 10, 12, 12, 12
11. 3, 1, 5
12. 2, 9, 4, 6, 7
13. 8, 3, 6, 5, 4
14. 10, 4, 12, 7, 15
15. 1, 8, 1, 6, 5
16. 7, 3, 4, 4, 7
17. 12, 19, 16, 3, 3
18. 2, 0, 1, 1, 4, 0, 2
19. 3, 8, 9, 2, 0, 9, 14
20. 5, 5, 5, 3, 7, 3, 3
21. 8, 10, 6, 2, 12, 14, 4
22. 1, 5, 3, 6, 7, 3, 8, 5, 1
23. 2, 9, 4, 4, 1, 8, 8, 9, 7
24. 4, 6, 6, 4, 6, 4, 4, 6, 4
25. 9, 12, 3, 6, 9, 18, 7, 3, 10

Mean, Median and Mode

Exercise 10

1. A football team scored the following goals in 9 matches:

2, 0, 1, 3, 2, 1, 4, 1, 3

Find:

(*a*) the median, (*b*) the mode.

2. The marks out of 10 obtained by 5 pupils were:

9, 4, 3, 5, 9

Find:

(*a*) the median, (*b*) the mode, (*c*) the mean.

3. The numbers of children in 7 families were:

$$3, 2, 2, 1, 3, 1, 2$$

Find:

(*a*) the median, (*b*) the mode.

4. Here are the shoe sizes of several people:

$$6, 4, 5, 6, 8, 7, 6, 8, 5, 7, 5, 7, 4, 6, 3$$

Find the mode.

5. Five people were weighed. Their masses were:

48 kg, 70 kg, 59 kg, 67 kg, 71 kg

Find:

(*a*) the median, (*b*) the mean.

Revision Exercises XX to XXVI

Revision Exercise XX

1. Find the missing numbers to make the statements correct:

(*a*) $\boxed{?} + 9 = 18$ (*d*) $\boxed{?} \times 4 = 48$

(*b*) $\boxed{?} - 9 = 12$ (*e*) $\dfrac{\boxed{?}}{4} = 4$

(*c*) $8 \times \boxed{?} = 32$

2. Solve the equations:

(*a*) $v + 4 = 15$ (*d*) $3g = 21$

(*b*) $q - 13 = 9$ (*e*) $\dfrac{n}{6} = 4$

(*c*) $12 + x = 20$

3. Brian thinks of a number.
He adds 9 and gets the answer 21. Find the number.

4. When a number is multiplied by 7 the answer is 35.
Find the number.

5. Sheena thinks of a number. She doubles it then adds 4.
The final answer is 10. Find the number.

Revision Exercise XXI

1. To change gallons into litres, multiply by 4.5.
Change into litres:
(*a*) 4 gal (*b*) 3 gal (*c*) 10 gal

2. To change kilograms into pounds, divide by 5 then multiply by 11. Change into pounds:
(*a*) 10 kg (*b*) 25 kg (*c*) 30 kg

3. Credit charge = HP price − cash price.
(*a*) Find the credit charge if the HP price is £145 and the cash price is £112.
(*b*) Find the HP price if the cash price is £74 and there is a credit charge of £25.
(*c*) Find the HP price if the cash price is £85.99 and there is a credit charge of £27.50.
(*d*) If the HP price is £75 and the credit charge is £21.65, find the cash price.

4. If the mass of an empty jar is e grams, the mass of the jar containing jam is f grams and the mass of the jam is j grams, then $f = e + j$.
Use the formula to find the mass of a full jar of jam if the empty jar has a mass of 226 g and the jam weighs 454 g.

Revision Exercise XXII

1. Draw a pair of axes as shown on p.291. The x-values should range from $^-10$ to $^+10$ (use a scale of 1 cm to 1 unit). The y-values should range from $^-8$ to $^+8$ (use a scale of 1 cm to 1 unit).
(*a*) Plot, then join the points A(1, 4), B(3, 7) and C(7, 4) to form a triangle. Find the area of the triangle.
(*b*) Plot and join the points D($^-1$, 3) and E(7, $^-1$). What are the co-ordinates of the mid-point of DE?
(*c*) Plot the points F($^-3$, $^-6$), G($^-1$, $^-3$) and H(5, $^-3$). Find a point I such that FGHI is a parallelogram.
Draw the parallelogram and give the co-ordinates of I.

(*d*) JKLM is a rectangle. Plot J($^-8$, $^-2$), K($^-4$, 2) and L($^-2$, 0) and give the co-ordinates of M.
Draw the rectangle.
Draw in the lines of symmetry.

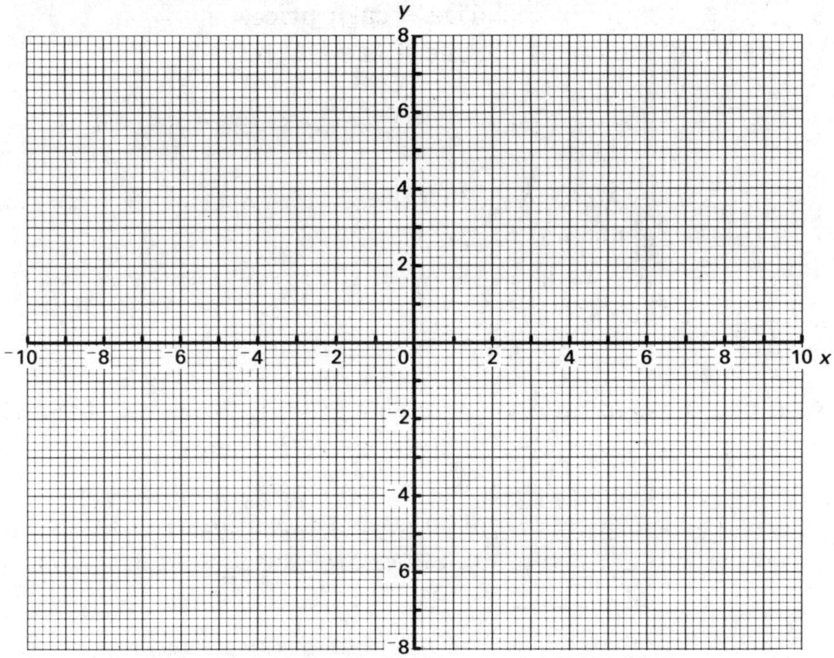

2. Mr Man's car can travel 13 km on each litre of petrol. The table below shows the distance travelled using the given number of litres of petrol:

Quantity of petrol used (ℓ)	0	1	2	3	4	5	6	7
Distance travelled (km)	0	13	26	39	52	65	78	91

(*a*) Draw a pair of axes as shown on p.292. Use a scale of 2 cm to 1 ℓ and a scale of 1 cm to 5 km.
(*b*) Draw a graph from the given table.

Use the graph to help you to find:
(*c*) The distance travelled on $4\frac{1}{2}$ ℓ of petrol.
(*d*) The distance travelled on 2.5 ℓ of petrol.

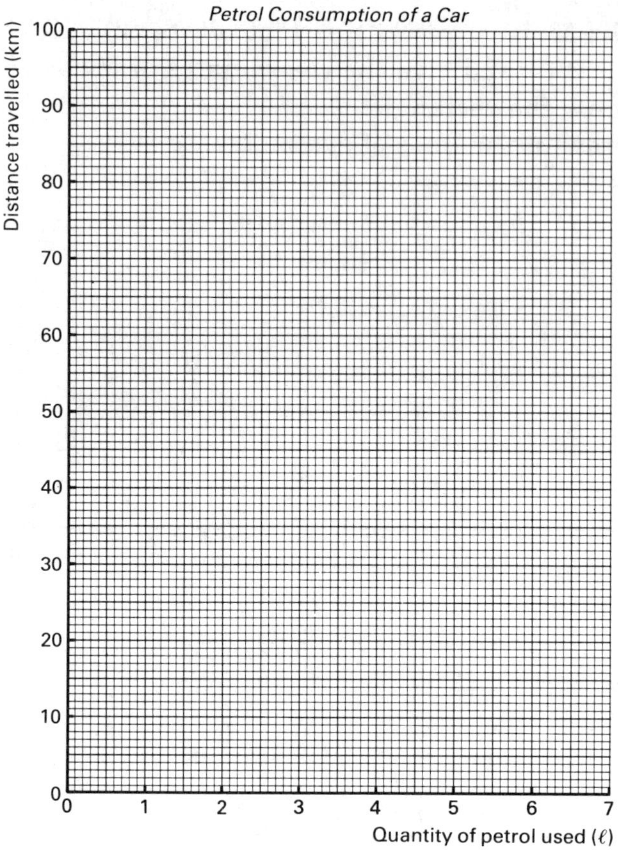

Revision Exercise XXIII

1. Susan sailed 20 km due north on a boat. She then sailed 9 km south-east. Darren started and finished at the same places as Susan but he walked directly from one place to the other following a straight route. Make a scale drawing. Use a scale of 1 cm to 2 km. Use your scale drawing to find the distance walked by Darren.

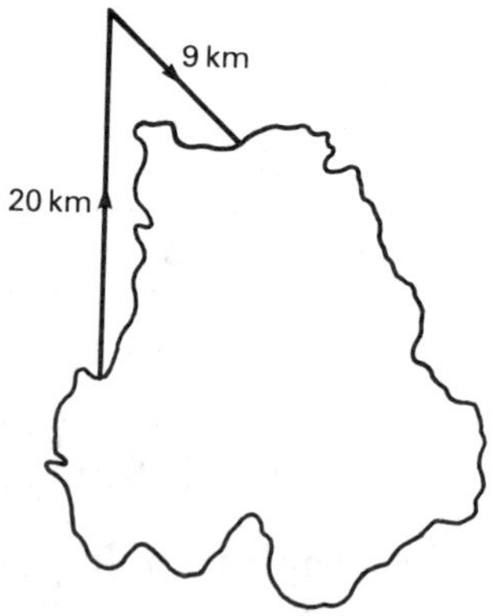

2. A rectangular swimming pool measures 25 m by 10 m. Make a scale drawing. Use a scale of 1 cm to 2 m.
If Sharon swam in a straight line from one corner of the pool to the opposite corner, use your scale drawing to find the distance swum.

Revision Exercise XXIV

For each question, write which two of the given shapes are similar:

1.

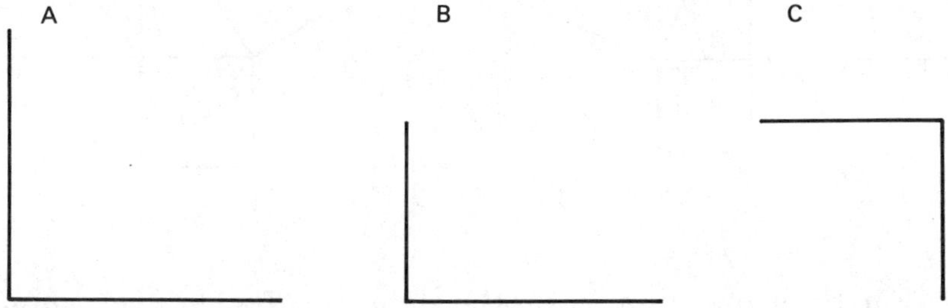

2.

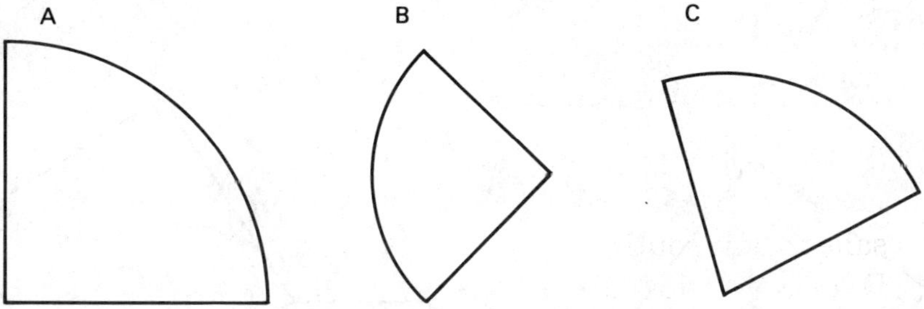

3.

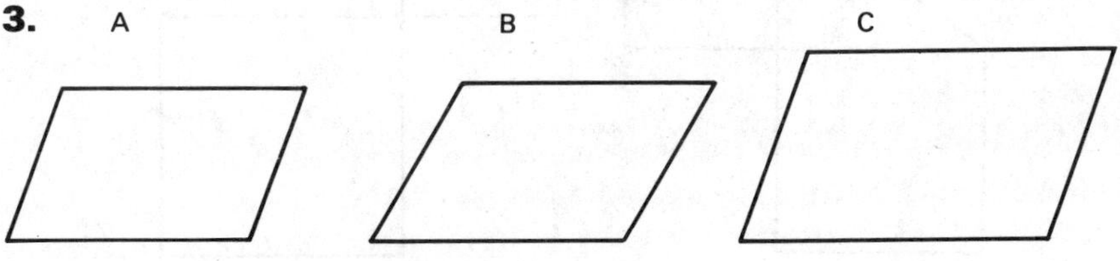

Revision Exercise XXV

For each question, the areas of two squares are given. Find the area of the third square:

1. (*a*)

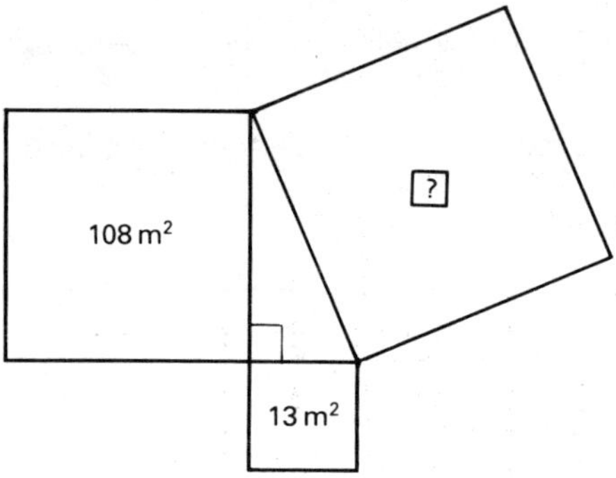

(*b*)

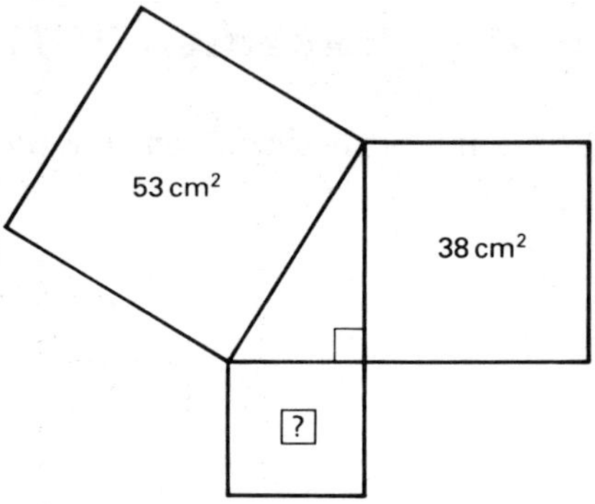

2. For each question, copy the diagram. Inside each square write a possible area. (Remember, in each diagram, the areas of the 3 squares are connected in some way.)

(*a*)

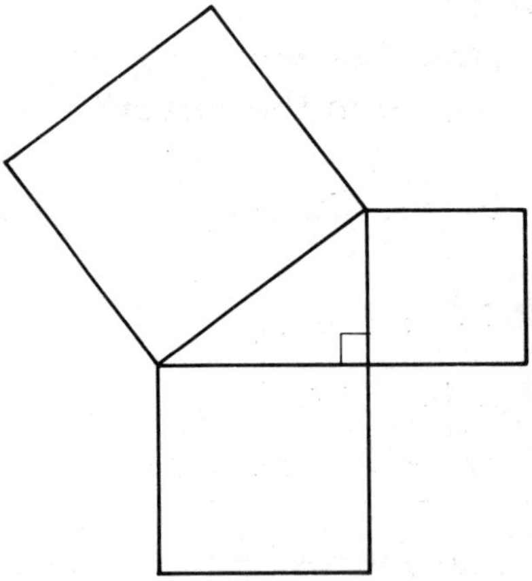

(*b*)

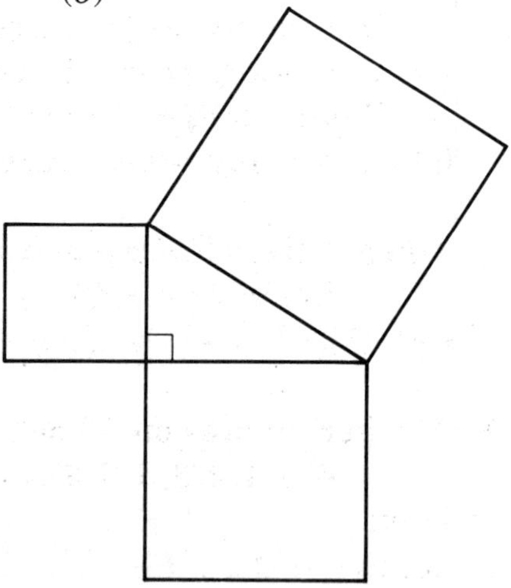

Revision Exercise XXVI

1. A survey was carried out to find the number of coins people normally carried in their pockets or purses. The given graph shows the results.

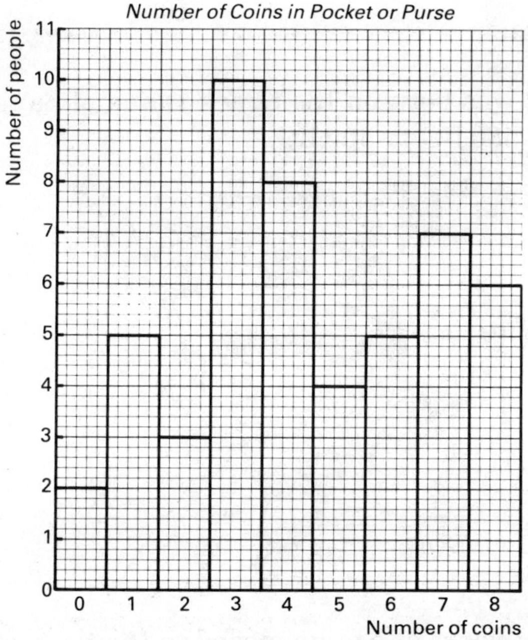

Answer the following questions using the graph:

(*a*) How many people carried 5 coins?

(*b*) How many people carried 8 coins?

(*c*) How many people had no coins?

(*d*) What number of coins was carried by most people?

(*e*) How many people were questioned in the survey?

2. Tim got the following marks in his exams:

52, 36, 56, 63, 43

Calculate his mean (average) mark.

3. A golfer's scores on 9 holes were:

4, 3, 4, 4, 3, 3, 5, 3, 3

Find:

(*a*) the mode, (*b*) the median.

Appendix: Pythagoras

Pythagoras

Pythagoras was born on the Greek island of Samos and lived from about 580 BC to 500 BC.

At about 50 years of age he moved to the town of Crotona in southern Italy where he formed a secret society called the Order of Pythagoreans. They had their own badge which was a star pentagram. (Note that there is a pentagon inside it.)

The Pythagoreans followed very strict rules among which were: they were not to wear wool, they were only to eat meat at religious ceremonies, they were not to tell non-members of the mathematical secrets they discovered (it is believed that the punishment for such a crime was death).

It is not certain what Pythagoras himself discovered, since, whatever was discovered by any members of the society, the credit was always given to Pythagoras, their founder and leader.

Not only did the Pythagoreans discover what is now called Pythagoras's theorem (which is dealt with in Chapter 25), they were the first to discover that the earth was not flat, they discovered the relation between the length of a string and the pitch of its note in early stringed instruments and they discovered the regular solids, the dodecahedron and the icosahedron shown below.

A Dodecahedron

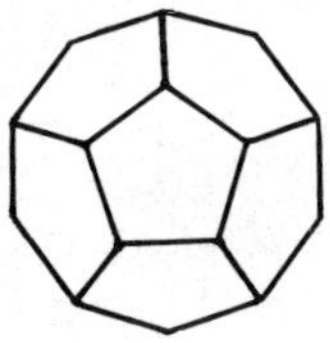

Net of a Dodecahedron

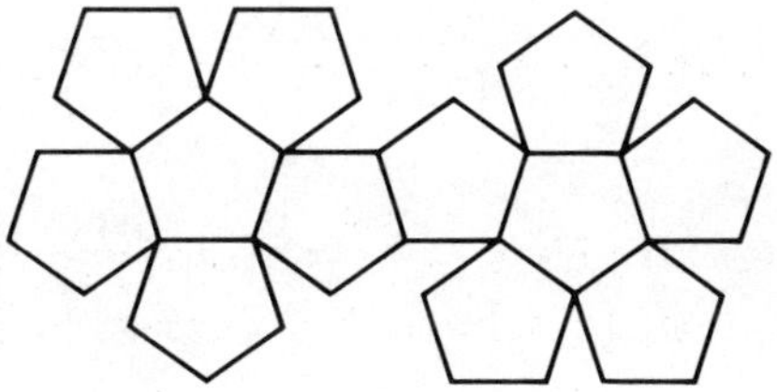

An Icosahedron

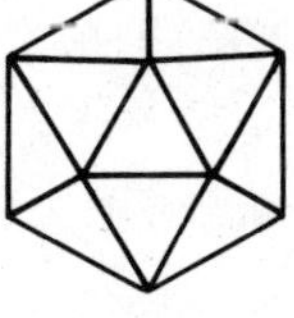

Net of an Icosahedron

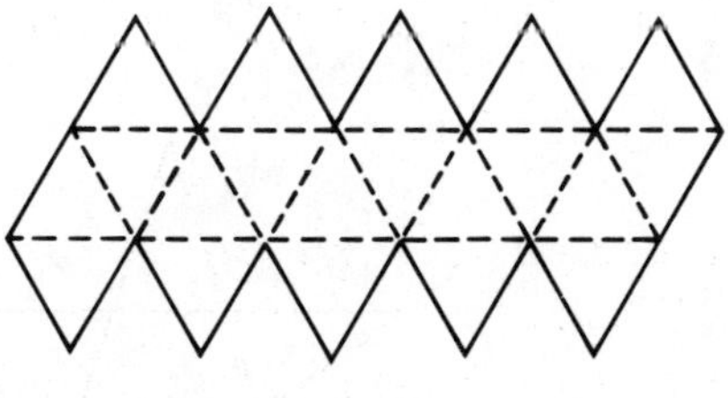

The Egyptians had already discovered the cube (6 faces, all squares), the tetrahedron (4 faces, equilateral triangles) and the octahedron (8 faces, equilateral triangles). These are the only five regular polyhedra. (Polyhedra are solids with flat faces. The faces are all polygons.)

The Pythagoreans also gave special meanings to numbers. The odd numbers were divine and were regarded as male. The even numbers were earthly and were female. 1 was regarded as the source of all numbers rather than as a number itself so 3 was the first true odd number and hence the first male number. The first female number was 2.

5 stood for marriage (since $3 + 2 = 5$, the combination of the first male and female numbers). The odd numbers were also regarded as lucky and the even numbers as unlucky and this belief was continued even in Shakespeare's day.

The Pythagoreans also used triangular numbers and square numbers. In fact, they called the triangular numbers 'Pythagorean numbers'. They made many other discoveries, particularly about numbers, too numerous to mention here.